AF323876

Coherent Doppler Wind Lidars in a Turbulent Atmosphere

Coherent Doppler Wind Lidars in a Turbulent Atmosphere

Viktor Banakh
Igor Smalikho

ARTECH HOUSE

BOSTON | LONDON
artechhouse.com

Library of Congress Cataloging-in-Publication Data
A catalog record for this book is available from the U.S. Library of Congress

British Library Cataloguing in Publication Data
A catalog record for this book is available from the British Library.

ISBN-13: 978-1-60807-667-3

Cover design by Vicki Kane

10 9 8 7 6 5 4 3 2 1

Contents

CHAPTER 1

Statistics of CDL Echo Signal   1

CHAPTER 2

Statistics of Lidar Estimates of the Radial Velocity and Doppler Spectrum Width   41

CHAPTER 5

Lidar Investigations of Aircraft Wake Vortices   179

Preface

It all started in 1988, when academician V. E. Zuev, director of the Institute of Atmospheric Optics (IAO) of the Siberian Branch of the USSR Academy of Sciences, obtained a resolution that the 15th International Laser Radar Conference would be held in the USSR, in Tomsk. As a result, in July 1990, about 100 internationally known specialists in laser sensing from all over the world visited Tomsk, which is located deep in what was then Soviet Siberia. As a result of this conference, hundreds of Soviet scientists had their first opportunity to communicate directly with their foreign colleagues. The conference gave rise to contacts, invitations, agreements, and so on.

One of the agreements for cooperation signed in 1991 was between IAO and the Lidar Group of the Institute of Optoelectronics (later called the Institute of Atmospheric Physics) of the German Aerospace Center (DLR), Oberpfaffenhofen, Germany. Dr. Christian Werner, head of the Lidar Group, cooperated with IAO on the study of the influence of turbulence on atmospheric Doppler anemometry through the International Bureau of the Ministry of Science of the Federal Republic of Germany. This agreement laid the foundation for many years of cooperation between IAO and the Lidar Group, whose scientific results have found their way into dozens of journal publications and form the subject of this book. During this period, Dr. Werner lent not only organizational, but also personal support to this cooperation, and felt deeply for Russian science, which was close to collapse in the 1990s.

Dr. Rod Frehlich, University of Colorado, who died before his time in 2012, made a tremendous contribution to our understanding of the operation of coherent Doppler wind lidars in the turbulent atmosphere. For many years, our studies in this field were parallel and complementary. There are few publications on coherent wind lidars that are without reference to Frehlich's publications. Frehlich's results have an important place in this book.

Finally, the investigations in the field that are discussed in this book were unlikely to have been carried out if not for another outstanding American scientist, Dr. Milton Huffaker. Huffaker conducted experiments on measurements of the wind velocity in the atmosphere based on coherent detection of laser echo signals in the early 1960s. Lidars based on solid-state lasers at the 2-μm wavelength were developed by Coherent Technology Inc. (CTI, now a part of Lockheed Martin), which was headed by Huffaker. These devices resulted in a true revolution in the practical use of coherent wind lidars in the 1990s. Many experimental investigations, whose results are reported in this book, were carried out with lidars developed and created at CTI.

We are related to these people by not only business relations, but also personal relations—and we really appreciate their attention and the help they rendered to us.

The results reported in this book were obtained in cooperation with many of our colleagues from DLR, ONERA, QinetiQ, NOAA, and IAO, namely, Ch. Werner, F. Köpp, S. Rahm, I. Leike, J. Streicher, R. Simmet, A. Dolfi, J.-P. Cariou, M. Harris, R. I. Young, Y. L. Pichugina, W. A. Brewer, N. N. Kerkis, N. P. Krivolutskii, A. V. Falits, and others. We are truly grateful to them.

We are also happy to express our gratitude to A. B. Gonchar, who agreed to translate this book into English, and E. V. Tereshchuk for help with manuscript preparation.

Introduction

The study of turbulent wind fields is one of the problems of modern atmospheric physics. Studies in this field have been ongoing for many decades. The fundamental regularities of turbulent flows were revealed by A. N. Kolmogorov and A. M. Obukhov as far back as the 1940s and 1950s [1–5]. Many principally important statistics and spectra for turbulent fluctuations in wind velocity and temperature of the atmospheric surface and boundary layers were obtained at the Institute of Atmospheric Physics of the Russian Academy of Science (RAS) in the 1960s [6–15]. In general, a great many papers and books are devoted to the problem of atmospheric turbulence. Among the books of note are those listed in [16–31], which generalize the various aspects of theoretical and experimental investigations of atmospheric turbulence.

In investigations of atmospheric turbulence, components of the wind velocity vector can be measured with the use of wind sensors; for example, cup or sonic anemometers installed on a meteorological tower. Substantial contributions to the study of wind turbulence in the atmospheric boundary layer were made by investigators at the Taifun Scientific Development and Production Center of the Russian Federal Service for Hydrometeorology and Environmental Monitoring (ROSHYDROMET), who used its unique 300m meteorological tower [29, 32–35]. Turbulence in the free atmosphere was studied from aircraft by investigators at the Central Aerological Observatory of ROSHYDROMET [23, 36].

Numerous experimental investigations in model turbulent flows and in the atmosphere show that turbulent flows can include organized quasideterministic structures in the form of vortices [37–40]. Along with the study of coherent structures of the velocity field in the atmosphere, of great practical interest is the study of vortical structures of technogenic origin, for example, aircraft wake vortices [41, 42]. The investigation of aircraft wakes is important both to reduce flight accidents and to increase the capacity of runways in airports.

In the past 40 years, many papers have been devoted to the study of aircraft wake vortices (see [43] and references therein (publications before 2002) and recent books [41, 42]). The results of investigation of the influence of various atmospheric factors on wake vortex parameters are published, in particular, in [44–57]. Aircraft wake vortices were studied with the use of numerical simulation, laboratory experiments, and field experiments by photographing smoked wakes behind aircraft.

At the same time, it is obvious that one promising direction for the development of tools for obtaining information about the dynamics of turbulent wind fields is the use of radiophysical methods. These methods allow for remote and real-time measurements and place fewer restrictions on the spatial and temporal resolution of obtained data than measurements with traditional sensors. In the 1950–1970s, investigators at the Institute of Atmospheric Physics RAS [20, 58–60], Physical Institute RAS [61], and Institute of Atmospheric Optics of the Siberian Branch (SB) of RAS [62–64] carried out basic research into the propagation of optical waves in the turbulent atmosphere. This research formed the physical basis for optical methods of atmospheric investigation. The investigations of atmospheric temperature turbulence by optical methods were started by A. S. Gurvich in the 1970s [59]. In recent years, Gurvich and coauthors have conducted extensive studies on the fine structure of the temperature field in the atmosphere using occultation techniques [65–72].

Radiophysical tools for investigation of atmospheric dynamics include sodars [73, 74], Doppler radars [75–86], and Doppler lidars [87–111]. Among these tools, coherent Doppler lidars are the best for remote measurement of parameters of wind turbulence in the lower troposphere, which is important not only for understanding exchange processes in the boundary layer, but also for applied aspects (aviation safety, diffusion of atmospheric admixtures, and so on). Coherent Doppler lidars significantly extend the possibilities of experimental investigation of not only wind turbulence, but also coherent structures, in particular, aircraft wake vortices.

Intense theoretical and experimental investigations on the development and manufacture of Doppler lidars, in particular, for wind measurements, go back to the 1970s. The fundamental contribution to this field was made by American scientists R. M. Huffaker, R. M. Hardesty, S. M. Hannon, S. Henderson, R. G. Frehlich, B. Ray, R. Menzies, and others [87, 88, 91, 93, 96–98, 101, 107, 110, 112–154]. In Europe, the wide use of coherent Doppler lidar systems would be impossible without the results of J. M. Vaughan from Great Britain [155–161, 166], P. H. Flamant and A. Dabas from France [102–105, 162–166], Ch. Werner, F. Köpp, S. Rahm, and O. Reitebuch from the German Aerospace Center (DLR) [90, 100, 103–105, 166–214], and many others. A multi-year scientific cooperative agreement between DLR (which has coherent Doppler lidars) and the Institute of Atmospheric Optics SB RAS has allowed the authors of this book to conduct field tests with the developed lidar measurement methods on wind velocity, atmospheric turbulence parameters, and aircraft wake vortices, as well as to conduct the studies, of which the results are reported here.

The operating principle of coherent Doppler lidars (CDLs) consists of launching laser radiation into the atmosphere and coherently detecting the signal backscattered by aerosol particles. The main elements of a CDL are as follows: lasers to generate probing and reference beams, a telescope, and detectors. A scanning device is used to change the angle of propagation direction of the probing beam. Information about the speed of particles moving with the airflow is extracted from the measured frequency shift of the backscattered optical wave, which is determined by the radial component (projection on the axis of the probing beam) of the wind velocity vector. The two main types of coherent lidars are continuous-wave (cw) and pulsed. Among all existing coherent lidars (see, for example, [87–111]), the continuous-wave CO_2 CDL [88–90, 99, 100] and pulsed 2-μm CDL developed by Coherent Technologies Inc. (now Lockheed Martin Coherent Technologies) [96, 97, 110] are used most widely both in scientific research and in practical applications. All experimental results discussed in this book were obtained just with these two types of lidars.

In developing CDL and lidar data processing procedures, it is necessary to know the statistical properties of the lidar echo signal (probability density, statistical characteristics, and characteristic timescales of the echo signal), estimates of the Doppler spectrum (lidar signal power spectrum), and its moments. Field experiments [140, 215] have revealed deviations of the one-dimensional probability density of the echo signal of continuous-wave CDLs from the Gauss distribution in the case of small sensing volume, in which the microstructure of aerosol particles begins to manifest itself. The question of how this influences the statistics of lidar estimates of radial velocity remained unanswered. The use of approximate methods [113, 115, 216, 217] for estimation of the influence of turbulent fluctuations of the refractive index on the statistical properties of echo signals imposes certain restrictions on the applicability of the results obtained and does not allow us to cover the whole range of problems with the investigation of possible CDL applications in wind measurements. In this connection, a problem arises: *that of studying the statistical properties of CDL echo signals and estimated moments of Doppler spectra for various atmospheric conditions and different optical-technical parameters of lidar systems. This book presents algorithms for computer simulation and processing of CDL signals, which allowed us to conduct numerical experiments simulating lidar operation for conditions maximally close to actual ones.*

Measurements by coherent Doppler lidars have some restrictions on the sounding range. If for continuous-wave CO_2 coherent Doppler lidar these restrictions are mostly caused by the worse spatial resolution at the increased focal length of the probing beam [88, 218], then in the case of pulsed lidars the decisive factor is the power of the backscattered signal. For pulsed CDLs, the signal-to-noise ratio (SNR), determined to be the ratio of the coherently detected echo signal to the power of noise in the selected detector bandwidth, is proportional to the energy of the probing pulse and the aerosol backscatter coefficient. As the probing pulse moves far away from the lidar, the level of the echo decreases. At high altitudes (in the free atmosphere), the aerosol concentration is low, and SNR can be very low (insufficient for wind measurements). Turbulent fluctuations of the refractive index of air also can significantly affect SNR values [113, 115, 128, 216, 217, 219, 220]. The study of CDL measurements under conditions with weak echo signals is especially urgent in connection with possible CDL use for measurement of wind distribution over the globe from space [138, 146, 149, 221–223].

Statistical properties of estimates of the radial wind velocity obtained from pulsed CDL measurements at low SNR are studied in [120–123, 125, 142, 144, 145, 147]. As SNR decreases, the estimate of the radial velocity is biased and its average value tends to zero regardless of the actual velocity value. To diminish this bias, it is proposed, before the estimate, to perform longer averaging of the power spectrum (or covariance function) of the lidar echo signal, from which the radial velocity is estimated; that is, to perform accumulation over probing pulses. *However, to determine the wind velocity vector from radial velocities measured by a coherent Doppler lidar, conical scanning is usually used, and therefore the application of the accumulation principle requires new (i.e., different from the traditionally used sine wave fitting [75, 169, 224]) techniques for processing of raw lidar data. That is why the book considers techniques for estimation of wind velocity and direction from CDL data obtained under conditions of weak echo signals.*

The idea of using CDL to study wind flow turbulence likely arose at the same time with the advent of the first lidar systems. It is known that in addition to estimates of the radial velocity obtained via Doppler lidar data, the width of the power spectrum of a lidar echo signal (Doppler spectrum) can also carry information about wind turbulence parameters (variance of wind velocity, dissipation rate of kinetic energy of turbulence, and outer scale of turbulence). In contrast to measurements that use point sensors, lidar measurements of wind are conducted in some spatial volume, and the problem of spatial averaging over the sensing volume inevitably arises in such a case. The error of estimates of the radial wind velocity and the width of the Doppler spectrum are important factors, which also should be necessarily taken into account when retrieving information about the dissipation rate of turbulence energy from statistical characteristics of wind velocity measured by CDLs.

The known methods for determination of the dissipation rate of kinetic energy of turbulence in the atmosphere are based on relations that follow from fundamental laws of transformation of the turbulent energy in the inertial range of scales of the wind flow discovered by A. N. Kolmogorov and A. M. Obukhov. Information about the dissipation rate in this case can be found from measurements of structure functions (or spectra) of wind velocity fluctuations, whose form in the inertial range is determined by the Kolmogorov-Obukhov two-thirds (−5/3 for spectra) law [1–3, 19]. In the case of time structure functions or spectra, the Taylor hypothesis of frozen turbulence is used [19, 29, 225].

For the case of cw CDLs, V. M. Gordienko and coauthors [99, 226] found a relationship between the mathematical expectation of the squared width of Doppler spectrum and the turbulence energy dissipation rate. Their results of simultaneous measurements of the dissipation rate with the cw CO_2 lidar and cup anemometers set at a meteorological tower are in good agreement. However, this approach can be applied only when the longitudinal dimension of the lidar sensing volume is smaller than the outer scale of turbulence. The longitudinal dimension of the sensing volume increases fast with the increase of the range (focal length of the beam) and at some instant it becomes longer than the outer scale of turbulence. Therefore, the technique for estimating the dissipation rate from the Doppler spectrum width (DSW) has some restrictions on the sounding range.

An alternative way is to retrieve the information about turbulence from the temporal spectrum (or structure function) of the radial wind velocity measured by cw CDLs. The justification of this approach free from restrictions of the DSW method required detailed studies, which would provide answers to the questions about the influence of averaging over the sensing volume, aerosol microstructure, and turbulent fluctuations of the refractive index of air on the high-frequency part of the velocity spectrum (corresponding to the inertial range of turbulence), within which the turbulence energy dissipation rate is estimated. The results of such studies, as well as analysis of the possibility of estimating the dissipation rate from scanning cw CDL data, are presented in this book.

Pulsed CDLs are most promising for the study of turbulence in the atmospheric boundary layer. In contrast to cw lidars, pulsed CDLs allow for direct (without the Taylor hypothesis of frozen turbulence) determination of the spatial structure function of wind velocity. Nevertheless, here also it is necessary to take into account the spatial averaging of the wind velocity over the sensing volume (determined by the probing pulse duration and the width of the time window [124]) and the error of lidar estimates of the radial velocity. Possibilities of measuring wind turbulence by pulsed CDL were studied in [124, 126, 127, 131, 135, 136, 227–230]. However, methodological problems associated with determination of turbulence parameters from data measured by lidar scanning in the vertical plane were not considered in those studies. The point is that scanning by the sounding beam in the vertical plane across an aircraft wake can give raw experimental data, which can carry the information about both wake vortices and wind turbulence. Thus, it becomes possible to use one device to study the influence of wind turbulence on the evolution of aircraft wake vortices. *Therefore, this book analyzes the possibilities of measuring wind turbulence with pulsed CDLs both in the case of conical scanning by probing beam and in the case of probing beam scanning in the vertical plane.*

The method for determination of wake vortex parameters (coordinates of the wake vortex axis and wake vortex circulation) from pulsed CDL measurements of Doppler spectral moments as functions of the scanning angle and the distance between the lidar and the sensing volume is proposed in [150–153, 231]. This method uses the model setting of wake vortices. However, such an approach does not allow us to always obtain an actual profile of the tangential wake vortex velocity, which can differ widely from the model one due to the wake vortex deformation under the effect of turbulent wind flow. The same is true for the method proposed in [134] for estimation of wake parameters from Doppler spectra by the maximum likelihood. *The book presents a method that yields actual profiles of the tangential wake vortex velocity and examines its efficiency.*

Numerous theoretical investigations [41–57] resulted in development of the wake vortex evolution models. However, before coherent Doppler lidars were applied to wake parameter measurements, theoretical conclusions could not be checked in a field experiment. This is of concern, in particular, to theoretical constructions describing regularities of wake vortex evolution depending on the turbulent state of the atmosphere. *In our book, this problem is considered with extensive experimental data obtained from lidar measurements and with application of the methods for estimation of wake and wind turbulence parameters.*

The book consists of five chapters. Chapter 1 describes the operating principles of cw and pulsed coherent Doppler lidars. With the use of the empirical model of atmospheric aerosol and numerical simulation of random realizations of cw CDL echo signals, the one-dimensional probability density function, variance, and temporal correlation function of echo signal power are calculated for different sizes of sensing volume. The results of investigations of the influence of turbulent fluctuations of the refractive index of air on the mean value, variance, and correlation coefficient of pulsed CDL echo signal power are presented.

Chapter 2 presents the relationships among the Doppler spectrum, spectral moments (echo signal power, radial velocity, and squared width of the Doppler spectrum), and weighting functions of spatial averaging over the sensing volume. The procedures for estimation of the Doppler spectrum and spectral moments from raw data measured by a coherent Doppler lidar are described, as well as algorithms for numerical simulation of random realizations of cw and pulsed CDL signals with allowance made for turbulent fluctuations of wind velocity and shot noise of the reference beam. Results of theoretical and experimental investigations of the influence of various factors (sensing volume size, atmospheric aerosol microstructure, wind and refractive turbulence, SNR) on statistical characteristics of lidar estimates of the radial velocity and width of the Doppler spectrum (variance, spectral and structure functions of radial velocity, mathematical expectation of the squared width of the Doppler spectrum) are presented.

Chapter 3 considers the procedures for estimating wind velocity and direction from measurement data of coherent Doppler lidars at conical scanning by a probing beam. The influence of measurement geometry, wind turbulence, and spatial and temporal averaging of wind velocity fluctuations on the accuracy of lidar estimates of the mean wind velocity and direction is analyzed. Techniques for estimation of the wind velocity and direction from pulsed CDL data under conditions of a weak echo signal are presented. Theoretically calculated errors of lidar estimation of the mean wind velocity are compared with results of the corresponding field experiments, as well as the results of simultaneous measurements of the wind velocity and direction by traditional wind sensors and pulsed CDL at very low SNRs. Vertical profiles of the wind velocity and direction retrieved from CDL data are presented. The possibility of measuring the global distribution of the wind field over the Earth by a coherent Doppler lidar from a satellite is studied numerically.

Chapter 4 studies the possibility of obtaining information about wind turbulence from data measured by cw and pulsed CDLs. Methods for estimation of turbulence parameters (wind velocity variance, dissipation rate of turbulence energy, and outer scale of turbulence) from lidar wind data using the width of the Doppler spectrum, from variance, temporal spectrum, temporal structure function, and longitudinal and transversal spatial structure functions of wind velocity fluctuations are considered. Ranges of applicability of these methods are determined. A comparative analysis of simultaneous measurements of the dissipation rate of turbulence energy by sonic anemometers and CDL is performed. Vertical profiles of wind turbulence parameters retrieved from lidar data are presented. The possibility of detecting turbulent zones in the clear sky by an airborne coherent Doppler lidar is analyzed numerically.

Chapter 5 presents the results of experimental investigations of the influence of atmospheric turbulence and wind on the wake vortex behavior and evolution. Methods for measurement of wake vortex parameters by cw CO_2 coherent lidars and pulsed 2-μm CDLs are considered. The results of simultaneous wake vortex measurements by two cw and one pulsed lidars are considered, and the measurement error is determined. Results of lidar investigations of wakes in atmospheric surface and boundary layers and in the free atmosphere are presented. These results are used to determine the empirical dependence of wake lifetime on the dissipation rate of atmospheric turbulence energy. Results of lidar investigation of atmosphere effect on a wind turbine wake are presented as well.

References

[1]   Kolmogorov, A.N., "Local structure of turbulence in incompressible viscous fluid at very large Reynolds numbers," *Doklady AN SSSR*, Vol. 30, No. 4, 1941, pp. 299–303.

[2]   Kolmogorov, A.N., "Scattering of energy at locally isotropic turbulence," *Doklady AN SSSR*, Vol. 32, No. 1, 1941, pp. 19–21.

[3]   Obukhov, A.M., "On the distribution of energy in the spectrum of turbulent flow," *Izvestiya AN SSSR, Ser. Geogr. i Geofiz.*, No. 4–5, 1941, pp. 453–463.

[4]   Obukhov, A.M., "Statistical description of continuous fields," *Trudy Geofiz. Inst. AN SSSR*, No. 24(151), 1954, pp. 3–42.

[5]   Obukhov, A.M., and Yaglom A.M., "Microstructure of turbulent flow," *Prikl. Matematika I Mekhanika*, Vol. 15, Issue 1, 1951, pp. 3–26.

[6]   Gurvich, A.S., "Spectra of pulsations of the vertical component of wind velocity and their relation with micrometeorological conditions," *Tr. IFA AN SSSR*, No. 4, 1962, pp. 101.

[7]   Gurvich, A.S., "Measurement of the asymmetry coefficient of distribution of velocity difference in the atmospheric surface layer," *DAN SSSR*, Vol. 134, No. 5, 1960, p.1073.

[8]   Gurvich, A.S., "Frequency spectra and probability distribution functions of the vertical wind component," *Izvestiya AN SSSR (Ser. Geofiz.)*, No. 7, 1960, p.1042.

[9]   Tsvang, L.R., "Some characteristics of spectra of temperature pulsations in the atmospheric boundary layer," *Izvestiya AN SSSR (Ser. Geofiz.)*, No. 10, 1963, p. 1594.

[10]   Tsvang, L.R., "Measurement of frequency spectra of temperature pulsations in the atmospheric surface layer," *Izvestiya AN SSSR (Ser. Geofiz.)*, No. 8, 1960, p. 1252.

[11]   Tsvang, L.R., "Measurements of turbulent heat fluxes and spectra of temperature fluctuations," *Trudy IFA AN SSSR Trudy IFA AN SSSR*, No. 4, 1962, p. 137.

[12]   Gurvich, A.S., and Kravchenko T.K., "On spectrum of temperature fluctuations in the range of small scales," *Trudy IFA AN SSSR*, No. 4, 1962, p. 144.

[13]   Tsvang, L.R., et al., "Measurement of some characteristics of turbulence in the lower 300-m atmospheric layer," *Izvestiya AN SSSR (Ser. Geofiz.)*, No. 5, 1963, p. 769.

[14]   Tsvang, L.R., "Measurement of temperature spectra in the free atmosphere," *Izvestiya AN SSSR (Ser. Geofiz.)*, No. 11, 1960, p. 76.

[15]   Zubkovskii, S.L., "Experimental investigation of spectra of pulsations of the vertical wind component in the free atmosphere," *Izvestiya AN SSSR (Ser. Geofiz.)*, No. 8. 1963, p. 1285.

[16]   Belinskii, V.A., *Dynamic Meteorology*, OGIZ, Moscow-Leningrad, 1948, p. 703.

[17]   Hinze, J.O., *Turbulence: An Introduction to Its Mechanism and Theory*, McGraw-Hill, New York, 1959.

[18]   Lumley, J.L., and Panofsky, H.A., *The Structure of Atmospheric Turbulence*, Interscience Publishers, New York, 1964.

[19]   Monin, A.S., and Yaglom, A.M., *Statistical Fluid Mechanics, Volume II: Mechanics of Turbulence*, MIT Press, Cambridge, MA, 1971.

[20]   Tatarskii, V.I., *Wave Propagation in a Turbulent Medium*, McGraw-Hill, New York, 1961.

[21]   Zilitinkevich, S.S., *Dynamics of Atmospheric Boundary Layer*, Gidrometeoizdat, Leningrad, 1970, p. 292.

[22]   Laikhtman, D.L., *Physics of Atmospheric Boundary Layer*, Gidrometeoizdat, Leningrad, 1970, p. 342.

[23]   Vinnichenko, N.K., et al., *Turbulence in the Free Atmosphere*, Gidrometeoizdat, Leningrad, 1976, p. 288.

[24]   Nieuwstadt, F.T.M., and Van Dop, H., *Atmospheric Turbulence and Air Pollution*, Hague, The Netherlands, 21–25 September 1981, p. 351.

[25]   Panofsky, H.A., and Dutton, J.A., *Atmospheric Turbulence*, John Wiley & Sons, New York, 1983, p. 397.

[26]   Stull, R.B., *An Introduction to Boundary Layer Meteorology*, Kluwer Academic Publishers. Dordrecht, 1988, p. 666.

[27]   Obukhov, A.M., *Turbulence and Dynamics of the Atmosphere*, Gidrometeoizdat, Leningrad, 1988, p. 413.

[28]   Monin, A.S., *Theoretical Principles of Geophysical Hydrodynamics*, Gidrometeoizdat, Leningrad, 1988, p. 424.

[29] Byzova, N.L., Ivanov, V.N., and Garger, E.K., *Turbulence in Atmospheric Boundary Layer*, Gidrometeoizdat, Leningrad, 1989. p. 263.

[30] Kaimal, J.C., and Finnigan, J.J., *Atmospheric Boundary Layer Flows. Their Structure and Measurement*, Oxford University Press, New York, 1994, p. 289.

[31] Kurbatskii, A.F., *Introduction to Simulation of Turbulence Transfer of Impulse and Scalar*, GEO, Novosibirsk, 2007, p. 327.

[32] Volkovitskaya, Z.I., and Ivanov, V.N., "Dissipation of turbulence energy in the atmospheric boundary layer," *Izv. AN SSSR. Fiz. Atmos. Okeana*, Vol. 6, No. 5, 1970, pp. 435–444.

[33] Byzova, N.L., Ivanov, V.N., and Matskevich, M.K., "Measurement of the vorticity components within the lower 300-m atmospheric layer," *Izvestiya, Atmospheric and Oceanic Physics*, Vol. 32, No. 3, 1996, pp. 298–302.

[34] Byzova, N.L., "Study of large vortices and dissipation of energy at neutral and weakly unstable stratification," in *Problems of Atmospheric Physics: Collection of Papers*, Gidrometeoizdat, St. Petersburg, 1998, pp. 227–246.

[35] Ivanov, V.N., "Peculiarities in conditions of appearance and structure of convective cells in the atmospheric boundary layer," in *Problems of Atmospheric Physics: Collection of Papers*, Gidrometeoizdat, St. Petersburg, 1998, pp. 467–487.

[36] Shur, G.N., et al., "Experience of studying stratospheric thermodynamics in high latitudes of the Northern Hemisphere from M-55 Geofizika Flying Laboratory," *Russian Meteorology and Hydrology*, No. 8, 2006, pp. 43–53.

[37] Shur, G.N., "Spectrum of continuous turbulence and coherent structures," *TsAGI Science Journal*, Vol. XXIV, Issue 3, 1993.

[38] Shur, G.N., and Volkov, V.V., "Coherent clusters in zones of intense atmospheric turbulence," *Russian Meteorology and Hydrology*, No. 1, 1999.

[39] Shur, G.N., et al., "Scales and energy of wind mesostructures in the low-latitude atmosphere," *Russian Meteorology and Hydrology*, No. 6, 2001, pp. 22–30.

[40] Shur, G.N., Lepukhov, B.N., and Sokolov, L.A., "Mesoscale structure of wind and temperature fields in the stratosphere of the Southern Hemisphere high latitudes," *Russian Meteorology and Hydrology*, No. 5, 2003, pp. 40–46.

[41] Babkin, V.I., et al., *Aircraft Flight Safety Systems*, Nauka, Moscow, 2008, p. 373.

[42] Ginevskii, A.S., and Zhelannikov, A.I., *Aircraft Wakes*, Fizmatlit, Moscow, 2008, p. 170.

[43] Gerz, T., Holzäpfel, F., and Darracq, D., "Commercial aircraft wake vortices," *Progress in Aerospace Sciences*, Vol. 38, 2002, pp. 181–208.

[44] Crow, S.C., "Stability theory for a pair of trailing vortices," *AIAA Journal*, Vol. 8, No. 12, 1970, pp. 2172–2179.

[45] Brashears, M.R., and Hallock, J.N., "Aircraft wake vortex transport model," *Journal of Aircraft*, Vol. 11, No. 5, 1974, pp. 256–272.

[46] Crow, S.C., and Bate, Jr., E.R., "Lifespan of trailing vortices in a turbulent atmosphere," *Journal of Aircraft*, Vol. 13, No. 7, 1976, pp. 476–482.

[47] Hecht, A.M., et al., "Turbulent vortices in stratified fluids," *AIAA Journal*, Vol. 18, No. 7, 1980, pp. 738–746.

[48] Greene, G.C., "An approximate model of vortex decay in the atmosphere," *Journal of Aircraft*, Vol. 23, No. 7, 1986, pp. 566–573.

[49] Sarpkaya, T., and Daly, J.J., "Effect of ambient turbulence on trailing vortices," *Journal of Aircraft*, Vol. 24, No. 61987, pp. 399–403.

[50] Robins, R.E., and Delisi, D.P., "Numerical study of vertical shear and stratification effect on the evolution of a vortex pair," *AIAA Journal*, Vol. 28, No. 4, 1990, pp. 661–669.

[51] Schilling, V., Siano, S., and Elting, D., "Dispersion of aircraft emissions due to wake vortices in stratified shear flows: A two-dimensional numerical study," *Journal of Geophysical Research*, Vol. 101, No. D15, 1996, pp. 20,965–20,974.

[52]   Hofbauer, T., and Gerz, T., "Effect of nonlinear shear on the dynamics of a counter-rotating vortex pair," in *Proceedings of the First International Symposium for Turbulence and Shear Flow Phenomena*, Santa Barbara, CA, USA, 12–15 September 1999.

[53]   Holzäpfel, F., "Probabilistic two-phase wake vortex decay and transport model," *Journal of Aircraft*, Vol. 40, No. 2, 2003, pp. 323–331.

[54]   Holzäpfel, F., et al., "Analysis of wake vortex decay mechanisms in the atmosphere," *Aerospace Science and Technology*, Vol. 7, No. 4, 2003, pp. 263–275.

[55]   Holzäpfel, F., and Robins, R.E., "Probabilistic two-phase aircraft wake-vortex model: Application and assessment," *Journal of Aircraft*, Vol. 41, No. 1, 2004, pp. 1–10.

[56]   Voevodin, A.V., et al., "Evolution of jet-vortical wake of passenger aircraft," *Aeromekhanika i Gazovaya Dinamika*, No. 4, 2004, pp. 23–31.

[57]   Vyshinskii, V.V., and Sudakov, G.G., *Aircraft Wake in the Turbulent Atmosphere*, TsAGI Proceedings, Moscow, Issue 2667, 2005, 155 p.

[58]   Klyatskin, V.I., *Stochastic Equations and Waves in Randomly Inhomogeneous Media*, Nauka, Moscow, 1980, 366 p.

[59]   Gurvich, A.S., et al., *Laser Radiation in the Turbulent Atmosphere*, Moscow: Nauka, 1976, 280 p.

[60]   Rytov, S.M., Kravtsov, Yu.A., and Tatarskii, V.I., *Introduction to Statistical Radiophysics. Part 2. Random Fields*, Nauka, Moscow, 1978, 464 p.

[61]   Gochelashvili, K.S., and Shishov, V.I., "Waves in randomly inhomogeneous media," *Itogi Nauki i Tekhniki. Radiphysics. Physical Principles of Electronics. Acoustics*, Vol. 1, VINITI, Moscow, 1981, 144 p.

[62]   Mironov, V.L., *Laser Beam Propagation in the Turbulent Atmosphere*, Nauka, Novosibirsk, 1981, 248 p.

[63]   Banakh, V.A., and Mironov, V.L., *Lidar in a Turbulent Atmosphere*, Artech House, Norwood, MA, 1987, 185 p.

[64]   Zuev, V.E., Banakh, V.A., and Pokasov, V.V., *Modern Problems of Atmospheric Optics. Part 5. Optics of the Turbulent Atmosphere*, Gidrometeoizdat, Leningrad, 1988, 272 p.

[65]   Grechko, G.M., et al., "Scintillations and random refraction during occultations by terrestrial atmosphere," *J. Opt. Soc. Am. A.*, Vol. 2, No. 12, 1985, pp. 2120–2123.

[66]   Aleksandrov, A.P., et al., "Spectra of temperature variations in the stratosphere from observations of star scintillations from space," *Izvestiya, Atmospheric and Oceanic Physics*, Vol. 26, No. 1, 1990, pp. 5–16.

[67]   Grechko, G.M., et al., "Observations of Atmospheric Turbulence at Altitudes of 20–70 km," *Trans. (Doklady) of the Russian Academy of Sciences/Earth Science Section*, Vol. 357A, No. 9, 1997, pp. 1382–1385.

[68]   Gurvich, A.S., et al., "Studying the turbulence and internal waves in the stratosphere from spacecraft observations of stellar scintillation: I. Experimental technique and analysis of the scintillation variance," *Izvestiya, Atmospheric and Oceanic Physics*, Vol. 37, No. 4, 2001, pp. 436–451.

[69]   Kan, V., Dalaudier, F., and Gurvich, A.S., "Chromatic refraction during star occultations with GOMOS. Part 2: Statistical properties of scintillations," *Appl. Opt.*, Vol. 40, No. 8, 2001, pp. 878–889.

[70]   Gurvich, A.S., and Kan, V., "Structure of air density irregularities in the stratosphere from spacecraft observations of stellar scintillation: 1. Three-dimensional spectrum model and recovery of its parameters," *Izvestiya, Atmospheric and Oceanic Physics*, Vol. 39, No. 3, 2003, pp. 300–310.

[71]   Gurvich, A.S., and Kan, V., "Structure of air density irregularities in the stratosphere from spacecraft observations of stellar scintillation: 2. Characteristic scales, structure characteristics, and kinetic energy dissipation," *Izvestiya, Atmospheric and Oceanic Physics*, Vol. 39, No. 3, 2003, pp. 311–321.

[72] Gurvich, A.S., and Fedorova, O.V., "Reconstruction of turbulence parameters under strong scintillation," *Atmos. Oceanic Opt.*, Vol. 21, No. 2, 2008, pp. 96–101.

[73] Granberg, I.G., et al., "A Study of the Spatial Structure of the Atmospheric Boundary Layer with a Doppler-Sodar Network," *Izvestiya, Atmospheric and Oceanic Physics*, Vol. 45, No. 5, 2009, pp. 541–548.

[74] Yushkov, V.P., et al., "Experience in Measuring the Wind-Velocity Profile in an Urban Environment with a Doppler Sodar," *Izvestiya, Atmospheric and Oceanic Physics*, Vol. 43, No. 2, 2007, pp. 168–180.

[75] Doviak, R.J., and Zrnic, D.S., *Doppler radar and weather observations*, Academic Press, San Diego, 1984, 458 p.

[76] Zrnic, D.S., "Estimation of spectral moments for weather echoes," *IEEE Trans. on Geoscience Electronics*, Vol. GE-17, No. 4, 1979, pp. 113–128.

[77] Maharatra, P.R., and Zrnic, D.S., "Practical algorithms for mean velocity estimation in pulse Doppler weather radars using a small numbers of samples," *IEEE Trans. on Geoscience Electronics*, Vol. GE-21, No. 4, 1983, pp. 491–501.

[78] May, P.T., and Strauch, R.G., "An examination of wind profiler signal processing algorithms," *Journal of Atmospheric and Oceanic Technology*, Vol. 6, 1989, pp. 731–735.

[79] Hildebrand, P.H., and Sekhon, R.S., "Objective determination of the noise level in Doppler spectra," *Journal of Applied Meteorology*, Vol. 13, No. 10, 1974, pp. 808–811.

[80] Zrnic, D.S., "Simulation of weather-like Doppler spectra and signals," *Journal of Applied Meteorology*, Vol. 14, No. 6, 1975, pp. 619–620.

[81] Naihanson, F.E., *Radar Design Principles*, McGraw-Hill, New York, 1980, 626 pp.

[82] Gorelik, A.G., and Sterlyadkin, V.V., "Wind sounding of the atmosphere with cw Doppler systems," *Izv. AN SSSR. Fiz. Atmos. Okeana*, Vol. 22, No. 7, 1986, pp. 720–727.

[83] Gorelik, A.G., and Sterlyadkin, V.V., "Doppler tomography in radar meteorology," *Izv. AN SSSR. Fiz. Atmos. Okeana*, Vol. 26, No. 1, 1990, pp. 47–54.

[84] Kononov, M.A., and Sterlyadkin, V.V., "Estimation of efficiency of calculated meteorological potential of wind Doppler radar station," in *Radiophysical Methods in Remote Sensing of Media: Collection of Papers of Fourth All-Russia Scientific School and Conference*, Murom, Russia, June 30–July 3, 2009, MIVl GU Press, 2009, pp. 93–97.

[85] Kononov, M.A., and Sterlyadkin, V.V., "Calculation of potential and estimation of capabilities of wind meteorological radar station of millimeter range," in *Nauchnyi Vestnik MGTU GA. Ser. Radiofiz. and Radiotekhn.*, MGTU GA, Moscow, No. 158, 2010.

[86] Kashcheev, B.L., Proshkin, E.G., and Lagutin, M.F. (Eds.), *Remote Methods and Tools for Investigation of Processes in the Earth's Atmosphere*, KhNURE, Biznes Inform, Kharkov, 2002, 426 p.

[87] Huffaker, R.M., "Laser Doppler detection systems for gas velocity measurement," *Applied Optics*, Vol. 9, No. 5, 1970, pp. 1026–1039.

[88] Lawrence, T.R., et al., "A laser velocimeter for remote wind sensing," *Review of Scientific Instruments*, Vol. 43, No. 3, 1972, pp. 512–518.

[89] Bilbro, J.M., "Atmospheric laser Doppler velocimetry: an overview," *Optical Engineering*, Vol. 20, No. 12, 1980, pp. 2048–2054.

[90] Köpp, F., Schwiesow, R.L., and Werner, Ch., "Remote measurements of boundary layer wind profiles using a cw Doppler lidar," *Journal of Climate Applied Meteorology*, Vol. 23, No. 1, 1984, pp. 148–158.

[91] Hall, F.F., et al., "Wind measurement accuracy of the NOAA pulsed infrared Doppler lidar," *Applied Optics*, Vol. 23, No. 15, 1984, pp. 2503–2506.

[92] Kane, T.J., et al., "Coherent laser radar at 1.06 μm using Nd:YAG lasers," *Optics Letters*, Vol. 12, 1987, pp. 232–241.

[93] Kavaya, M.J., et al., "Remote wind profiling with a solid-state Nd:YAG coherent lidar systems," *Optics Letters*, Vol. 14, 1989, pp. 776–778.

[94] Chanin, M.L., et al., "Doppler lidar for measuring winds in the middle atmosphere," *Geophys. Re. Lett.*, Vol. 16, 1989, pp. 1273–1276.

[95] Rees, D., and McDermid, I.S., "Doppler lidar atmospheric wind sensor: Reevaluation of a 355-nm incoherent Doppler lidar," *Applied Optics*, Vol. 29, 1990, pp. 4133–4157.

[96] Henderson, S.W., et al., "Eye-safe coherent laser radar system at 2 μm using Tm. Ho: YAG lasers," *Optics Letters*, Vol. 16, 1991, pp. 773–775.

[97] Henderson, S.W., et al., "Coherent laser radar at 2 μm using solid-state lasers," *IEEE Trans. Geosci. Remote Sens.*, Vol. 31, No. 1, 1993, pp. 4–15.

[98] Hawley, J.G., et al., "Coherent launch-site atmospheric wind sounder: theory and experiment," *Applied Optics*, Vol. 32, 1993, pp. 4557–4567.

[99] Gordienko, V.M., et al., "Coherent CO_2 lidars for measuring wind velocity and atmospheric turbulence," *Optical Engineering*, Vol. 33, No. 10, 1994, pp. 3206–3213.

[100] Rahm, S., "Measurement of a wind field with an airborne continuous-wave lidar," *Optics Letters*, Vol. 20, 1995, pp. 581–599.

[101] Huffaker, R.M., and Hardesty, R.M., "Remote sensing of atmospheric wind velocities using solid-state and CO_2 coherent laser systems," *Proc. IEEE*, Vol. 84, 1996, pp. 181–204.

[102] Drobinski, Ph., Dabas, A., and Flamant, P.H., "Remote measurements of turbulent wind spectra by heterodyne Doppler lidar technique," *Journal of Applied Meteorology*, Vol. 39, 2000, pp. 2434–2451.

[103] Werner, Ch., et al., "Wind Infrared Doppler Lidar Instrument," *Optical Engineering*, Vol. 40, No. 1, 2001, pp. 115–125.

[104] Reitebuch, O., et al., "Experimental validation of wind profiling performed by the Airborne 10 μm heterodyne Doppler lidar WIND," *Journal of Atmospheric and Oceanic Technology*, Vol. 18, No. 8, 2001, pp. 331–1344.

[105] Reitebuch, O., et al., "Determination of air flow across the alpine ridge by a combination of airborne Doppler lidar, routine radiosounding and numerical simulation," *Quart. J. Roy. Meteorol. Soc.*, Vol. 129, 2003, pp. 715–728.

[106] Ando, T., et al., "Development of low cost all coherent Doppler LIDAR (CDL) system," in *Proceedings of the 13th Coherent Laser Radar Conference*, Kamakura, Japan, 2005, pp. 170–173.

[107] Hannon, S.M., Pelk, J.V., and Henderson, S.W., "Recent wind and aerosol measurements using WindTracer," in *Proceedings of the 13th Coherent Laser Radar Conference*, Kamakura, Japan, 2005, pp. 84–87.

[108] Ando, T., Kameyama, S., and Hirano, Y., "All-fiber coherent Doppler LIDAR technologies at Mitsubishi Electric Corporation," in *Proceedings of the 14th International Symposium for the Advancement of Boundary Layer Remote Sensing*, IOP Conference Series, *Earth and Environmental Science*, Vol. 1, 2008, doi:10.1088/1755-1307/1/1/012011.

[109] Davis, J.C., et al., "Doppler lidar measurements of boundary layer winds and sensible heat flux," in *Proceedings of the 14th International Symposium for the Advancement of Boundary Layer Remote Sensing*, IOP Conference Series, *Earth and Environmental Science*, Vol. 1, 2008, doi:10.1088/1755-1307/1/1/012029.

[110] Hannon, S.M., "Wind resource assessment using long range pulsed Doppler lidar," in *Proceedings of the 15th Coherent Laser Radar Conference*, Toulouse, France, 2009, pp. 59–62.

[111] Parmentier, R., et al., "Windcube™ pulsed lidar compact wind profiler: Overview on more than two years of comparison with calibrated sensors at different location," in *Proceedings of the 15th Coherent Laser Radar Conference*, Toulouse, France, 2009, pp. 267–270.

[112] Frehlich, R.G., "Conditions for optimal performance of monostatic coherent laser radar," *Optics Letter*, Vol. 15. 1990, pp. 643–645.

[113] Frehlich, R.G., and Kavaya, M.J., "Coherent laser radar performance for general atmospheric turbulence," *Applied Optics*, Vol. 30, 1991, pp. 5325–5337.

[114] Rye, B.J., and Frehlich, R.G., "Optimal truncation and optical efficiency of an apertured coherent lidar focused on an incoherent backscatter target," *Applied Optics*, Vol. 31, No. 15, 1992, pp. 2891–2899.

[115] Frehlich, R.G., "Effects of refractive turbulence on coherent laser radar," *Applied Optics*, Vol. 32, No. 12, 1993, pp. 2122–2139.

[116] Frehlich, R.G., "Optimal local oscillator field for a Monostatic coherent laser radar with a circular aperture," *Applied Optics*, Vol. 32, No. 24, 1993, pp. 4569–4577.

[117] Frehlich, R.G., "Cramer-Rao bound for Gaussian random processes and applications to radar processing of atmospheric signals," *IEEE Trans. on Geoscience and Remote Sensing*, Vol. 31, No. 6, 1993, pp. 1123–1131.

[118] Frehlich, R.G., "Coherent Doppler lidar signal covariance including wind shear and wind turbulence," *Applied Optics*, Vol. 33, No. 27, 1994, pp. 6472–6481.

[119] Frehlich, R.G., "Heterodyne efficiency for a coherent laser radar with diffuse or aerosol targets," *Journal of Modern Optics*, Vol. 41, No. 11, 1994, pp. 2115–2129.

[120] Frehlich, R.G., and Yadlowsky, M.J., "Performance of mean-frequency estimators for Doppler radar and lidar," *Journal of Atmospheric and Oceanic Technology*, Vol. 11, No. 5, 1994, pp. 1217–1230.

[121] Frehlich, R.G., Hannon, S.M., and Henderson, S.W., "Performance of a 2-μm coherent Doppler lidar for wind measurements," *Journal of Atmospheric and Oceanic Technology*, Vol. 11, No. 6, 1994, pp. 1517–1528.

[122] Frehlich, R.G., "Comparison of 2- and 10-μm coherent Doppler lidar performance," *Journal of Atmospheric and Oceanic Technology*, Vol. 12, No. 2, 1995, pp. 415–420.

[123] Frehlich, R.G., "Simulation of coherent Doppler lidar performance in the weak-signal regime," *Journal of Atmospheric and Oceanic Technology*, Vol. 13, No. 6, 1996, pp. 646–658.

[124] Frehlich, R.G., "Effect of wind turbulence on coherent Doppler lidar performance," *Journal of Atmospheric and Oceanic Technology*, Vol. 14, No. 2, 1997, pp. 54–75.

[125] Frehlich, R.G., Hannon, S.M., and Henderson, S.W., "Coherent Doppler lidar measurements of winds in the weak signal regime," *Applied Optics*, Vol. 36, No. 15, 1997, pp. 3491–3499.

[126] Frehlich, R.G., Hannon, S.M., and Henderson, S.W., "Coherent Doppler lidar measurements of wind field statistics," *Boundary-Layer Meteorology*, Vol. 86, No. 1, 1998, pp. 223–256.

[127] Frehlich, R.G., and Conman, L.B., "Coherent Doppler lidar signal spectrum with wind turbulence," *Applied Optics*, Vol. 38, No. 36, 1999, pp. 7456–7466.

[128] Frehlich, R.G., "Effect of refractive turbulence on ground-based verification of coherent Doppler lidar performance," *Applied Optics*, Vol. 39, No. 24, 2000, pp. 4237–4246.

[129] Frehlich, R.G., "Estimation of velocity error for Doppler lidar measurements," *Journal of Atmospheric and Oceanic Technology*, Vol. 18, No. 10, 2001, pp. 1628–1639.

[130] Frehlich, R.G., "Errors for space-based Doppler lidar wind measurements: Definition, performance and verification," *Journal of Atmospheric and Oceanic Technology*, Vol. 18, No. 11, 2001, pp. 1749–1772.

[131] Frehlich, R.G., and Cornman, L.B., "Estimating spatial velocity statistics with coherent Doppler lidar," *Journal of Atmospheric and Oceanic Technology*, Vol. 19, No. 3, 2002, pp. 355–366.

[132] Frehlich, R.G., and Kavaya, M.J., "Comment on "Heterodyne lidar returns in the turbulent atmosphere: Performance evaluation of simulated systems," *Applied Optics*, Vol. 41, No. 9, 2002, pp. 1595–1600.

[133] Frehlich, R.G., "Velocity error for coherent Doppler lidar with pulse accumulation," *Journal of Atmospheric and Oceanic Technology*, Vol. 21, No. 6, 2004, pp. 905–920.

[134] Frehlich, R.G., and Sharman R., "Maximum likelihood estimates of vortex parameters from simulated coherent Doppler lidar data," *Journal of Atmospheric and Oceanic Technology*, Vol. 22, No. 2, 2005, pp. 117–129.

[135] Frehlich, R.G., et al., "Measurements of boundary layer profiles in urban environment," *Journal of Applied Meteorology and Climatology*, Vol. 45, No. 6, 2006, pp. 821–837.

[136] Frehlich, R.G., and Kelley, N., "Measurements of wind and turbulence profiles with scanning Doppler lidar for wind energy applications," *IEEE Journal of Selected Topics in Applied Earth Observations and Remote Sensing (J-STARS)*, Vol. 1, 2008, pp. 42–47.

[137] Frehlich, R.G., and Kelley, N., "Coherent Doppler lidar for wind energy research," *Proceedings of the 15th Coherent Laser Radar Conference*, Toulouse, France, 2009.

[138] Huffaker, R.M., et al., "Feasibility studies for global wind measuring satellite system (Windsat): Analysis of simulated performance," *Applied Optics*, Vol. 23, 1984, pp. 2523–2536.

[139] Targ, R., et al., "Coherent lidar airborne windshear sensor: Performance evaluation," *Applied Optics*, Vol. 30, 1991, pp. 2013–2026.

[140] Post, M.J., et al., "Calibration of coherent lidar target," *Applied Optics*, Vol. 19, No. 16, 1980, pp. 2828–2832.

[141] Hardesty, R.M., et al., "Characteristics of coherent lidar returns from calibration targets and aerosols," *Applied Optics*, Vol. 20, No. 21, 1981, pp. 3763–3769.

[142] Hardesty, R.M., "Performance of a discrete spectral peak frequency estimator for Doppler wind velocity measurements," *IEEE Trans. on Geoscience and Remote Sensing*, Vol. GE-24, No. 5, 1986, pp. 777–783.

[143] Zhao, Y., Post, M.J., and Hardesty, R.M., "Receiving efficiency of monostatic pulsed coherent lidars. 1: Theory," *Applied Optics*, Vol. 29, No. 28, 1990, pp. 4111–4119.

[144] Ray, B.J., and Hardesty, R.M., "Discrete spectral peak estimation in incoherent backscatter heterodyne lidar. I: Spectral accumulation and Cramer-Rao lower bound," *IEEE Trans. on Geoscience and Remote Sensing*, Vol. 31, No. 1, 1993, pp. 16–27.

[145] Ray, B.J., and Hardesty, R.M., "Discrete spectral peak estimation in incoherent backscatter heterodyne lidar. II: Correlogram accumulation," *IEEE Trans. on Geoscience and Remote Sensing*, Vol. 31, No. 1, 1993, pp. 28–35.

[146] Baker, W.F., et al., "Lidar-Measured Wind from Space: A Key Component for Weather and Climate Prediction," *Bull. Amer. Meteorol. Soc.*, Vol. 76, No. 6, 1995, pp. 869–888.

[147] Ray, B.J., and Hardesty, R.M., "Detecting techniques for validating Doppler estimates in heterodyne lidar," *Applied Optics*, Vol. 36, No. 9, 1997, pp. 1940–1951.

[148] Wulfmeyer, V., et al., "2-μm Doppler lidar transmitter with high frequency stability and low chirp," *Optics Letters*, Vol. 27, No. 17, 2000, pp. 1228–1230.

[149] Hardesty, R.M., et al., "A potential NPOESS winds mission," *Proceedings of the 13th Coherent Laser Radar Conference*, 2005, pp. 45–48.

[150] Hannon, S.M., and Thomson, J.A., "Aircraft wake vortex detection and measurement with pulsed solid-state coherent laser radar," *Journal of Modern Optics*, Vol. 41, 1994, pp. 2175–2196.

[151] Hannon, S.M., et al., "Windshear, turbulence and wake vortex characterization using pulsed solid-state coherent lidar," in *SPIE Proc. Air Traffic Control Technology*, Orlando, FL, USA, 18–19 April 1995, Vol. 2464, pp. 94–102.

[152] Hannon, S.M., and Henderson, S.W., "Wind measurement applications of coherent lidar," *Review of Laser Engineering*, Vol. 23, 1995, pp. 124–130.

[153] Hannon, S.M., and Thomson, J.A., "Real time wake vortex detection, tracking and strength estimation with pulsed coherent lidar," *Proceedings of the 9th Coherent Laser Radar Conference*, Linköping, Sweden, 23–27 June 1997, pp. 202–205.

[154] Hannon, S.M., "Autonomous infrared Doppler radar: Airport surveillance applications," *Physics and Chemistry of the Earth*, Vol. 25, No. 10, 2000, pp. 1005–1011.

[155] Alldritt, M., et al., "The processing of digital signals by a surface acoustic wave spectrum analyzer," *Journal of Physics E: Scientific Instruments*, Vol. 11, 1978, pp. 1–4.

[156] Woodfield, A.A., and Vaughan, J.M., "Airspeed and wind shear measurements with an airborne CO_2 cw laser," *International Journal of Aviation Safety*, Vol. 1, No. 2, 1983, pp. 207–224.

[157] Foord, R., et al., "Precise comparison of experimental and theoretical SNRs in CO_2 laser heterodyne systems," *Applied Optics*, Vol. 23, No. 23, 1983, pp. 3787–3795.

[158] Keeler, R.J., et al., "An airborne air motion sensing system. Pat I: Concept and preliminary experiment," *Journal of Atmospheric and Oceanic Technology*, Vol. 4, No. 3, 1987, pp. 113–127.

[159] Constant, G., et al., "Coherent laser radar and the problem of aircraft wakes," *Journal of Modern Optics*, Vol. 41, No. 11, 1994, pp. 2153–2173.

[160] Harris, M., et al., "Aircraft wake vortices: a comparison of wind-tunnel data with field-trial measurements by laser radar," *Aerospace Science and Technology*, Vol. 4, No. 5, 2000, pp. 363–370.

[161] Vaughan, J.M., and Harris, M., "Lidar measurement of B747 wakes: Observation of a vortex within a vortex," *Aerospace Science and Technology*, Vol. 5, No. 6, 2001, pp. 409–411.

[162] Salamitou, P., Darde, F., and Flamant, P.H., "A semi-analytic approach for coherent laser radar system efficiency, the nearest-Gaussian approximation," *Journal of Modern Optics*, Vol. 41, 1994, pp. 2101–2113.

[163] Salamitou, P., Dabas, A., and Flamant, P.H., "Simulation in the time domain for heterodyne coherent laser radar," *Applied Optics*, Vol. 34, 1995, pp. 499–506.

[164] Dabas, A.M., Drobinski, P., and Flamant, P.H., "Chirp-induced bias in velocity measurements by a coherent Doppler CO_2 lidar," *Journal of Atmospheric and Oceanic Technology*, Vol. 15, 1998, pp. 407–415.

[165] Dabas, A.M., Drobinski, P., and Flamant, P.H., "Velocity biases of adaptive filter estimates in heterodyne Doppler lidar measurements," *Journal of Atmospheric and Oceanic Technology*, Vol. 17, 2000, pp. 1189–1202.

[166] Vaughan, J.M., et al., "Coherent Laser Radar in Europe," *Proceedings of IEEE*, Vol. 84, 1996, pp. 205–226.

[167] Werner, Ch., Köpp, F., and Schwiesow, R.L., "Influence of clouds and fog on LDA wind measurements," *Applied Optics*, Vol. 23, 1984, pp. 2482–2487.

[168] Köpp, F., Bachstein, F., and Werner, Ch., "On-line data system for a cw laser Doppler anemometer," *Applied Optics*, Vol. 23, 1984, pp. 2488–2491.

[169] Werner, Ch., "Fast sector scan and pattern recognition for a cw laser Doppler anemometer," *Applied Optics*, Vol. 24, No. 21, 1985, pp. 3557–3564.

[170] Schwiesow, R.L., Köpp, F., and Werner, Ch., "Comparison of cw-lidar-measured wind values by full conical scan, conical sector scan and two-point techniques," *Journal of Atmospheric and Oceanic Technology*, Vol. 2, No. 1, 1985, pp. 3–14.

[171] Biselli, E., and Werner, Ch., "Determination of the direction of motion on the basis of cw-homodyne laser Doppler radar," *Applied Optics*, Vol. 28, No. 5, 1989, pp. 915–920.

[172] Reitebuch, O., et al., "Airborne Doppler lidar observations and numerical simulations of 'alpine pumping' mesoscale flow," in *Proceedings of the 12th Coherent Laser Radar Conf.*, Bar Harbor, Maine, USA, 2003, pp. 102–105.

[173] Banakh, V., et al., "Wind turbulence parameters measurements over sea by Doppler lidar and radar," in *Proceedings of the 12th Coherent Laser Radar Conference*, Bar Harbor, Maine, USA, 2003, pp. 191–196.

[174] Werner, Ch., et al., "Wind and turbulence measurements over sea by Doppler lidar and SAR," *Atmos. Oceanic Opt.*, Vol. 17, No. 8, 2004, pp. 576–583.

[175] Banakh, V.A., and Werner, Ch., "Computer simulation of coherent Doppler lidar measurement of wind velocity and retrieval of turbulent wind statistics," *Optical Engineering*, Vol. 44, No. 7, 2005, pp. 1–19.

[176] Köpp, F., "Doppler lidar investigation of wake vortex transport between closely spaced runways," *AIAA Journal*, Vol. 32, No. 4, 1994, pp. 805–810.

[177] Köpp, F., "Wake-vortex characteristics of military-type aircraft measured at airport Oberpfaffenhofen using the DLR laser Doppler anemometer," *Aerospace Science and Technology*, Vol. 3, 1999, pp. 191–199.

[178] Harris, M., et al., "Wake vortex detection and monitoring," *Aerospace Science and Technology*, Vol. 6, 2002, pp. 325–331.

[179] Keane, M., et al., "Axial detection of aircraft wake vortices using Doppler lidar," *Journal of Aircraft*, Vol. 39, No. 5, 2002, pp. 850–861.

[180] Holzäpfel F., et al., "Strategies for circulation evaluation of aircraft wake vortices measured by lidar," *Journal of Atmospheric and Oceanic Technology*, Vol. 20, No. 8, 2003, pp. 1183–1195.

[181] Rahm, S., "Precursor experiment for an active true airspeed sensor," *Optics Letters*, 2001, Vol. 26, No. 6, pp. 319–321.

[182] Rahm, S., Simmet, R., and Wirth, M., "Airborne two micron coherent lidar wind profiles," in *Proceedings of the 12th Coherent Laser Radar Conference*, Bar Harbor, Maine, USA, 15–20 June 2003, pp. 94–97.

[183] Weissmann, M., et al., "Targeted observations with an airborne wind lidar," *Journal of Atmospheric and Oceanic Technology*, Vol. 22, No. 10, 2005, pp. 1706–1719.

[184] Banakh, V.A., et al., "Effect of dynamic turbulence of the atmospheric boundary layer on the accuracy of Doppler lidar wind-velocity measurements," *Atmos. Oceanic Opt.*, Vol. 6, No. 11, 1993, pp. 786–793.

[185] Köpp, F., et al., "Laser Doppler wind measurements in the planetary boundary layer," *Contributions to Atmospheric Physics*, Vol. 67, No. 4, 1994, pp. 269–286.

[186] Banakh, V.A., et al., "Representativeness of wind measurements with a cw Doppler lidar in the atmospheric boundary layer," *Applied Optics*, Vol. 34, No. 12, 1995, pp. 2055–2067.

[187] Banakh, V.A., et al., "Turbulence measurements with a cw Doppler lidar in the atmospheric boundary layer," *Atmos. Oceanic Opt.*, Vol. 8, No. 12, 1995, pp. 955–959.

[188] Banakh, V.A., et al., "Measurement of the turbulent energy dissipation rate with a scanning Doppler lidar," *Atmos. Oceanic Opt.*, Vol. 9, No. 10, 1996, pp. 849–853.

[189] Banakh, V.A., et al., "Fluctuation spectra of wind velocity measured with a Doppler lidar," *Atmos. Oceanic Opt.*, Vol. 10, No. 3, 1997, pp. 202–208.

[190] Werner, Ch., et al., "Intercomparison of laser Doppler wind measurements with other methods and forecast model," *Journal of Optics A: Pure and Applied Optics*, Vol. 7, No. 12, 1998, pp. 1473–1487.

[191] Banakh, V.A., et al., "Computer simulation of cw Doppler wind lidar operation in the turbulent atmosphere," *Atmos. Oceanic Opt.*, Vol. 12, No. 10, 1999, pp. 905–911.

[192] Banakh, V.A., et al., "Measurements of turbulent energy dissipation rate with a cw Doppler lidar in the atmospheric boundary layer," *Journal of Atmospheric and Oceanic Technology*, Vol. 16, No. 8, 1999, pp. 1044–1061.

[193] Werner, Ch., et al., "Update on the results of DLR/DWD/IAO impact study," presented at *NOAA-Working Group Meeting on Space-Based Lidar Winds*, Key West, FL, USA, 19–22 January 1999.

[194] Banakh, V.A., Werner, Ch., and Smalikho, I.N., "The effect of aerosol microstructure on the error in estimating wind velocity with a Doppler lidar," *Atmos. Oceanic Opt.*, Vol. 13, No. 8, 2000, pp. 685–691.

[195] Banakh, V.A., Werner, Ch., and Smalikho, I.N., "Effect of turbulent fluctuations of refractive index on the time spectrum of wind velocity measured by Doppler lidar," *Atmos. Oceanic Opt.*, Vol. 13, No. 9, 2000, pp. 741–746.

[196] Banakh, V.A., et al., "Multi-aperture coherent reception in the turbulent atmosphere," *Atmos. Oceanic Opt.*, Vol. 13, No. 10, 2000, pp. 850–854.

[197] Banakh, V.A., Smalikho, I.N., and Werner, Ch., "Effect of aerosol particle microstructure on statistics of cw Doppler lidar signal," *Applied Optics*, Vol. 39, No. 30, 2000, pp. 5393–5402.

[198] Banakh, V.A., Smalikho, I.N., and Werner, Ch., "Numerical simulation of effect of refractive turbulence on the statistics of a coherent lidar return in the atmosphere," *Applied Optics*, Vol. 39, No. 30, 2000, pp. 5403–5414.

[199] Banakh, V.A., et al., "Modeling of wind reconstruction from measurements with a space-borne coherent Doppler lidar," *Atmos. Oceanic Opt.*, Vol. 14, No. 10, 2001, pp. 848–855.

[200] Banakh, V.A., Werner, Ch., and Smalikho, I.N., "Remote sensing of clear sky turbulence using Doppler lidar. Numerical simulation," *Atmos. Oceanic Opt.*, Vol. 14, No. 10, 2001, pp. 856–863.

[201] Leike, I., et al., "Virtual Doppler lidar instrument," *Journal of Atmospheric and Oceanic Technology*, Vol. 18, No. 9, 2001, pp. 1447–1456.

[202] Smalikho, I.N., et al., "Laser remote sensing of the mean wind," *Atmos. Oceanic Opt.*, Vol. 15, No. 8, 2002, pp. 607–614.

[203] Banakh, V.A., et al., "Accuracy of estimates by the method of variational spectrum accumulation of the wind velocity in turbulent atmosphere from lidar data," *Atmos. Oceanic Opt.*, Vol. 16, No. 8, 2003, pp. 658–662.

[204] Köpp, F., et al., "Characterization of aircraft wake vortices by multiple-lidar triangulation," *AIAA Journal*, Vol. 41, No. 6, 2003, pp. 1081–1088.

[205] Köpp, F., Rahm, S., and Smalikho, I.N., "Characterization of aircraft wake vortices by 2-μm pulsed Doppler lidar," *Journal of Atmospheric and Oceanic Technology*, Vol. 21, No. 2, 2004, pp. 194–206.

[206] Banakh, V.A., et al., "Estimates of turbulence parameters from measurements of wind velocity with a pulsed coherent Doppler lidar," *Atmos. Oceanic Opt.*, Vol. 18, No. 12, 2005, pp. 955–957.

[207] Köpp, F., et al., "Comparison of wake-vortex parameters measured by pulsed and continuous-wave lidars," *Journal of Aircraft*, Vol. 42, No. 4, 2005, pp. 916–923.

[208] Smalikho, I.N., Köpp, F., and Rahm, S., "Measurement of atmospheric turbulence by 2-μm Doppler lidar," *Journal of Atmospheric and Oceanic Technology*, Vol. 22, No. 11, 2005, pp. 1733–1747.

[209] Rahm, S., Smalikho, I.N., and Köpp, F., "Characterization of aircraft wake vortices by airborne coherent Doppler lidar," *Journal of Aircraft*, Vol. 44, No. 3, 2007, pp. 799–805.

[210] Banakh, V.A., et al., "Measurement of atmospheric turbulence parameters by vertically-scanning pulsed coherent lidar," *Atmos. Oceanic Opt.*, Vol. 20, No. 12, 2007, pp. 1019–1023.

[211] Rahm, S., and Smalikho, I.N., "Aircraft wake vortex measurement with airborne coherent Doppler lidar," *Journal of Aircraft*, Vol. 45, No. 4, 2008, pp. 1148–1155.

[212] Smalikho, I.N., and Rahm, S., "Measurements of aircraft wake vortex parameters with a coherent Doppler lidar," *Atmos. Oceanic Opt.*, Vol. 21, No. 11, 2008, pp. 854–868.

[213] Smalikho, I.N., and Rahm, S., "Lidar investigation of the effect of wind and atmospheric turbulence on aircraft wake vortices," *Atmos. Oceanic Opt.*, Vol. 22, No. 12, 2009, pp. 1160–1169.

[214] Holzäpfel, F., et al., "The wake vortex prediction and monitoring system WSVBS—Part I: Design," *Air Traffic Control Quarterly*, Vol. 17, No. 4, 2009, pp. 301–322.

[215] Harris, M., et al., "Single-particle laser Doppler anemometry at 1.55 μm," *Applied Optics*, Vol. 40, No. 6, 2001, pp. 969–973.

[216] Yura, H.T., "Signal-to-noise ratio of heterodyne lidar systems in the presence of atmospheric turbulence," *Optica Acta*, Vol. 26, No. 5, 1979, 627–644.

[217] Clifford, S.F., and Wandzura, S., "Monostatic heterodyne lidar performance: The effect of the turbulent atmosphere," *Applied Optics*, Vol. 20, No. 3, 1981, pp. 514–516.

[218] Sonnenschein, C.M., and Horrigan, F.A., "Signal-to-noise relationship for coaxial systems that heterodyne backscatter from the atmosphere," *Applied Optics*, Vol. 10, No. 7, 1971, pp. 1600–1604.

[219] Belmonte, A., and Rye, B.J., "Heterodyne lidar returns in the turbulent atmosphere: Performance evaluation of simulated systems," *Applied Optics*, Vol. 39, No. 15, 2000, pp. 2401–2411.

[220] Belmonte, A., "Feasibility study for the simulation of beam propagation: Consideration of coherent lidar performance," *Applied Optics*, Vol. 39, No. 30, 2000, pp. 5426–5445.

[221] Menzies, R.T., "Doppler lidar atmospheric wind sensors: A comparative performance evaluation for global measurement applications from earth orbit," *Applied Optics*, Vol. 25, 1986, pp. 2546–2553.

[222] Petheram, J.C., Frohbeiter, G., and Rosenberg, A., "Carbon dioxide Doppler lidar wind sensor on a space station polar platform," *Applied Optics*, Vol. 28, 1989, pp. 834–839.

[223] Endemann, M., and Ingmann, P., "The European spaceborne Doppler wind lidar ADM-aeolus," in *Proceedings 13th Coherent Laser Radar Conference*, 2005, pp. 49–52.

[224] Lhermitte, R.M., and Atlas, D., "Precipitation motion by pulse Doppler," in *Proceedings of the 9th Weather Radar Conference*, Kansas City, MO, USA, 1961, pp. 218–223.

[225] Gurvich, A.S., "Influence of temporal evolution of turbulent inhomogeneities on frequency spectra," *Izv. AN SSSR. ser. Fiz. Atmos. Okeana*, Vol. 16, No. 4, 1980, pp. 345–354.

[226] Byzova, N.L., et al., "Joint measurements of wind velocity by Doppler lidar and high-tower anemometers," *Meteorol. Gidrol.*, No. 3, 1991, pp. 114–117.

[227] Eberhard, W.L., Cupp, R.E., and Healy, K.R., "Doppler lidar measurements of profiles of turbulence and momentum flux," *Journal of Atmospheric and Oceanic Technology*, Vol. 6, 1989, pp. 809–819.

[228] Gal-Chen, T., Xu, M., and Eberhard, W.L., "Estimations of atmospheric boundary layer fluxes and other turbulence parameters from Doppler lidar data," *Journal of Geophysical Research*, Vol. 97, No. D17, 1992, pp. 18,409–18,423.

[229] Davies, F., et al., "Doppler lidar measurements of turbulent structure function over an urban area," *Journal of Atmospheric and Oceanic Technology*, Vol. 21, No. 5, 2004, pp. 753–761.

[230] Banakh, V.A., and Falits, A.V., "Estimating the atmospheric turbulence parameters from the wind velocity measured with a pulsed coherent CO_2 Doppler lidar," *Atmos. Oceanic Opt.*, Vol. 17, No. 4, 2004, pp. 260–267.

[231] Thomson, J.A., and Hannon, S.M., "Wake vortex modeling for airborne and ground-based measurements using a coherent lidar," in *SPIE Proc. Air Traffic Control Technology*, Orlando, FL, USA, 18–19 April 1995. Vol. 2464, pp. 63–78.

CHAPTER 1

Statistics of CDL Echo Signal

1.1 Introduction

Laser Doppler techniques of liquid and gas flow velocimetry are based on detection of radiation scattered in a flow and estimation of the Doppler frequency shift caused by the motion of scatterers entrained by the flow. The range of laboratory measurements by laser Doppler anemometers covers velocities from 10^{-3} m/s to supersonic ones with a spatial resolution up to 1 mm. The accuracy of measurements can be so high that the relative error does not exceed 1% [1–3].

For measurement of wind fields, Doppler laser radars (or lidars) have been developed. There are two types of lidars: (1) incoherent Doppler lidar (IDL) and (2) coherent Doppler lidar (CDL). In the former, the radiation scattered by aerosol particles or air molecules is collected by a telescope and, after passage through an interferometer, directly detected by a photodetector in the chosen spectral channels. The Doppler frequency shift is determined from the measured spectrum [4–9]. In the case of CDL, the probing laser radiation backscattered by aerosol particles passes through the aperture of a telescope, mixes with the reference laser beam, and comes to the sensitive plate of the detector. The signal (echo signal) carrying the information about the Doppler frequency shift is separated by a narrowband frequency filter from measurements of the photocurrent arising in the detector circuit. Thus, CDL uses laser heterodyning [10]. The necessary condition of this heterodyning is the fulfillment of strict requirements on matching of wavefronts of the scattered radiation and the reference beam in the plane of the detector's sensitive plate. The heterodyne efficiency depends on the coherence area of the scattered radiation with respect to the area of the reference beam cross section [10–17].

The probing laser beam can be both continuous-wave or pulsed; that is, coherent Doppler lidars can be divided into two groups: (1) continuous-wave (cw) CDL and (2) pulsed CDL. In the case of cw CDL, the sensing volume is formed by focusing the laser beam to a preset distance. As the focal length increases, along with an increase in the measurement range, the sensing volume increases [18, 19]. For pulsed CDL, the sensing volume is determined by the probing pulse duration and the transverse dimension of the laser beam. In this case, the distance from a lidar to the center of the sensing volume is defined by the time interval between launching of the laser pulse into the atmosphere and the moment of registration of the sample of the echo signal.

The signal measured by a coherent Doppler lidar is a sum of the useful component (echo signal) and the noise component, whose level is mostly determined by the reference beam power. Since the accuracy of estimation of the Doppler frequency shift depends on the echo signal and noise power, the signal-to-noise ratio (SNR) is an important characteristic in terms of the possibility of sensing the atmosphere using a

coherent Doppler lidar. In the development of lidars and the procedures for processing raw data, it is also important to know the statistical properties of a lidar signal.

It has been shown [18, 19] with no regard for the atmospheric turbulence that a
cw CDL SNR in the near diffraction zone of the probing beam is independent of the
measurement range. The influence of turbulent fluctuations of the refractive index
of air on the SNR in the case of pulsed CDL was studied theoretically with the use
of the Huygens-Kirchhoff method [12, 14, 20]. In addition, the SNR was estimated
in the bistatic approximation, in which the correlation of optical waves propagating in the forward and backward directions is ignored [12, 21, 22]. However, the
accuracy of the approximations used remained unstudied.

Experimentally, fluctuations of the echo signal power of cw CDLs were studied in
[23, 24]. In particular, it was shown that if a lidar forms a small sensing volume, the
echo signal has non-Gaussian statistics, whereas for the probability density function
of a pulsed CDL's echo signal it was usually assumed *a priori* that it is described by
the normal distribution law [25, 26]. The influence of the sensing volume size and
turbulent fluctuations of the refractive index on the statistical properties of the lidar
echo signal in the case of cw and pulsed CDL was studied theoretically in [27–36].

This chapter presents the results of calculations of echo signal power characteristics: mean value (signal-to-noise ratio), relative variance, and time correlation
coefficient. Numerical simulation is used to take into account correctly the influence
of correlation between counterpropagating waves in the turbulent atmosphere on
statistical characteristics of the echo signal power in the case of pulsed CDL, as well
as to analyze the influence of the microstructure of aerosol particles on the probability density function of the echo signal amplitude and power in the case of cw
CDL. This chapter is based on the results reported in [27–36].

1.2   Coherent Detection and Governing Equations for CDL Echo Signals

Figure 1.1 shows the monostatic scheme of coherent detection of the radiation scattered
in the atmosphere. The elements are shown in the Cartesian coordinate system $\{z, \boldsymbol{\rho}\}$,
where $\boldsymbol{\rho} = \{x, y\}$ with an origin at the optical axis $z = 0$ in the plane of the transmit/
receive aperture. Vectors in the planes transverse to the optical axis and coinciding,
respectively, with the plane of the detector sensitive plate and the plane in front of and
behind the transmit/receive aperture of the telescope are denoted as $\boldsymbol{\rho}'''$, $\boldsymbol{\rho}''$, and $\boldsymbol{\rho}'$.

The complex amplitude of the probing beam at the point $\{z_i, \boldsymbol{\rho}_i\}$ and time t with
ignored extinction due to absorption and scattering can be written as [37]:

$$U_p(z_i, \boldsymbol{\rho}_i, t) = \int_{-\infty}^{+\infty} d^2\rho' A_t(\boldsymbol{\rho}') U_{0P}(\boldsymbol{\rho}', t - z_i/c) G(0, \boldsymbol{\rho}'; z_i, \boldsymbol{\rho}_i; t) \qquad (1.1)$$

where $\int d^2\rho$ denotes $\int dx \int dy$ and

$$A_t(\boldsymbol{\rho}) = \Pi(\boldsymbol{\rho}) \exp\left[-j\frac{\pi}{\lambda F}\rho^2\right] \qquad (1.2)$$

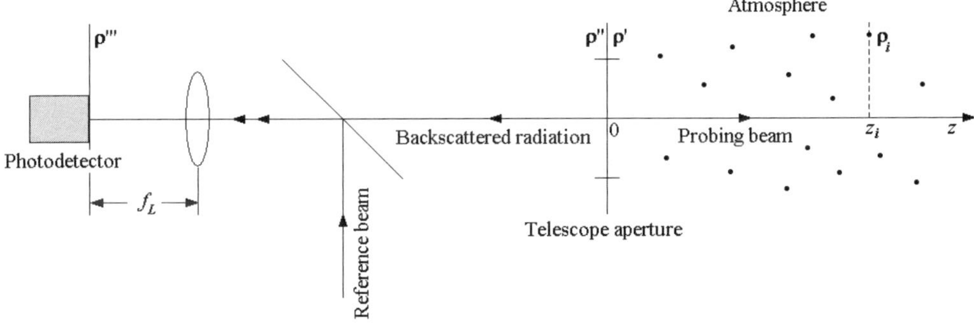

Figure 1.1   Coherent detection of the laser radiation backscattered in the atmosphere.

$\Pi(\rho)$ is the pupil function of the transmit/receive aperture of the telescope, $j = \sqrt{-1}$, λ is the optical wavelength, F is the focal length of the probing beam, $\rho^2 = x^2 + y^2$, $U_{0P}(\rho, t)$ is the field of the laser radiation at the aperture, c is the speed of light, and $G(0,\rho_a;z,\rho_b;t)$ is the Green's function describing the propagation of a spherical wave from point $\{0,\rho_a\}$ to point $\{z,\rho_b\}$ and satisfying the reciprocity relation

$$G(0,\rho_a;z,\rho_b) = G(z,\rho_b;0,\rho_a) \tag{1.3}$$

The complex amplitude of the wave field $U_B(\rho'',t)$ backscattered by N_s aerosol particles in the single scattering approximation [38–44] is described by the equation

$$U_B(\rho'',t) = A_t(\rho'')\lambda\sum_{i=1}^{N_s}\alpha_i U_P(z_i,\rho_i,t - z_i/c)G(z_i,\rho_i;0,\rho'';t) \tag{1.4}$$

where α_i is the backscatter amplitude of the ith particle.

In the case of heterodyning, the principle of superposition of the fields of backscattered $E_B(\rho'',t) = U_B(\rho'',t)\exp[-2\pi jvt]$ and reference $E_L(\rho'',t) = U_{0L}(\rho'')\exp[-2\pi j(v + f_I)t]$ radiation is true: $E(\rho'',t) = E_B(\rho'',t) + E_L(\rho'',t)$, where $v = c/\lambda$, f_I is an intermediate frequency, and $U_{0L}(\rho'')$ is the complex amplitude of the field of the reference cw radiation beam (local oscillator field). Then the complex amplitude of the resultant field $U(\rho'',t) = E(\rho'',t)\exp[2\pi jvt]$ can be written in the form

$$U(\rho'',t) = U_B(\rho'',t) + U_{0L}(\rho'')\exp[-2\pi jf_I t] \tag{1.5}$$

After the mixing, the radiation is focused by a lens with the focal length f_L to the sensitive plate of the photodetector. In this case, for the complex amplitude of the field, we have:

$$U(\rho''',t) = \frac{1}{j\lambda f_L}\exp\left[j\frac{\pi}{\lambda f_L}\rho''^2\right]\int_{-\infty}^{+\infty}d^2\rho''U(\rho'',t)\exp\left[-j\frac{2\pi}{\lambda f_L}\rho''\rho'''\right] \tag{1.6}$$

where $\rho_a\rho_b = x_a x_b + y_a y_b$. The power of radiation within the sensitive plate is determined as follows:

$$P_D(t) = \int_D d^2\rho''' I(\rho''', t) \tag{1.7}$$

where $I(\rho, t) = U(\rho, t)U^*(\rho, t) \equiv |U(\rho)|^2$ is the radiation intensity. The integration is performed over the sensitive plate. The plate dimensions exceed the transverse dimensions of the optical beam coming to the photodetector (full interception), and that is why in (1.6) we can substitute $\int_D$ with $\int_{-\infty}^{+\infty}$. Then, from (1.5) through (1.7) after the integration with respect to ρ''', we obtain

$$P_D(t) = P_L + P_B(t) + P_C(t) \tag{1.8}$$

where $P_L = \int_{-\infty}^{+\infty} d^2\rho'' |U_{0L}(\rho'')|^2$ is the reference beam power, $P_B(t) = \int_{-\infty}^{+\infty} d^2\rho'' |U_B(\rho'', t)|^2$ is the power of scattered radiation at incoherent its detection, and

$$P_C(t) = 2\,\mathrm{Re}\left\{ \exp[2\pi\, jf_L t] \int_{-\infty}^{+\infty} d^2\rho'' U_{0L}^*(\rho'') U_B(\rho'', t) \right\} \tag{1.9}$$

is the coherent component of the power, which is a result of the interference of the reference and backscattered waves and carries the information about the velocities of the backscattering particles. Because of the random character of probing beam propagation and scattering in the atmosphere, the photocurrent components P_B and P_C are random values.

On substituting (1.4) into (1.9) with an allowance made for (1.1) and (1.3), we have:

$$P_C(t) = 2\,\mathrm{Re}\left\{ \exp[2\pi\, jf_L t]\lambda \sum_{i=1}^{N_s} \alpha_i \tilde{U}_P(z_i, \rho_i, t)\tilde{U}_L(z_i, \rho_i, t) \right\} \tag{1.10}$$

where

$$\tilde{U}_P(z_i, \rho_i, t) = \int_{-\infty}^{+\infty} d^2\rho A_t(\rho) U_{0P}(\rho, t - 2z_i/c) G(0, \rho; z_i, \rho_i, t) \tag{1.11}$$

is the complex amplitude of the probing beam and

$$\tilde{U}_L(z_i, \rho_i, t) = \int_{-\infty}^{+\infty} d^2\rho A_t(\rho) U_{0L}^*(\rho) G(0, \rho; z_i, \rho_i, t) \tag{1.12}$$

is the complex amplitude of the equivalent reference beam propagating from the lidar to the atmosphere (or the backpropagated local oscillator field).

Since the airflow transfers scattering particles, their coordinates are functions of time ($\{z_i(t), \boldsymbol{\rho}_i(t)\}$). While a scattering particle stays within the sensing volume, its velocity is nearly constant. Taking into account that the longitudinal (along the optical axis) dimension of the lidar sensing volume far exceeds the transverse dimension, we can neglect the velocity dependence on transverse coordinates. Therefore, the coordinates $\{z_i(t), \boldsymbol{\rho}_i(t)\}$ can be represented in the form

$$\begin{cases} z_i(t) = z_i + (t - t_0)V_r(z_i) \\ \boldsymbol{\rho}_i(t) = \boldsymbol{\rho}_i + (t - t_0)\mathbf{V}_\perp(z_i) \end{cases} \tag{1.13}$$

where $V_r(z_i)$ is the radial velocity (projection of the velocity vector on the axis of the probing beam) and $\mathbf{V}_\perp(z_i)$ is the transverse component of the velocity vector of the ith particle at time t_0. The Green's function in a homogeneous medium has the form [37]

$$G\left(0, \boldsymbol{\rho}; z_i, \boldsymbol{\rho}_i\right) = \frac{1}{j\lambda z_i} \exp\left[j2\pi z_i/\lambda + j\frac{\pi}{\lambda z_i}\left(\boldsymbol{\rho} - \boldsymbol{\rho}_i\right)^2 \right] \tag{1.14}$$

Except for the first term inside the square brackets, in other terms the time dependence of z_i can be neglected, and we can take $t_0 = 0$ when using (1.13) and represent the Green's function in the form

$$G\left(0, \boldsymbol{\rho}; z_i(t), \boldsymbol{\rho}_i(t); t\right) = \exp\left[2\pi\, jt V_r(z_i)/\lambda \right]\tilde{G}\left(0, \boldsymbol{\rho}; z_i, \boldsymbol{\rho}_i(t); t\right) \tag{1.15}$$

where $\tilde{G}$ is the Green's function, in which the coordinate z_i is independent of time. After the substitution of (1.15) into (1.11) and (1.12), for the coherent component of the power with allowance made for (1.10), we have

$$P_C(t) = 2\,\mathrm{Re}\left\{ \lambda \sum_{i=1}^{N_s} \alpha_i \exp\left[2\pi\, j(f_I + f_{ri})t \right]\tilde{U}_P(z_i, \boldsymbol{\rho}_i(t), t)\tilde{U}_L(z_i, \boldsymbol{\rho}_i(t), t) \right\} \tag{1.16}$$

The motion of the ith particle causes a shift of the frequency f_{ri} defined by the Doppler equation

$$f_{ri} = 2V_r(z_i)/\lambda \tag{1.17}$$

Because the radiation is incident on the sensitive area of the detector, the electric current arises in the receiving system circuit due to the photo effect. Neglecting all noises except for the shot noise, the photocurrent J measured for the time τ_D can be written in the form

$$J = en/\tau_D \tag{1.18}$$

where e is the electron charge and n is the number of photons captured by electrons for time τ_D. The probability of the number of photoelectrons P_n obeys the Poisson statistics:

$$P_n = \frac{\langle n \rangle_p^n}{n!} \exp\left(-\langle n \rangle_p\right) \tag{1.19}$$

where

$$\langle n \rangle_p = \sum_{n=0}^{\infty} n P_n = \eta \frac{P_D \tau_D}{h\nu} \tag{1.20}$$

is the mean number of photoelectrons, η is the quantum efficiency (probability of photon capture by an electron), $h\nu$ is the photon energy, and h is Plank's constant. Let the photocurrent J be written as a sum

$$J = J_S + J_N \tag{1.21}$$

where, with allowance for (1.18) and (1.20),

$$J_S = \langle J \rangle_P = e\langle n \rangle_p / \tau_D = \frac{e\eta}{h\nu} P_D \tag{1.22}$$

is the signal component of the photocurrent and

$$J_N = J - \langle J \rangle_P = e(n - \langle n \rangle_p)/\tau_D \tag{1.23}$$

is the noise component of the photocurrent (shot noise). The power of the radiation incident on the photodetector sensitive area is determined primarily by the power of the reference beam P_L, and for time τ_D many photoelectrons n are always generated. Consequently, the noise probability density is Gaussian with zero mean and variance equal to the noise power $P_{JN} = \langle J_N^2 \rangle_P = e^2(\langle n^2 \rangle_P - \langle n \rangle_P^2)/\tau_D^2$ For the noise power in the passband B_D determined by the Nyquist frequency

$$B_D = 1/(2\tau_D) \tag{1.24}$$

with allowance for $\langle n^2 \rangle_p = \sum_{n=0}^{\infty} n^2 P_n = \langle n \rangle_p^2 + \langle n \rangle_p$, (1.20), and the approximate equality $P_D \approx P_L$, we can write [45]

$$P_{JN} = 2\frac{e^2 \eta B_D}{h\nu} P_L \tag{1.25}$$

By analogy with (1.8), we write the signal component of the photocurrent in the form $J_S = J_L + J_B + J_C$, where each of the terms is described by (1.22) with the corresponding substitution of the subscript D with L, B, and C. Since the coherent component of the photocurrent J_C bears the information about velocities of scattering particles, it can be separated through filtering in the passband B_F (for example, in the vicinity of the intermediate frequency f_I), and the result is the complex signal

$$J_F(t) = J_C(t) + J_{NF}(t) \tag{1.26}$$

where, according to (1.16),

$$J_C(t) = 2\frac{e\eta\lambda}{h\nu}\sum_{i=1}^{N_S}\alpha_i\exp[j\psi_i + 2\pi j(f_I + f_{ri})t]I_P^{1/2}(z_i,\rho_i(t),t)I_L^{1/2}(z_i,\rho_i(t),t) \tag{1.27}$$

and $\psi_i = \arg[\tilde{U}_P(z_i,\rho_i(t),t)] + \arg[\tilde{U}_L(z_i,\rho_i(t),t)]$, $I_P = |\tilde{U}_P|^2$ and $I_L = |\tilde{U}_L|^2$. The component $J_C(t)$ is proportional to the complex amplitude of the scattered wave, and in what follows it is referred to as an *echo signal*. The noise component of the photocurrent (or simply the noise) J_{NF} is also a complex value with zero mean and statistically independent real and imaginary parts, whose variances are identical and defined by (1.25) with the substitution of B_D with B_F.

In processing of raw lidar measurement data, the normalization of a signal to noise is often used. Multiplying $J_F(t)$ by $\exp(-2\pi jf_I t)/\sqrt{P_{JN}}$, we pass to the normalized complex signal $Z(t)$, which can be represented in the form

$$Z(t) = Z_S(t) + Z_N(t) \tag{1.28}$$

where, according to (1.26), (1.27), and (1.25) (in the last equation, B_D should be replaced with B_F),

$$Z_S(t) = \lambda\sqrt{\frac{2\eta}{h\nu B_F P_L}}\sum_{i=1}^{N_S}\alpha_i\exp[j\psi_i + 2\pi j f_{ri}t]I_P^{1/2}(z_i,\rho_i(t),t)I_L^{1/2}(z_i,\rho_i(t),t) \tag{1.29}$$

is the normalized echo signal, and $Z_N(t)$ is the normalized noise. With (1.28), we can write the power of the normalized signal $P(t) = (1/2)|Z(t)|^2$ measured at time t in the form

$$P(t) = P_S(t) + P_N(t) + \mathrm{Re}\{Z_S(t)Z_N^*(t)\} \tag{1.30}$$

where

$$P_S(t) = (1/2)|Z_S(t)|^2 \tag{1.31}$$

is the normalized power of the echo signal and

$$P_N(t) = (1/2)\left|Z_N(t)\right|^2 \tag{1.32}$$

is the normalized power of noise.

Since the probability density of phase ψ_i has a uniform distribution in the interval $[0, 2\pi]$, the mean value of the lidar signal is equal to zero ($\langle Z \rangle = 0$). Taking into account the normalization and statistical independence of Z_S and Z_N^*, from (1.30) we obtain for the mean normalized signal power $\langle P(t) \rangle$ the following equation:

$$\langle P(t) \rangle = \langle P_S(t) \rangle + 1 \tag{1.33}$$

where angle brackets refer to ensemble averaging.

In the averaging of $|Z_S(t)|^2$ with the use of (1.29), one should take into account the statistical independence of the phases ψ_i and ψ_k (when $i \neq k$) and can use the following equation [42]:

$$\left\langle \sum_{i=1}^{N_S} |\alpha_i|^2 \, F(z_i, \rho_i) \right\rangle = \int\limits_0^\infty dz \int\limits_{-\infty}^{+\infty} d^2\rho \, \beta_\pi(z) \langle F(z, \rho) \rangle \tag{1.34}$$

where

$$\beta_\pi = \rho_0 \langle \sigma_\pi \rangle \tag{1.35}$$

is the backscatter coefficient, ρ_0 is the concentration of aerosol particles (mean number of particles in the unit volume),

$$\langle \sigma_\pi \rangle = \int\limits_0^\infty da \, f_s(a) \sigma_\pi(a) \tag{1.36}$$

$f_s(a)$ is the aerosol particle size distribution function, and $\sigma_\pi(a) = |\alpha(a)|^2$ is the differential backscatter cross section of a particle with radius a. The equation we seek for the mean normalized power of the echo signal can be written in the form

$$\langle P_S(t) \rangle = \frac{\eta \lambda^2}{h v B_F P_L} \int\limits_0^\infty dz \int\limits_{-\infty}^{+\infty} d^2\rho \, \beta_\pi(z) \langle I_P(z, \rho, t) I_L(z, \rho) \rangle \tag{1.37}$$

In the general case, in (1.37) we can take into account the atmospheric transmission

$$T_A(z) = \exp\left[-\int\limits_0^z dz' \beta_t(z') \right] \tag{1.38}$$

where β_t is the radiation extinction coefficient due to absorption and scattering by air molecules and aerosol particles, by multiplying the integrand by $T_A^2(z)$.

The signal-to-noise ratio is an important characteristic for lidar sensing of the atmosphere. We define SNR as a ratio of the mean echo signal power $(1/2) \langle |J_C(t)|^2 \rangle$ to the mean noise power P_{JN} in the frequency band B_F. Then

$$\mathrm{SNR} = \langle P_S(t) \rangle \tag{1.39}$$

where $\langle P_S(t) \rangle$ is described by (1.37).

In the analysis of statistical properties of the echo signal, it is important to know what sensing volume is formed by a lidar and what is the concentration of the particles that are major contributors to the power of the echo signal $\langle P_S(t) \rangle$. According to (1.37), the echo signal power is proportional to the spatial integral of the function $F_I(z, \rho) = \beta_\pi(z) \langle I_P(z, \rho, t) I_L(z, \rho) \rangle \geq 0$. We determine the effective lidar sensing volume as follows:

$$V_{eff} = \int\limits_0^\infty dz \int\limits_{-\infty}^{+\infty} d^2\rho \, \frac{\beta_\pi(z)\langle I_P(z,\rho,t) I_L(z,\rho)\rangle}{\beta_\pi(R)\langle I_P(R,0,t) I_L(R,0)\rangle} \tag{1.40}$$

where $\{R,0\} = \{z_{max}, \rho_{max}\}$ is the maximum point of the function $F_I(z, \rho)$.

The mean number N_p of aerosol particles in the sensing volume is calculated as

$$N_p = \rho_0 V_{eff} \tag{1.41}$$

Scattering particles may have different sizes (backscatter amplitudes). Therefore, the contributions to the lidar echo signal power from different particles differ. The larger the particle, the larger the contribution. The particle size distribution function satisfying the normalization condition $\int\limits_0^\infty da \, f_s(a) = 1$ can be represented in the form

$$f_s(a) = -\frac{1}{\rho_0} \cdot \frac{d\rho_s(a)}{da} \tag{1.42}$$

where $\rho_s(a)$ is the concentration of particles with radii larger than a. In this case

$$\rho_s(a) = \rho_0 \int\limits_a^\infty da' \, f_s(a') \tag{1.43}$$

If we know the function $f_s(a)$, the total concentration $\rho_0 = \rho_s(0)$, and the complex refractive index $m = n + j\kappa$ of the particulate matter (m depends on wavelength λ), then we can calculate the concentration of particles with radii $a \geq a_m$; that is, $\rho_s(a_m)$, the major contributors to the mean echo signal power $\langle P_s(t) \rangle$. The particle radius a_m can be found in the following way [31]. Let $\beta_\pi(z) \approx \beta_\pi(R)$ within the sensing volume.

Then, according to (1.35) through (1.37), (1.40), and (1.41), the mean echo signal power can be written as

$$\langle P_S(t) \rangle = \mu_e N_p \langle \sigma_\pi \rangle \left[\int_{a_m}^{\infty} da \, f_s(a)\sigma_\pi(a)/\langle \sigma_\pi \rangle + \eta_s(a_m) \right] \tag{1.44}$$

where $\mu_e = \dfrac{\lambda^2 \eta}{h v B_F P_L} \cdot \langle I_P(R,0,t)I_L(R,0) \rangle$ and

$$\eta_s(a_m) = \int_{0}^{a_m} da \, f_s(a)\sigma_\pi(a)/\langle \sigma_\pi \rangle \tag{1.45}$$

According to (1.36), (1.37), and (1.45), the sum of terms in the square brackets in (1.44) is equal to unity. Particles with radii $a \geq a_m$ are major contributors to the mean echo signal power only provided that the first term in (1.44) is far greater than the second one. Taking, for example, $\eta_s = 0.1$, we find that particles with radii $a \geq a_m$ 90% determine the mean echo power.

By analogy with (1.41), we determine the mean number N_{eff} of efficiently scattering particles (mean number of particles, which 90% determine the mean echo signal power) as [31]

$$N_{eff} = \rho_s(a_m)V_{eff} \tag{1.46}$$

For the case of radiation scattering by an ensemble of N_p identical particles, Ref. [46] shows that under the condition $N_p \gg 1$ the one-dimensional probability density function (ODPDF) of the field of the scattered wave has a normal distribution by virtue of the central limit theorem. If the condition $N_p \gg 1$ is not fulfilled, then the statistics of the scattered wave are non-Gaussian [46]. Since in our case particles have different sizes, the ODPDF of the lidar echo signal has the Gaussian distribution only under the condition $N_{eff} \gg 1$ (that is, N_p is replaced with N_{eff}). The estimates of SNR, V_{eff}, and N_{eff} can be found in the following sections of this chapter for both cw and pulsed CDLs.

1.3   Echo Signal Statistics for Continuous-Wave CDLs

Figure 1.2 depicts the principal diagram of cw CDL operation. The main lidar elements are a cw laser, a receive/transmit telescope, and a photodetector. The laser output radiation is separated into two beams: a probing beam and a reference beam. If an acousto-optical modulator (AOM) is used, the frequency of the reference beam v changes for $v + f_I$ (heterodyne system). Otherwise ($f_I = 0$), we have a homodyne system. The telescope focuses the probing beam to a preset distance and collects the backscattered radiation. The received radiation together with the reference beam comes to the sensitive area of the photodetector. Nonsmoothed estimates of the lidar

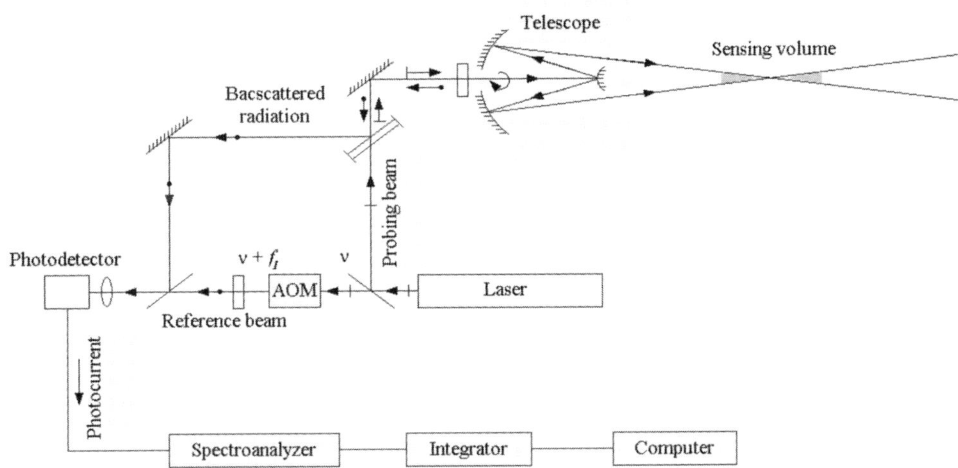

Figure 1.2 Diagram of cw CDL operation.

signal spectrum are obtained from measurements of the photocurrent in the circuit of the lidar receiving system and then accumulated (averaged) in the integrator.

The amplitudes of the fields of the probing and equivalent reference beams with plane phase fronts have the Gauss distribution in the plane of the exit aperture of the telescope:

$$U_{0P}(\boldsymbol{\rho}) = \sqrt{\frac{P_P}{\pi a_0^2}} \exp\left[-\frac{\rho^2}{2a_0^2}\right] \tag{1.47}$$

$$U_{0L}(\boldsymbol{\rho}) = \sqrt{\frac{P_L}{\pi a_0^2}} \exp\left[-\frac{\rho^2}{2a_0^2}\right] \tag{1.48}$$

where P_P is the probing beam power and a_0 is the beam radius determined at the e^{-1} intensity level. Let the telescope radius be $a_t = 2a_0$. Then in (1.11) and (1.12) for the function $A_t(\boldsymbol{\rho})$ determined by (1.2), we can take approximately $\Pi(\boldsymbol{\rho}) = 1$. As a result, for the normalized complex echo signal $Z_S(t)$ from (1.47), (1.48), (1.11) through (1.13), (1.17), and (1.28), we have

$$Z_S(t) = \frac{\lambda}{\pi a_0^2} \sqrt{\frac{2\eta P_P}{h\nu B_F}} \sum_{i=1}^{N_S} \alpha_i \exp[j\psi_i + 4\pi j V_r(z_i)t/\lambda] I_{PN}(z_i, \boldsymbol{\rho}_i + tV_\perp(z_i), t) \tag{1.49}$$

where

$$I_{PN}(z, \boldsymbol{\rho}, t) = \left| \int_{-\infty}^{+\infty} d^2\rho' \exp\left[-\frac{\rho'^2}{2a_0^2} - j\frac{\pi}{\lambda F}\rho'^2\right] G(0, \boldsymbol{\rho}'; z, \boldsymbol{\rho}; t) \right|^2 \tag{1.50}$$

is the normalized (dimensionless) intensity of the probing beam.

Replacing $\langle I_P(z, \rho, t)I_L(z, \rho)\rangle$ with $\langle I_{PN}^2(z,\rho)\rangle P_P P_L /(\pi a_0^2)^2$ in (1.37) and (1.40), and taking into account (1.39), we find for the SNR and for the sensing volume V_{eff}:

$$\text{SNR} = \frac{\eta \lambda^2 P_P}{h\nu B_F \pi^2 a_0^4} \int\limits_0^{\infty} dz \beta_\pi(z) \int\limits_{-\infty}^{+\infty} d^2\rho \langle I_{PN}^2(z,\rho)\rangle \tag{1.51}$$

$$V_{eff} = \int\limits_0^{\infty} dz \int\limits_{-\infty}^{+\infty} d^2\rho \, \frac{\beta_\pi(z)\langle I_{PN}^2(z,\rho)\rangle}{\beta_\pi(R)\langle I_{PN}^2(R,0)\rangle} \tag{1.52}$$

where $R = z_{\max}$ is the point of maximum of the integrand function, that is, the sensing range.

As a laser beam propagates in the atmosphere, its amplitude and phase distort due to turbulent inhomogeneities of the refractive index of air [43, 47–52], and this fact affects SNR and V_{eff}. It is shown in [29] for the case of cw lidar with $\lambda = 10.6 \; \mu m$ that at moderate turbulence the longitudinal dimension of the sensing volume is nearly the same as in the absence of turbulence. Therefore, to estimate the sensing volume based on (1.52), we assume that the beam propagates in a homogeneous medium; that is, in (1.50) for G we use (1.14), and then for the normalized intensity we have the well-known equation

$$I_{PN}(z,\rho) = \exp[-\rho^2/(g(z)a_0^2)]/g(z) \tag{1.53}$$

where $g(z) = (1 - z/F)^2 + (z/L_d)^2$, and $L_d = 2\pi a_0^2/\lambda$ is the diffraction length. Assuming that $\beta_\pi(z) \approx \beta_\pi(R)$ within the sensing volume, from the maximum of the function $\int\limits_{-\infty}^{+\infty} d^2\rho \langle I_{PN}^2(z,\rho)\rangle$ with allowance for (1.53) we find that the sensing range R depends on the focal length F as

$$R = \frac{F}{1 + (F/L_d)^2} \tag{1.54}$$

For the volume V_{eff} after the integration in (1.52) we obtain

$$V_{eff} = \frac{1}{2} S_R \Delta z \tag{1.55}$$

where $S_R = \pi a_R^2$ is the effective area in the cross section of the beam with the minimal radius

$$a_R = \frac{a_0 F/L_d}{\sqrt{1 + (F/L_d)^2}} \tag{1.56}$$

and

$$\Delta z = \frac{F^2/L_d}{1 + (F/L_d)^2}\left[\frac{\pi}{2} + \arctan\left(\frac{L_d}{F}\right)\right] \tag{1.57}$$

is the effective longitudinal dimension of the sensing volume [31].

For a typical laser with $\lambda = 10.6\ \mu m$ and $a_0 = 7.5$ cm ($L_d = 3334$m), the focal length of the sensing beam does not exceed 1 km. According to (1.54) and (1.57), in this case $\delta R = F - R = 82$m and $\Delta z = 855$m. When the beam is focused to the near diffraction zone and $\delta R < 1$m ($R \approx F$), the simpler equations can be derived from (1.55) through (1.57):

$$a_R = \frac{\lambda F}{2\pi a_0}, \ \Delta z = \frac{\lambda F^2}{2a_0^2}, \ \text{and}\ V_{eff} = \frac{\lambda^3}{16\pi}\frac{F^4}{a_0^4} \tag{1.58}$$

It follows from these equations that in the absence of turbulent distortions of the sensing beam we can achieve extremely small values of Δz and V_{eff} (see Table 1.1).

From (1.51) and (1.53) provided that $\beta_\pi(z) = $ const and $F \ll L_d$, we obtain [18, 19]

$$\text{SNR} = \frac{\pi\lambda\beta_\pi\eta P_P}{h\nu B_F} \tag{1.59}$$

It follows then that the SNR, in contrast to V_{eff}, is independent of the sensing range. To calculate SNR in this way, it is necessary not only to know the lidar parameters, but also to have information about the aerosol backscatter coefficient β_π determined by the concentration of particles ρ_0, the particle size distribution function $f_s(a)$, and the differential backscatter cross section $\sigma_\pi(a)$ of a particle with radius a. We assume particles to be spherical with the identical complex refractive index $m = n + j\kappa$ of the particulate matter. Then the backscatter cross section, which is a function of a, λ, and m, can be calculated using Mie theory [42, 53, 54].

We use the empirical model for $\rho_s(a)$ proposed in [53] for the atmospheric boundary layer. The data of this model shown by dots in Figure 1.3 are well approximated by the following equation [31]:

$$\rho_s(a) = \rho_0\left(\frac{c_1}{1 + \tilde{\rho}_1 a^3} + \frac{c_2}{1 + \tilde{\rho}_2 a^3}\right) \tag{1.60}$$

where $\rho_0 = 1430$ cm^{-3}, $c_1 = 1 - c_2$, $c_2 = 2.5 \cdot 10^{-4}$, $\tilde{\rho}_1 = 7.15 \cdot 10^3\ \mu m^{-3}$, and $\tilde{\rho}_2 = 0.286\ \mu m^{-3}$. In (1.60) the particle radius a is in micrometers. Curve 1 in Figure 1.3 is a result of calculation by this equation. From (1.60) and (1.42) we obtain for the particle size distribution function $f_s(a)$

$$f_s(a) = 3a^2\left[\frac{c_1\tilde{\rho}_1}{(1 + \tilde{\rho}_1 a^3)^2} + \frac{c_2\tilde{\rho}_2}{(1 + \tilde{\rho}_2 a^3)^2}\right] \tag{1.61}$$

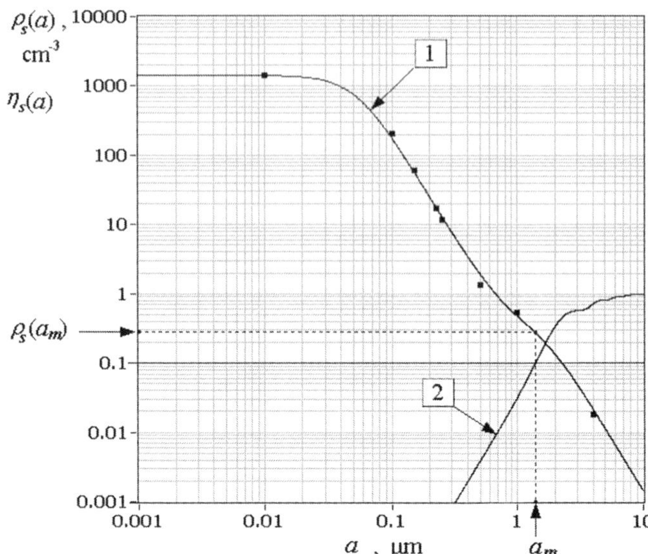

Figure 1.3 Model for particle size distribution function (dots and curve 1) and parameter $\eta_s(a)$ (curve 2). (© 2000 Optical Society of America. From [31].)

In calculations by (1.35), (1.36), (1.43), and (1.61), we take $m = 1.4 + j0.08$ [53]. Then, at $\lambda = 10.6$ μm $\beta_\pi = 0.55 \cdot 10^{-7}$ (m sr)$^{-1}$. For cw CO_2 lidar with $P_P = 4$ W, $\eta = 0.4$, and $B_F = 5$ MHz [55], from (1.59) we find SNR = 29. This rather large number of coherently detected photons coming from the atmosphere allows us to neglect the noise component of the signal when interpreting experimental data.

The calculated values of the total number of particles being in the sensing volume $N_p = \rho_0 V_{eff}$ are summarized in Table 1.1. If we assume that scattering amplitudes (sizes) of all particles are identical, then under the condition $N_p \gg 1$ the one-dimensional probability density of the real and imaginary parts Z_S has the Gauss distribution (in virtue of the central limit theorem). According to the data of Table 1.1, this condition is fulfilled even at $R = 10$m. However, most particles of the total number N_p have small size (see Figure 1.3), and they are optically inactive; that is, larger particles are major contributors to the power of the scattered radiation. The

function $\eta_s(a) = \int_0^a da' f_s(a') \sigma_\pi(a)/\langle \sigma_\pi \rangle$ is shown in Figure 1.3, from which it follows

that the particle radius a_m determined from solution of the equation $\eta_s(a_m) = 0.1$ equals 1.39 μm and $\rho_s(a_m) = 0.28$ cm^{-3}. The calculated mean number of efficiently scattering particles N_{eff} at different dimensions of the sensing volume is given in

Table 1.1 Mean Number of Scattering Particles at Different Focal Lengths

F [m]	R [m]	a_R [mm]	Δz [m]	V_{eff} [cm^3]	N_p	N_{eff}
10	10	0.23	0.09	$7.5 \cdot 10^{-3}$	11	$2 \cdot 10^{-3}$
25	25	0.56	0.59	0.3	430	$8 \cdot 10^{-2}$
50	50	1.12	2.35	4.7	$6.72 \cdot 10^3$	1.3
100	100	2.25	9.42	74.9	$1.07 \cdot 10^5$	21
500	489	11.2	219.5	$4.4 \cdot 10^4$	$6.26 \cdot 10^7$	$1.2 \cdot 10^4$

Table 1.1, from which it follows that that the one-dimensional probability density of the lidar echo signal is strictly Gaussian only at the maximal ($F = 500$m in the Table 1.1) sensing range.

1.3.1 Statistical Characteristics of Echo Signal of Continuous-Wave CDLs

To calculate the statistical characteristics of the echo signal from (1.49), we assume the following. Intensity fluctuations of the sensing beam caused by turbulent variations of the refractive index of air can be neglected and, therefore, we can use (1.53). Scattering particles are fully entrained by the turbulent wind flow, which is stationary, isotropic, and statistically homogeneous. The probability density of the radial wind velocity has the form

$$p(V_r) = \exp\left[-(V_r - \langle V_r \rangle)^2/(2\sigma_V^2)\right]/(\sqrt{2\pi}\sigma_V) \qquad (1.62)$$

where $\langle V_r \rangle$ is the mean velocity and $\sigma_V^2 = \langle V_r^2 \rangle - \langle V_r \rangle^2$ is the velocity variance. For the cross wind, the condition $|\langle V_\perp \rangle| \gg \sigma_V$ is true, which allows us to replace $V_\perp(z)$ with $\langle V_\perp \rangle$ in (1.49).

For the covariance function of the normalized complex echo signal

$$C_S(\tau) = \frac{1}{2}\langle Z_S(t + \tau)Z_S^*(t)\rangle \qquad (1.63)$$

from (1.49) with allowance made for the above assumptions after the averaging [similar to that described in the previous section for the derivation of the equation for $\langle P_S(t) \rangle$, (1.37), (1.44)] under the condition $F \ll L_d$, we obtain

$$C_S(\tau) = \langle P_S \rangle \exp\left[j\frac{4\pi}{\lambda}\langle V_r \rangle\tau - \frac{1}{2}\left(\frac{4\pi}{\lambda}\sigma_V\tau\right)^2\right]F_1(\langle V_\perp \rangle\tau/a_R) \qquad (1.64)$$

where $\langle P_S \rangle \equiv$ SNR is given by (1.59),

$$F_1(x) = \frac{2}{\pi}\int_0^\infty \frac{d\xi}{1 + \xi^2}\exp\left[-\frac{x^2}{2(1 + \xi^2)}\right]$$

and the radius a_R is defined by the first equation in (1.58). Since in the used lidar systems the telescope diameter usually does not exceed 30 cm, the condition

$$2\sigma_V \gg |\langle V_\perp \rangle|a_0/F \qquad (1.65)$$

is always true in the atmosphere, even if, for example, $|\langle V_\perp \rangle| = 10$ m/s and $F = 10$m. Therefore, in (1.64) we can take $F_1(x) = 1$, and thus we come to the well-known equation for the covariance function [10], from which one can obtain estimates of the mean radial velocity and variance of velocities of scattering particles. Most often these estimates are found from the spectral power density of the lidar echo signal $S_S(f)$, which can be written in the form [56]

$$S_S(f) = \int\limits_{-\infty}^{+\infty} d\tau\, C_S(\tau)\exp(-2\pi\, j\, f\tau) \tag{1.66}$$

After the substitution of (1.64) [at $F_1(x) = 1$] in (1.66) and integration, we come to the well-known relation [10] between the spectral power density of the echo signal and the probability density function of the radial wind velocity, (1.62):

$$S_S(f) = \langle P_S \rangle (\lambda/2)p(\lambda f/2) \tag{1.67}$$

To obtain stable estimates of the covariance function and the power spectrum of the echo signal from raw lidar data, it is necessary to know the value and characteristic timescales of fluctuations of the echo signal power. The relative variance σ_{PS}^2 and the correlation coefficient $K_{PS}(\tau)$ of echo signal power fluctuations are determined as

$$\sigma_{PS}^2 = \frac{\langle P_S^2 \rangle - \langle P_S \rangle^2}{\langle P_S \rangle^2} \tag{1.68}$$

and

$$K_{PS}(\tau) = \frac{\langle P_S(t+\tau)P_S(t)\rangle - \langle P_S \rangle^2}{\langle P_S^2 \rangle - \langle P_S \rangle^2} \tag{1.69}$$

Within the framework of assumptions drawn in the derivation of (1.64), the following equations for σ_{PS}^2 and K_{PS} are true based on (1.49) at $F \ll L_d$ [27]

$$\sigma_{PS}^2 = 1 + \frac{3}{4}\frac{\gamma_\pi}{N_{eff}} \tag{1.70}$$

and

$$K_{PS}(\tau) = \frac{\exp[-(\tau V_\perp/a_R)^2 I_1(\tau) + \dfrac{3}{4}\dfrac{\gamma_\pi}{N_{eff}} I_2(\tau)]}{1 + \dfrac{3}{4}\dfrac{\gamma_\pi}{N_{eff}}} \tag{1.71}$$

where $\gamma_\pi = \dfrac{\rho_s(a_m)}{\rho_0}\dfrac{\langle \sigma_\pi^2 \rangle}{\langle \sigma_\pi \rangle^2}$, $N_{eff} = \rho_s(a_m)V_{eff}$ is the number of efficiently scattering particles [see (1.46)],

$$I_1(\tau) = \frac{2}{\pi} \int_0^\infty \frac{d\xi}{1 + \xi^2} \exp\left\{ -\left(\frac{4\pi}{\lambda}\sigma_V\tau\right)^2 \left[1 - K_V\left(\frac{2}{\pi}\Delta z \xi\right)\right]\right\} \tag{1.72}$$

where

$$K_V(r) = [\langle V_r(z + r)V_r(z)\rangle - \langle V_r\rangle^2]/\sigma_V^2 \tag{1.73}$$

is the correlation coefficient of wind velocity fluctuations, and

$$I_2(\tau) = \frac{16}{3\pi} \int_0^\infty \frac{d\xi}{(1 + \xi^2)^3} \exp\left[-\frac{(\tau V_\perp / a_R)^2}{1 + \xi^2} \right] \tag{1.74}$$

In (1.69) through (1.74), the parameters a_R, Δz, and V_{eff} are determined by (1.55) through (1.57).

To calculate the function $I_1(\tau)$, it is necessary to know the correlation coefficient $K_V(r)$. If the longitudinal dimension of the sensing volume Δz is an order of magnitude larger than the integral correlation scale of wind velocity fluctuations L_V

$$L_V = \int_0^\infty dr\, K_V(r) \tag{1.75}$$

(or the outer scale of turbulence, which can be determined as $L_o = 1.34L_V$ [57]), then in (1.72) we can take $K_V = 0$. If the condition $\Delta z \ll L_V$ is true, then the correlation coefficient K_V can be described by the following equation [58]:

$$K_V(r) = 1 - C_K \varepsilon^{2/3} r^{2/3} / (2\sigma_V^2) \tag{1.76}$$

for the inertial interval of turbulence $l_V \ll r \ll L_V$, where $l_V \sim 1$ mm is the inner scale of wind turbulence, $C_K \approx 2$ is the Kolmogorov constant, and ε is the dissipation rate of the kinetic energy of turbulence.

The parameters σ_{PS}^2 and $K_{PS}(\tau)$ were calculated for a lidar with $\lambda = 10.6\ \mu$m and $a_0 = 7.5$ cm with the use of model (1.60), (1.70) through (1.74), and, on the assumption that the condition $\Delta z \ll L_V$ is true, (1.76), at $\varepsilon = 10^{-3}$ m^2/s^3 and $|\langle V_\perp\rangle| = 5$ m/s.

The bold curve in Figure 1.4 shows the calculated dependence of the relative variance of the echo signal power σ_{PS}^2 on the focal length of the probing beam F. This figure shows the dependence of N_{eff} on F as well. One can see that at $F \geq 150$m, when the mean number of efficiently scattering particles $N_{eff} > 100$, the relative variance is $\sigma_{PS}^2 = 1$. Consequently, the one-dimensional probability density function of the echo signal should be described by the normal distribution. As the focal length F decreases, the sensing volume V_{eff} and the number N_{eff} also decrease. As this takes place, the relative variance σ_{PS}^2 increases and already at $F = 50$m, when $N_{eff} \sim 1$, becomes roughly equal to 2.5. At $F = 10$m, σ_{PS}^2 is close to 1,000.

Figure 1.5 shows the calculated correlation coefficient of the echo signal power $K_{PS}(\tau)$ at different focal lengths of the sensing beam F. Since $N_{eff} \ll 1$ at $F = 20$m

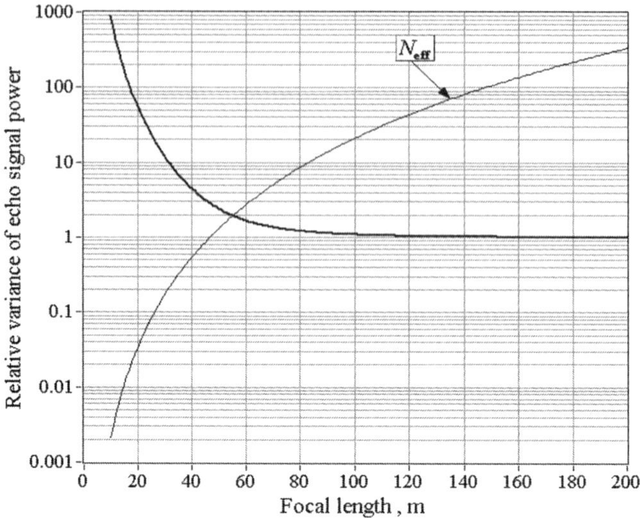

Figure 1.4 Relative variance of the echo signal power σ^2_{PS} and the number of efficiently scattering particles N_{eff}.

according to Figure 1.4, the first terms in the numerator and denominator of (1.71) can be neglected. Then $K_{PS}(\tau) \approx I_2(\tau)$ and, consequently, the characteristic time correlation scale is determined by the time of transfer of one or several large particles by the cross wind to a distance $\sim a_R$. As the focal length increases, the contribution of the first terms in the numerator and denominator of (1.71) to the correlation coefficient increases, and, as can be seen in Figure 1.5, $K_{PS}(\tau)$ has two characteristic timescales, τ_{C1} and τ_{C2}, which was discovered earlier in the field experiment [23]. According to

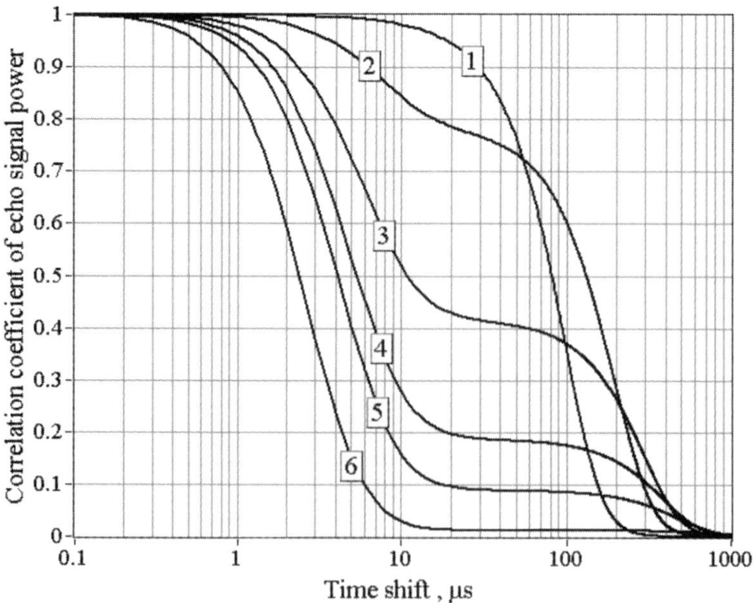

Figure 1.5 Correlation coefficient of the echo signal power $K_{PS}(\tau)$ at $F = 20$m (1), 40m (2), 60m (3), 80m (4), 100m (5), and 200m (6).

the performed calculations, the two-scale character of the correlation function of the echo signal power is most pronounced for the focal lengths from 30m to 150m.

Define the scales τ_{C1} and τ_{C2} as $\tau_{C1} = \int_0^{\infty} d\tau I_i(\tau)$. Then from (1.72), (1.74), and (1.76) we have [27]

$$\tau_{C1} = \frac{\lambda}{8\sqrt{\pi}\tilde{\sigma}_V(\Delta z)} \tag{1.77}$$

where

$$\tilde{\sigma}_V(\Delta z) = \sigma_V \left\{ \frac{2}{\pi} \int_0^{\infty} \frac{d\xi}{1+\xi^2} \left[1 - K_V\left(\frac{2}{\pi}\Delta z\xi\right)\right]^{-1/2} \right\}^{-1} \tag{1.78}$$

and

$$\tau_{C2} = \frac{16}{9\sqrt{\pi}} \frac{a_R}{|V_{\perp}|} \tag{1.79}$$

The integral in (1.78) can be taken analytically only in two limit cases: $\Delta z \gg L_V$ and $\Delta z \ll L_V$. At $\Delta z \gg L_V$, taking $K_V = 0$ in (1.77), we obtain $\tilde{\sigma}_V(\Delta z) = \sigma_V$. In the case $\Delta z \ll L_V$, using (1.76) in (1.77), after integration we have [27] $\tilde{\sigma}_V(\Delta z) = [(3/8)(2/\pi)^{2/3} C_K (\varepsilon \Delta z)^{2/3}]^{1/2}$. The estimation of τ_{C1} and τ_{C2} for a lidar with $\lambda = 10.6\ \mu m$ and a telescope diameter of 30 cm at the focusing of the sensing beam to a distance $F \leq 200$ m has shown that τ_{C1} does not take values smaller than ~0.5 μs, and τ_{C2} does not exceed 5 ms.

Thus, when the probing beam is focused to short distances, when $N_{eff} \ll 1$, the correlation time of echo signal power fluctuations is determined by the time of transfer of individual large particles by the cross wind through the beam cross section in the waist zone. In another limit case, when the sensing volume houses a very large number of efficiently scattering particles $N_{eff} > 100$, the correlation time of the echo signal power is determined by the time, during which the distance between scattering particles transported by a turbulent wind flow changes, on average, approximately over λ. In this case, the more intense the turbulence, the shorter the correlation time. If in the sensing volume $0.2 < N_{eff} < 100$, then the time correlation function of echo signal power fluctuations has a two-scale character.

1.3.2 One-Dimensional Probability Density Functions of the Amplitude and Power of the Echo Signal of Continuous-Wave CDLs

The probability density functions of the echo signal amplitude and power have been calculated with the use of numerical simulation data. The following algorithm of numerical simulation of a complex echo signal was proposed in [31]. We assume that fluctuations of the number of particles in the sensing volume V_{eff} can be neglected, and then at $F \ll L_d$ according to (1.49) and (1.59) (SNR $\equiv \langle P_s \rangle$) the complex echo signal can be written as

$$Z_S = \mu_p \sum_{i=1}^{N_p} \sigma_\pi^{1/2}(a_i) \exp(j\psi_i) \tag{1.80}$$

where $\mu_p = \sqrt{\dfrac{\pi\lambda\eta P_P}{h\nu B_F V_{eff}}}$ and $N_p = [\rho_0 V_{eff}]$. The phase ψ is simulated with the use of a generator of pseudorandom numbers ξ'' uniformly distributed in the $[0,2\pi]$ interval.

Let ξ' be a random value uniformly distributed in the interval $[0,1]$. The random value a with the probability density function $f_s(a)$ ($a \in [0,\infty]$) is related to ξ' through the differential equation

$$da\, f_s(a) = d\xi' \tag{1.81}$$

After the substitution of (1.81) into (1.42) and integration, we obtain

$$1 - \rho_s(a)/\rho_0 = \xi' \tag{1.82}$$

From (1.60) and (1.82), we can find the relation between a and ξ' in the form

$$a = \left[\frac{\beta + (\beta^2 + 4\gamma\xi')^{1/2}}{2\gamma} \right]^{1/3} \tag{1.83}$$

where $\beta = c_1\tilde{\rho}_2 + c_2\tilde{\rho}_1 - (1 - \xi)(\tilde{\rho}_1 + \tilde{\rho}_2)$ and $\gamma = (1 - \xi')\tilde{\rho}_1\tilde{\rho}_2$. The pseudorandom number generator generates a value of ξ'. Equation (1.83) is used to determine the particle radius a, and the backscattering cross section $\sigma_\pi(a)$ for a particle with this radius is calculated using Mie theory. Then a random realization of Z_S is calculated by (1.80).

With an increase of the focal length F, the sensing volume ($V_{eff} \sim F^4$) increases fast according to (1.58) and, consequently, the number of particles $N_p = [\rho_0 V_{eff}]$ increases as well. As a result, computer realization of the simulation algorithm becomes difficult. In [31], the following approach was proposed. The sensing volume is divided into N elementary volumes each containing n particles. In this case, $N_p = N \cdot n$. Then Z_S can be written as

$$Z_S = \mu_p \sum_{k=1}^{N} \left[\sum_{l=1}^{n} \sigma_\pi^{1/2}(a_{kl}) \exp(j\psi_{kl}) \right] \tag{1.84}$$

It is obvious that at relatively small n one large particle with the radius $a = a_{\max}$ can contribute decisively to $\sum_{l=1}^{n} \sigma_\pi^{1/2}(a_{kl}) \exp(j\psi_{kl})$ and, correspondingly, the contribution from other particles can be neglected through the replacement of the expression in square brackets in (1.84) with $\sigma_\pi^{1/2}(a_{\max}^{(k)}) \exp(j\psi_k)$.

The probability density function $f_s^{(n)}(a)$ of the maximal particle size distribution $a = a_{\max}$ in the volume containing n particles is described by the following equation [59]:

$$f_s^{(n)}(a) = f_s(a)\sum_{i=1}^{n}\prod_{k\neq 1}^{n-1}\int_0^a da_k f_s(a_k) \tag{1.85}$$

After the substitution of (1.42) into (1.85) and integration over all a_k, we obtain

$$f_s^{(n)}(a) = \frac{n}{\rho_0}\frac{d\rho_s(a)}{da}[1 - \rho_s(a)/\rho_0]^{n-1} \tag{1.86}$$

According to (1.43), the particle concentration $\rho_s^{(n)}(a)$ is calculated as
$\rho_s^{(n)}(a) = \rho_s^{(n)}(a)\int_a^\infty da' f_s^{(n)}(a')$, from which we find

$$\rho_s^{(n)}(a) = \rho_s^{(n)}(0)\{1 - [1 - \rho_s(a)/\rho_0]^n\} \tag{1.87}$$

It is obvious that the total concentration $\rho_s^{(n)}(0)$ is n times lower than ρ_0; that is, $\rho_s^{(n)}(0) = \rho_0/n$. It follows from (1.87) that, in the simulation of $\sigma_\pi^{1/2}(a_{max}^{(k)})\exp(j\psi_k)$, ξ' in (1.83) should be replaced with $(\xi')^{1/n}$. The performed analysis shows that for the use of (1.84) [with replacement of the expression in square brackets with $\sigma_\pi^{1/2}(a_{max}^{(k)})\exp(j\psi_k)$] it is necessary for the number N to be no less than 300 and n to take integer values 1, 2, 3, ... This is not true as a rule (for example, for $F = 10m$, according to Table 1.1, $N_p = 11$ is much less than 300, for $F = 25m$ $N_p = 430$, and the ratio $N_p/N \approx 1.5$ is not an integer number n). Therefore, in the simulation for the case of $N_p \leq 300$, it was assumed that $N = N_p$ and $n = 1$, and for $N_p > 300$ the parameters N and n were considered equal to, respectively, 300 and $N_p/300 \equiv \rho_0 V_{eff}/300$ (this ratio should not necessarily be an integer number).

In [31], $c_2 = 0$ was taken in (1.83), and the data of numerical simulation of Z_S (with 10^6 independent realizations obtained as described above) were used to calculate one-dimensional probability density functions for the amplitude $p(A_S)$, phase $p(\psi_S)$, and power $p(P_S)$ of the echo signal, where $A_S = |Z_S|$, $\psi_S = \arg(Z_S)$, and $P_S = (1/2)|Z_S|^2$. As expected, the obtained function $p(\psi_S)$ has a uniform distribution in the interval $[0, 2\pi]$ for any dimensions of the sensing volume.

The solid curves in Figure 1.6 are the probability density functions $p(A_S)$ (Figure 1.6(a)) and $p(P_S)$ (Figure 1.6(b)) calculated from the data of numerical simulation with allowance for the aerosol microstructure at different focal lengths of the sensing beam F. The dashed curves are the results for the Rayleigh distribution of $p(A_S)$

$$p(A_S) = (\pi/2)(A_S/\langle A_S\rangle^2)\exp[-(\pi/4)A_S^2/\langle A_S\rangle^2] \tag{1.88}$$

and for the exponential distribution of $p(P_S)$

$$p(P_S) = \langle P_S\rangle^{-1}\exp(-P_S/\langle P_S\rangle) \tag{1.89}$$

It is well known (see, for example, [42]) that if the probability density functions of the real and imaginary parts of Z_S are distributed by the normal law, then $p(A_S)$

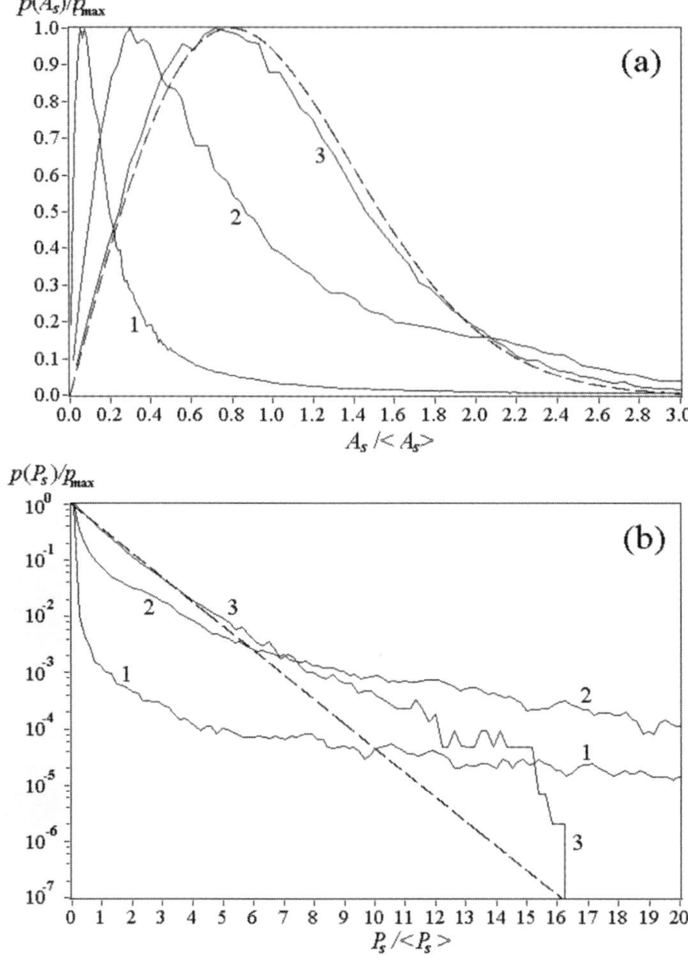

Figure 1.6 Normalized probability density functions of the (a) amplitude and (b) power of the lidar echo signal. Curves 1, 2, and 3 correspond to the cases of focusing of the sensing beam to a distance of 10m, 50m, and 100m, respectively. The (a) Rayleigh distribution and (b) exponential distribution are shown by dashed curves. (© 2000 Optical Society of America. From [31].)

is described by the Rayleigh distribution, while $p(P_S)$ is described by the exponential distribution. One can see from Figure 1.6 that even at $F = 100\text{m}$ the probability densities $p(A_S)$ and $p(P_S)$ differ from the Rayleigh distribution and the exponential distribution. That is, the one-dimensional probability density of the signal Z_S becomes non-Gaussian with a decrease in the sensing volume. The difference of the Z_S signal statistics from the Gaussian one at small focal lengths F was also found in the field experiment for the focusing of the sensing beam to short distances in the case of a cw CDL with $\lambda = 10.6\ \mu\text{m}$ [23] and $\lambda = 1.55\ \mu\text{m}$ [24].

Note that even under the condition $N_{eff} \gg 1$ the two-dimensional probability density $p(Z_{S1}, Z_{S2})$, where Z_{S1} and Z_{S2} are the real (or imaginary) parts of, respectively, $Z_S(t_1)$ and $Z_S(t_2)$, differs from the two-dimensional probability density of a Gaussian process due to the random inhomogeneity of the radial velocity $V_r(z)$ (wind flow turbulence) [46].

1.4   Echo Signal Statistics for Pulsed CDLs

Among existing pulsed CDLs [21, 60–76], 2-μm lidars [72, 75] are more efficient for atmospheric studies. Below we give a brief description of the 2-μm lidar used in the experiments discussed in this book. Figure 1.7 shows the principal block diagram of this lidar operation.

The pulsed 2-μm CDL is based on the transceiver unit MAG-1 from CLR Photonics [63, 64]. It comprises two Tm:LuAG lasers with photodiode pumping and a wavelength of $\lambda = 2,022$ nm. A master laser generates a cw beam. This beam is divided into two beams. One of them is injected into the slave (pulsed) laser after passage through an acousto-optical modulator (frequency shift $f_I \sim 100$ MHz); another is used as a reference beam (local oscillator). The energy of the generated pulse is $E_P = 2$ mJ, and its duration (determined by the power drop down to the half-maximum level to the right and to the left from the point of maximum) is $\tau_P = 400$ ns. The generated pulse is divided into the probing and control beams. Probing pulses (pulse repetition frequency can be up to 500 Hz) are emitted through the telescope (aperture diameter of 108 mm) into the atmosphere. The radiation backscattered in the atmosphere is collected by the telescope and, after mixing with the reference beam, comes to the photodetector. To control the intermediate frequency f_I, the control pulsed beam is also mixed with the reference beam and detected by another photodetector. The detected (backscatter and control) signals are digitized with a frequency of 500 MHz.

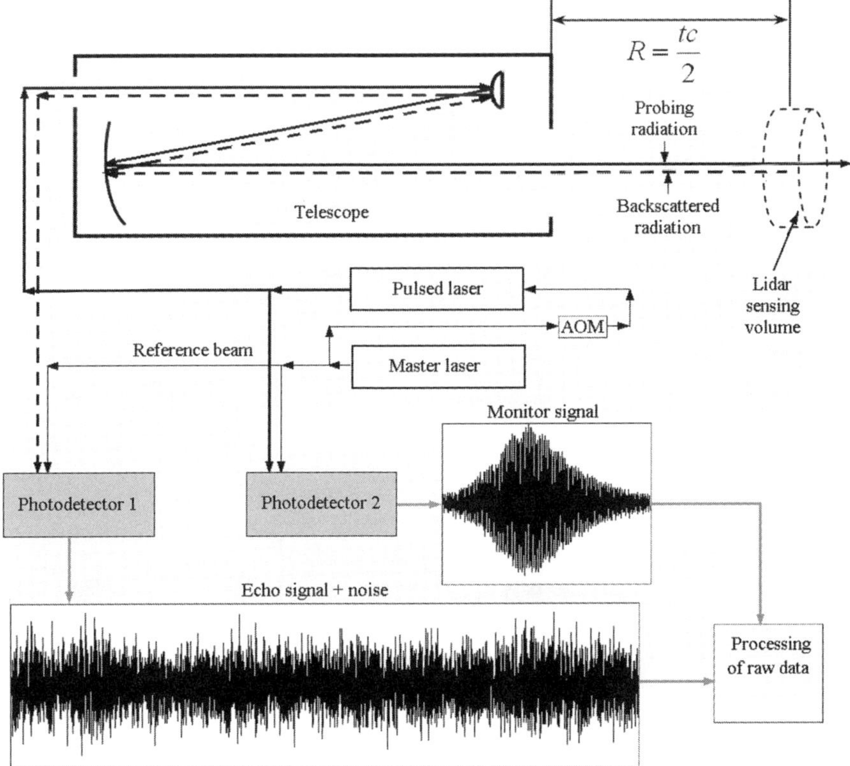

Figure 1.7   Block diagram of pulsed CDL operation.

The issue of optimization of the relationship among the telescope radius, the initial radius of the sensing beam, and the initial radius of the equivalent reference beam is considered in detail in [13, 15–17]. We assume that the initial radii of the probing and equivalent reference beams are identical, and the normalized intensity $I_{PN}(z,\rho)$ can be written in the form of (1.50). In the case of the pulsed 2-μm CDL, the power of the probing beam as a function of time for a single pulse is well described by the equation

$$P_P(t') = \frac{E_P}{\sqrt{\pi}\sigma_P}\exp\left(-\frac{t'^2}{\sigma_P^2}\right) \tag{1.90}$$

where $E_P = \int\limits_{-\infty}^{+\infty} dt' P_P(t')$ is the pulse energy and $2\sigma_P = \tau_P/\sqrt{\ln 2}$ is the pulse duration determined from the power drop to the e^{-1} level to the left and to the right from the point of maximum. Then, from (1.11), (1.12), (1.29), (1.47), (1.48), and (1.90), we obtain for the normalized complex echo signal $Z_S(t)$

$$Z_S(t) = \frac{\lambda}{\pi a_0^2}\sqrt{\frac{2\eta E_P}{h\nu B_F\sqrt{\pi}\sigma_P}}\sum_{i=1}^{N_S}\alpha_i\exp\left[-\frac{(t-2z_i/c)^2}{2\sigma_P^2} + j\psi_i + 2\pi j\frac{2V_r(z_i)}{\lambda}t\right] \tag{1.91}$$
$$I_{PN}(z_i,\rho_i,t)$$

where the normalized intensity of the sensing beam $I_{PN}(z,\rho)$ is described by (1.50). In contrast to (1.49), (1.91) ignores the transfer of scattering particles by the cross wind, because it is negligible for the sensing pulse durations used in lidars.

As a sensing pulse is launched in the atmosphere, at the time $t \in [0,\infty]$ the lidar measures the echo signal that comes from the sensing volume with its center at a distance $R = ct/2$ from the lidar. According to (1.91), we can substitute $\langle I_{PN}^2(z,\rho)\rangle P_L E_P\exp\{-(R-z)^2/\Delta p^2\}/[(\pi a_0^2)^2\sqrt{\pi}\sigma_P]$, where $\Delta p = \sigma_P c/2$, for $\langle I_P(z,\rho,t) I_L(z,\rho)\rangle$ in (1.37) and (1.40). As a result, for the SNR with allowance for (1.39) and for the sensing volume V_{eff} we have:

$$\mathrm{SNR} = \frac{\eta\lambda^2 E_P/(\sqrt{\pi}\sigma_P)}{h\nu B_F\pi^2 a_0^4}\int\limits_0^\infty dz\beta_\pi(z)\exp\{-(R-z)^2/\Delta p^2\}\int\limits_{-\infty}^{+\infty} d^2\rho\langle I_{PN}^2(z,\rho)\rangle \tag{1.92}$$

and

$$V_{eff} = \int\limits_0^\infty dz\int\limits_{-\infty}^{+\infty} d^2\rho\frac{\beta_\pi(z)\exp\{-(R-z)^2/\Delta p^2\}\langle I_{PN}^2(z,\rho)\rangle}{\beta_\pi(R)\langle I_{PN}^2(R,0)\rangle} \tag{1.93}$$

where it is assumed that the point of maximum of the integrand function in (1.93) is $z_{max} \approx R$ (or $|z_{max} - R| \ll \Delta p$).

We assume that the condition $R \gg \Delta p$ is true and the functions $\beta_\pi(z)$ and $\langle I_{PN}^2(z,\rho)\rangle$ weakly depend on z inside a layer with a thickness $4\Delta p$ and a center at the point $z = R$. Then in (1.92) and (1.93) we can take $z = R$ and, after the substitution $z \to z' + R\left(\int\limits_0^\infty dz \to \int\limits_{-\infty}^{+\infty} dz'\right)$, perform the integration with respect to the variable z'. As a result, taking into account $\Delta p = \sigma_P c/2$ and atmospheric transmission, (1.38), we obtain

$$\text{SNR} = \frac{\eta E_P c}{2h\nu B_F} \beta_\pi(R) T_A^2(R) \left(\frac{\lambda}{\pi a_0^2}\right)^2 \int\limits_{-\infty}^{+\infty} d^2\rho \langle I_{PN}^2(R,\rho)\rangle \tag{1.94}$$

and

$$V_{eff} = \sqrt{\pi}\Delta p \int\limits_{-\infty}^{+\infty} d^2\rho \frac{\langle I_{PN}^2(R,\rho)\rangle}{\langle I_{PN}^2(R,0)\rangle} \tag{1.95}$$

When turbulent distortions of the probing beam can be neglected, (1.53) is applicable for the normalized intensity $I_{PN}(z,\rho)$. In this case, from (1.95) we have

$$V_{eff} = \frac{1}{2}[\sqrt{\pi}\Delta p] \cdot \pi a_0^2 \left[\left(1 - \frac{R}{F}\right)^2 + \left(\frac{\lambda R}{2\pi a_0^2}\right)^2\right] \tag{1.96}$$

For wind measurements in the atmospheric boundary layer by a ground-based 2-μm CDL, the sensing beam is usually focused to a distance $F = 1.5$ km [77, 78]. The calculation of the minimal volume V_{eff} (in the waist zone of the sensing beam) at $F = 1.5$ km, pulse duration $\tau_P = 0.4$ μs ($\Delta p = 36$m), and $a_0 = 2.5$ cm by (1.96) gives a value of $2.3 \cdot 10^4$ cm^3. The calculation using Mie theory (for $\lambda = 2$ μm the complex refractive index of the particulate matter is $m = 1.36 + j2.5 \cdot 10^{-3}$ [53]) and the model (1.60) yields that the concentration of optically active particles $\rho_s(a_m)$ is roughly equal to 1 cm^{-3}. Then the mean number of efficiently scattering particles in the sensing volume is $N_{eff} = V_{eff}\rho_s(a_m) \approx 2.3 \cdot 10^4$. Even if the concentration $\rho_s(a_m)$ would be two to three orders of magnitude smaller, at a volume $V_{eff} = 2.3 \cdot 10^4$ cm^3, the one-dimensional probability density of the real (and imaginary) part Z_S would have the Gaussian or close to the Gaussian distribution.

For the above case, when (1.53) can be used, from (1.94) we obtain [12]

$$\text{SNR} = \frac{\eta E_P c}{h\nu B_F} T_A^2(R)\beta_\pi(R)\frac{S_c(R)}{R^2} \tag{1.97}$$

where $S_c(R) = \pi a_c^2 = \pi a_0^2 (R/L_d)^2[(1 - R/F)^2 + (R/L_d)^2]^{-1}$ is the area of the coherent plate (a_c is the coherence radius of the backscattered wave in the telescope plane) and $L_d = 2\pi a_0^2/\lambda$. With an increase of the range R, the value of $S_c(R)$ increases (consequently, the efficiency of heterodyning increases), achieving the maximum at a point $R = F$ and saturating under the condition $R \gg L_d$. Thus, in the far diffraction zone, the SNR decreases proportionally to $T_A^2(R)\beta_\pi(R)R^{-2}$ with an increase of the range R. The backscattering coefficient β_π for $\lambda = 2\ \mu m$ calculated by model (1.60) is $\beta_\pi = 1.3 \cdot 10^{-7}(m\ sr)^{-1}$. Then, taking $\eta = 0.8$, $E_P = 2$ mJ, $B_F = 50$ MHz, $T_A = 1$, $a_0 = 2.5$ cm, and $R = F = 1.5$ km in (1.97), we obtain the estimate of SNR = 11. Turbulent inhomogeneities of the refractive index of air in the atmosphere deteriorate the probing beam spatial coherence and the spatial coherence of the scattered wave in the telescope plane and, consequently, cause an additional decrease of the SNR.

1.4.1 Statistical Characteristics of the Echo Signals of Pulsed CDLs

For the case of pulsed CDLs, we consider the covariance function of the normalized complex echo signal $C_S(\tau)$ from (1.63), the spectral density of the lidar echo signal $S_S(f)$ from (1.66), normalized mean power $\langle P_S \rangle \equiv$ SNR from (1.94), relative variance σ_{PS}^2 from (1.68), and the correlation coefficient $K_{PS}(\tau)$ from (1.69) of the echo power.

After the substitution of (1.91) into (1.63) and averaging with the use of (1.62), we obtain [10]

$$C_S(\tau) = \langle P_S \rangle \exp\left[j\frac{4\pi}{\lambda}\langle V_r \rangle \tau - \frac{1}{2}\left(\frac{4\pi}{\lambda}\sigma_V \tau\right)^2 - \left(\frac{\tau}{2\sigma_P}\right)^2 \right] \tag{1.98}$$

It follows from (1.98) that, as in the case of cw CDLs, the covariance function $C_S(\tau)$ contains the information about the mean value and variance of radial wind velocity. Equation (1.98) for the ratio $C_S(\tau)/\langle P_S \rangle$ differs from (1.64), where we can take $F_1(x) = 1$, only by the factor $\exp[-\tau^2/(4\sigma_P^2)]$.

From (1.66) and (1.98), we have for the spectral power density of the echo signal [10]

$$S_S(f) = \frac{\langle P_S \rangle (\lambda/2)}{\sqrt{2\pi[\sigma_V^2 + (\lambda\sigma_{f1}/2)^2]}} \exp\left\{ \frac{(\lambda f/2 - \langle V_r \rangle)^2}{2[\sigma_V^2 + (\lambda\sigma_{f1}/2)^2]} \right\} \tag{1.99}$$

where $\sigma_{f1} = (2\sqrt{2}\pi\sigma_P)^{-1}$. Equation (1.99) is obtained at $\tau \in [-\infty, +\infty]$ without multiplication of $C_S(\tau)$ by the function of a temporal window. To conduct measurements with a pulsed CDL having the required spatial and frequency resolution, a temporal window is used, with which the width of the obtained spectrum can by significantly larger than that given by (1.99).

After the substitution of (1.91) into (1.68) and (1.69) and the averaging, we have the following equations for the relative variance σ_{PS}^2 and the correlation coefficient $K_{PS}(\tau)$ of the echo signal power [32, 79]:

$$\sigma_{PS}^2 = 1 + 2\sigma_T^2 \tag{1.100}$$

and

$$K_{PS}(\tau) = \frac{(1 + \sigma_T^2)\mu(\tau)\exp[-\tau^2/(2\sigma_P^2)] + B_T(\tau)}{1 + 2\sigma_T^2} \tag{1.101}$$

where

$$\sigma_T^2 = \langle \bar{P}_S^2 \rangle / \langle P_S \rangle^2 - 1 \tag{1.102}$$

$$B_T(\tau) = \langle \bar{P}_S(t + \tau)\bar{P}_S(t) \rangle / \langle P_S \rangle^2 - 1 \tag{1.103}$$

$$\bar{P}_S(t) = \frac{\eta E_P c}{2 h \nu B_F} \beta_\pi(R) T_A^2(R) \left(\frac{\lambda}{\pi a_0^2} \right)^2 \int\limits_{-\infty}^{+\infty} d^2\rho I_{PN}^2(R,\rho,t) \tag{1.104}$$

is the normalized power of the echo signal averaged over microphysical parameters of the scattering medium, ($\langle \bar{P}_S \rangle \equiv \langle P_S \rangle \equiv \mathrm{SNR}$),

$$\mu(\tau) = (\sqrt{\pi}\Delta p)^{-1} \int\limits_{-\infty}^{+\infty} dz' \exp[-(z'/\Delta p)^2 - 2(2\pi/\lambda)^2 D_V(z')\tau^2] \tag{1.105}$$

$D_V(z_1 - z_2) = \langle [V_r'(z_1) - V_r'(z_2)]^2 \rangle$ is the structure function of the radial wind velocity, and $V_r' = V_r - \langle V_r \rangle$. It follows from (1.100) that due to turbulent fluctuations of the refractive index in the atmosphere, the one-dimensional probability density function of the real (imaginary) part of the complex echo signal of the pulsed lidar is not Gaussian even at $N_{eff} \gg 1$, since $\sigma_{PS}^2 \neq 1$.

First, we consider the case that turbulent distortions of the probing beam are very small and (1.53) is applicable in (1.104) for the normalized intensity $I_{PN}(z,\rho)$. Then $\sigma_T^2 = B_T(\tau) = 0$ and, according to (1.100) and (1.101), $\sigma_{PS}^2 = 1$,

$$K_{PS}(\tau) = \mu(\tau)\exp[-\tau^2/(2\sigma_P^2)] \tag{1.106}$$

It follows from (1.106) that the correlation time of the echo signal power τ_C determined at the e^{-1} level of $K_{PS}(\tau)$ depends on the pulse duration and the structure function of wind turbulence. If the condition $(4\pi\sigma_P/\lambda)^2 D_V(\Delta p) \ll 1$ is true, we can take $\mu(\tau) = 1$ in (1.106), and then $\tau_C = \sqrt{2}\sigma_P$. The process is Gaussian, because the factorization condition is fulfilled ($K_{PS}(\tau) = [|C_S(\tau)|/\langle P_S \rangle]^2$) [46]. For the pulse duration $\tau_P = 0.4 \; \mu s$ ($\sigma_P = 0.24 \; \mu s$), τ_C equals $0.34 \; \mu s$. The intensification of the wind turbulence leads only to a decrease of τ_C.

1.4.2 Influence of Turbulent Fluctuations of the Refractive Index of Air on Echo Signal Power Statistics

To study the influence of turbulent fluctuations of the refractive index of air on the statistical characteristics of the echo signal power, it is necessary to know the second and fourth intensity moments of the probing beam [see (1.94) and (1.100) through (1.104)], analytical expressions for which are obtained only for the turbulent conditions characterized as the regimes of "weak" and "strong" intensity fluctuations [47–50, 52]. The bistatic approximation is used sometimes to estimate SNR; that is, $\langle I_{PN}^2(R,\rho) \rangle$ in (1.94) is replaced with $\langle I_{PN}(R,\rho) \rangle^2$. The expression for the normalized mean intensity of the probing beam propagating in the turbulent atmosphere has the form [48, 49, 52]

$$\langle I_{PN}(R,\rho)\rangle = \frac{a_0^2}{a_T^2(R)}\exp\left[-\frac{\rho^2}{a_T^2(R)}\right] \tag{1.107}$$

where

$$a_T(R) = a_0\left[\left(1-\frac{R}{F}\right)^2 + \left(1+\frac{4a_0^2}{r_s^2(R)}\right)\left(\frac{\lambda R}{2\pi a_0^2}\right)^2\right]^{1/2} \tag{1.108}$$

is the effective laser beam radius,

$$r_s(R) = \left[1.45(2\pi/\lambda)^2\int\limits_0^R dz\, C_n^2(z)(1-z/R)^{5/3}\right]^{-3/5} \tag{1.109}$$

is the coherence radius of the spherical wave propagating from the point $z = R$ to the point $z = 0$ (under the condition that $r_s \gg l_n$), C_n^2 is the structure characteristic of the refractive index, and l_n is the inner scale of refractive turbulence. After the substitution of (1.107) into (1.94) ($\langle P_S\rangle \equiv$ SNR), we have for the normalized mean power $\langle P_S\rangle_M$ in the bistatic approximation [12, 21, 22]

$$\langle P_S\rangle_M = \frac{\eta E_p c}{4h\nu B_F}\beta_\pi(R)T_A^2(R)\frac{\lambda^2}{\pi a_T^2(R)} \tag{1.110}$$

In [32], a rigorous method was developed for the calculation of $\langle P_S\rangle$, σ_T^2, and $B_T(\tau)$ with the use of numerical simulation of random realizations of $\tilde{P}_S(t)$ [see (1.104)]. Below we describe the numerical simulation algorithm.

As an optical wave propagates in the atmosphere, turbulent inhomogeneities of the refractive index induce fluctuations of the wave amplitude and phase. Let us assume that an optical plane wave passes through a thin layer of atmospheric turbulence. Once a plane wave passes through a thin layer with the thickness δz, it acquires the random phase change $\Psi(\rho)$, where $\rho = \{x, y\}$. To simulate the random phase $\Psi(\rho)$(a random phase screen), we use the model of the spectrum of phase fluctuations of the plane wave $S_\Psi(\kappa) = \int\limits_{-\infty}^{+\infty} d^2\rho\, K_\Psi(\rho)\exp[-2\pi j\,\kappa\rho]$ $[K_\Psi(\rho) = \langle\Psi(\rho)$ $\Psi(0)\rangle$ is the correlation function of the phase and $\kappa = \{\kappa_x, \kappa_y\}$ is the two-dimensional spatial frequency] in the form [47–52]

$$S_\Psi(\kappa) = 0.265\sigma_\Psi^2\frac{(8.42L_n)^2}{[1 + (8.42L_n)^2\kappa^2]^{11/6}} \tag{1.111}$$

where $\sigma_\Psi^2 = 1.273C_n^2 L_n^{5/3}(2\pi/\lambda)^2\delta z$ is the variance of the phase and L_n is the integral (outer) scale of turbulent fluctuations of the refractive index. At $(8.42L_n)^2\kappa^2 \gg 1$, we have from (1.111) the spectrum

$$S_\Psi(\kappa) = 0.382 C_n^2 \lambda^{-2} \delta z |\kappa|^{-11/3} \tag{1.112}$$

which corresponds to the Kolmogorov-Obukhov model of developed turbulence for the air refractive index in the atmosphere.

If we use (1.112), then for the structure function of the phase $D_\Psi(\rho) = [\langle\Psi(\rho) - \Psi(0)]^2\rangle = 2\int_{-\infty}^{+\infty} d^2\kappa S_\Psi(\kappa)[1 - \exp(2\pi j\kappa\rho)]$, we obtain the equation

$$D_\Psi(\rho) = 2.92 C_n^2 (2\pi/\lambda)^2 \delta z |\rho|^{5/3} \tag{1.113}$$

which coincides with the expression for the phase structure function of a plane wave that passes a path of length δz in a turbulent atmosphere [47, 48].

A random phase screen is simulated with the use of the two-dimensional fast Fourier transform (FFT) as [32, 80, 81]

$$\Psi(z, n_x\delta x, n_y\delta x) = \mathrm{Re}\left\{ \sum_{k_x=0}^{N_x-1} \sum_{k_y=0}^{N_y-1} \xi(k_x, k_y) \left[\frac{1}{(\delta x)^2 N_x N_y} S_\Psi\left(\frac{k_x'}{\delta x N_x}, \frac{k_y'}{\delta x N_y} \right) \right]^{1/2} \right.$$
$$\left. \times \exp\left[-2\pi j\left(\frac{k_x n_x}{N_x} + \frac{k_y n_y}{N_y} \right) \right] \right\} \tag{1.114}$$

where δx is the step of the computational grid; $n_{x,y} = 0, 1, 2, \ldots, N_{x,y}$; $\xi(k_x, k_y)$ are complex pseudorandom numbers with the Gaussian statistics and $\langle\xi\rangle = \langle\xi^2\rangle = 0$, $\langle|\xi|^2\rangle = 1$, $\langle\xi(k)\xi(l)\rangle = \delta(|k-l|)$ [$k = \{k_x, k_y\}$, $\delta(|k|)$ is the Dirac function, $\delta(|k|) = 1$ at $|k| = 0$ and $\delta(|k|) = 0$ at $|k| \neq 0$]; $k_{x,y}' = k_{x,y}$ at $k_{x,y} < N_{x,y}/2$, and $k_{x,y}' = N_{x,y} - k_{x,y}$ at $k_{x,y} \geq N_{x,y}/2$.

If the scale L_n exceeds the minimal size of the computational grid $\delta x N_x$ or $\delta x N_y$, then in (1.114) we should use (1.113) taking $S_\Psi(0,0) = S\left(\frac{1}{\delta x N_x}, \frac{1}{\delta x N_y} \right)$. In this case, the outer scale $L_n \sim \delta x N_x$ (if $N_y = N_x$). For the statistics of the simulated screens to correspond to (1.113) at any $|\rho|$ (within the computational grid), one can use the method described in detail in [80, 82]. If the inner scale l_n should be taken into account, then we can use the models of the spectrum reported, for example, in [48]. In addition, in the case (1.111) or (1.112) the size of a cell of the computational grid δx can be considered as an analog of the scale l_n.

As a wave propagates through one thin screen, phase distortions occur, which gradually transform into amplitude distortions during the propagation in a homogeneous medium. Therefore, to simulate the propagation of a laser beam in the turbulent atmosphere, we can divide the entire path with a length R into N_z layers each having a thickness of δz, set a phase screen in front of each layer, and calculate only diffraction of a beam in a homogeneous medium inside a layer before the next phase screen. This approach is referred to as a separation with respect to physical factors [81]. The complex amplitude of the wave field $U(z,\rho)$ [the factor $\exp(-2\pi jz/\lambda)$ is omitted] at the transition from one layer to another is calculated with the use of the two-dimensional direct and inverse FFT by the following algorithm:

$$U((n_z + 1)\delta z, n_x \delta x, n_y \delta x) = N^{-2} \sum_{k_z=0}^{N-1} \sum_{k_y=0}^{N-1} F_U(n_z, k_x, k_y) \exp\left\{-j \frac{\pi \lambda \delta z}{(N \delta x)^2}\right.$$
$$\left. [(k'_x)^2 + (k'_y)^2] + 2\pi j \frac{k_x n_x + k_y n_y}{N}\right\} \tag{1.115}$$

where

$$F_U(n_z, k_x, k_y) = \sum_{n_z=0}^{N-1} \sum_{n_y=0}^{N-1} U(n_z \delta z, n_x \delta x, n_y \delta x)$$
$$\exp\left\{j\Psi(n_z \delta z, n_x \delta x, n_y \delta x) - 2\pi j \frac{k_x n_x + k_y n_y}{N}\right\} \tag{1.116}$$

$n_z = 0, 1, 2, 3, \ldots, N_z - 1$, $N \le N_x$, $N \le N_y$ and $\langle \Psi(n_z) \Psi(n_z + k) \rangle = 0$ at any integer $k \ne 0$.

Note that the parameters δz, N_z, δx, and N should be chosen so that the values of statistical moments of the simulated field $U(R, \boldsymbol{\rho})$ coincide with the well-known analytical solutions [47–52], in particular, with the results of analytical calculations of the mean intensity $\langle I_{PN}(R, \boldsymbol{\rho}) \rangle = \langle |U(R, \boldsymbol{\rho})|^2 \rangle$ [48, 52].

For the simulation of random changes in the intensity of the probing beam in time, the computational grids of phase screens are shifted along one of the axes in the plane transverse to the beam axis (that is, we assume that the Taylor's hypothesis of frozen turbulence is true [58, 47, 83] and the cross wind does not vary along the propagation path). According to (1.104) the simulated array $I_{PN}(R, n_x \delta x, n_y \delta x, t)$ is summed over n_x, n_y, and we obtain one realization $\overline{P}_S(t)$. Then, $\langle P_S \rangle$ and $B_T(\tau)$ $[\sigma_T^2 = B_T(0)]$ are calculated from at least 1,000 independent realizations of $\overline{P}_S(t)$. Below in this section, we present the results of calculations with this algorithm for pulsed CDL with $\lambda = 2$ μm and $a_0 = 2.5$ cm on the assumption that C_n^2 and β_π do not change with the range R (homogeneous path) and $T_A = 1$.

Figure 1.8 shows $\langle P_S(R) \rangle / P_0$ calculated from simulated data (bold curves), $\langle P_S(R) \rangle_M / P_0$ calculated by (1.110) (thin curves), and the ratio $\langle P_S(R) \rangle / \langle P_S(R) \rangle_M$ at moderate $[C_n^2 = 10^{-14} \text{ m}^{-2/3} (1)]$ and strong $[C_n^2 = 10^{-13} \text{ m}^{-2/3} (2)]$ turbulence. The dashed curve shows $\langle P_S(R) \rangle / P_0$ calculated in the absence of turbulence ($C_n^2 = 0$). According to (1.110), the value of $P_0 = \langle P_S(0) \rangle$ is described as

$$P_0 = \frac{\eta E_P c \beta_\pi}{4 h \nu B_F} \frac{\lambda^2}{\pi a_0^2} \tag{1.117}$$

Since the probing beam and the backscattered wave pass through the same turbulent inhomogeneities of the refractive index of air, the lidar echo signal power is always higher than the lidar echo signal power in the case of the bistatic sounding scheme because of the backscatter amplification effect [20, 50]. Moreover, as follows from Figure 1.8, at some R the value of $\langle P_S(R) \rangle / P_0$ can exceed by a small amount even the value corresponding to the nonturbid atmosphere (see, for example, bold curve 2 and dashed curve at $R < 0.7$ km). As follows from the data depicted for the

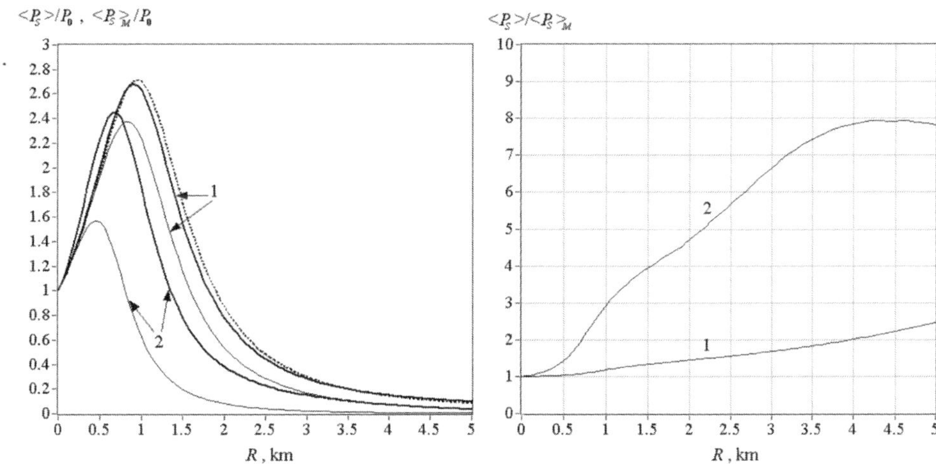

Figure 1.8 Normalized mean power of the lidar echo signal $\langle P_S \rangle/P_0$ (bold curves), $\langle P_S \rangle_M/P_0$ (thin curves), and ratio $\langle P_S \rangle/\langle P_S \rangle_M$ as functions of the range R at $C_n^2 = 10^{-14}\,\mathrm{m}^{-2/3}$ (1) and $C_n^2 = 10^{-13}\,\mathrm{m}^{-2/3}$ (2). The dashed curve is the calculation at $C_n^2 = 0$.

ratio $\langle P_S(R) \rangle/\langle P_S(R) \rangle_M$, the use of the bistatic approximation in SNR calculations can lead to the result being underestimated by nearly an order of magnitude. To check the correctness of the numerical algorithm being used, the ratio $\langle P_S(R) \rangle_M/P_0$, for which an analytical representation is available, was calculated at $R = 5$ km and $C_n^2 = 10^{-13}\,\mathrm{m}^{-2/3}$ with the use of 10,000 independent realizations of $I_{PN}(R,\rho)$. The results obtained coincided with the calculations of (1.110) and (1.117).

The influence of the effect of backscattering amplification on the ratio $\langle P_S(R) \rangle/\langle P_S(R) \rangle_M$ was analyzed on the assumption of full (ideal) matching of the transverse dimensions, axes, and wavefronts of the reference and sensing beams. The influence of mismatching of axes and beam wavefronts on the CDL operation is considered in [14], where the phase approximation of the Huygens-Kirchhoff method was used. Figure 1.9 shows the ratio $\langle P_S(R) \rangle/\langle P_S(R) \rangle_M$ calculated at different angles of mismatch of the reference and probing beams $\Delta\theta_s$ based on numerical simulation. For this purpose, $I_{PN}^2(R,\rho)$ was replaced with $I_{PN}(R,\rho)I_{LN}(R,\rho)$ in (1.104), where $I_{LN}(R,\rho)$ is the normalized intensity of the equivalent reference beam propagating in the turbulent atmosphere at an angle $\Delta\theta_s$ to the axis of the sensing beam.

The calculations have been performed for a homogeneous path of different length R and constant $C_n^2 = 10^{-13}\,\mathrm{m}^{-2/3}$. The value of $\langle P_S \rangle_M$ was calculated by (1.110) with the wavefront mismatch ($\Delta\theta_s = 0$) neglected. One can see that at $C_n^2 = 10^{-13}\,\mathrm{m}^{-2/3}$ the effect of backscattering amplification becomes marked at mismatch angles $\Delta\theta_s \leq 20\,\mu\mathrm{rad}$.

The parameter σ_T^2 in (1.100) is the relative variance [see (1.102)] of the echo signal power $\overline{P}_S$ averaged over microphysical parameters of the scattering medium. The value of σ_T^2 is nonzero due to the presence of turbulent inhomogeneities of the refractive index in the propagation path of the probing beam ($C_n^2 \neq 0$). Figure 1.10 shows the calculated dependence of σ_T on the distance R at different C_n^2. One can see that the value of σ_T does not exceed 0.5. Consequently, the relative variance of the echo signal power σ_{PS}^2 [see (1.100)] does not exceed 1.5.

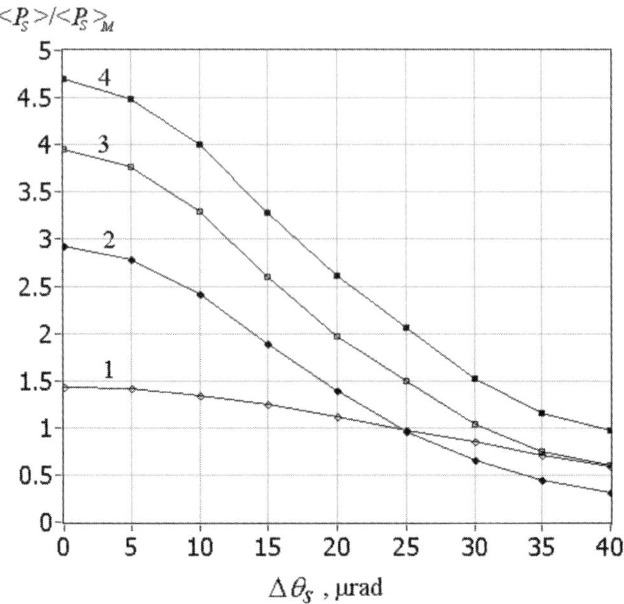

Figure 1.9 Ratio $\langle P_S \rangle / \langle P_S \rangle_M$ as a function of the mismatch angle of wavefronts of the sensing and reference beams at $C_n^2 = 10^{-13}\,\text{m}^{-2/3}$ and $R = 0.5$ km (1), 1 km (2), 1.5 km (3), and 2 km (4).

In [33], one can find the results of the asymptotic analysis of the variance of power fluctuations σ_{PS}^2 for the conditions of weak optical turbulence and the processing of experimental data confirming this conclusion.

The parameter $B_T(\tau)$ in (1.101) is the temporal correlation function [see (1.103)] of the normalized echo signal power $\overline{P}_S / \langle P_S \rangle$. Figure 1.11 shows the correlation coefficient $K_T(\tau) = B_T(\tau)/\sigma_T^2$ calculated at $C_n^2 = 10^{-13}\,\text{m}^{-2/3}$ and $R = 2$ km. It can be

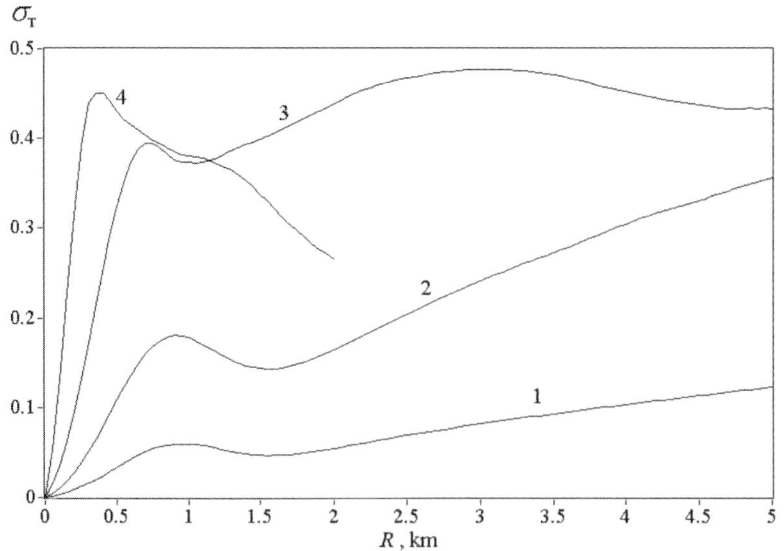

Figure 1.10 Dependence of σ_T on the range R at $C_n^2 = 10^{-15}\,\text{m}^{-2/3}$ (1), $C_n^2 = 10^{-14}\,\text{m}^{2/3}$ (2), $C_n^2 = 10^{-13}\,\text{m}^{-2/3}$ (3), and $C_n^2 = 10^{-12}\,\text{m}^{-2/3}$ (4).

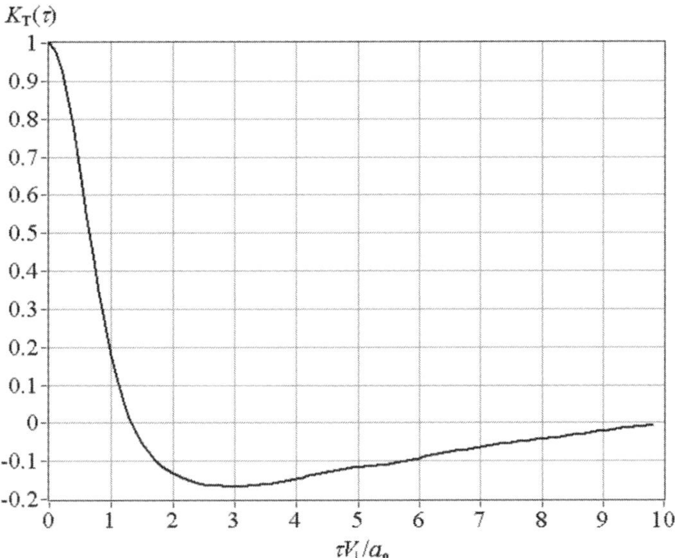

Figure 1.11 Temporal correlation coefficient of fluctuations of the echo signal power averaged over microphysical parameters of the scattering medium.

seen that the characteristic time correlation scale of echo signal power fluctuations caused by the refractive turbulence coincides in the order of magnitude with the time, for which refractive index inhomogeneities are transferred by the cross wind with the velocity $V_\perp$ to a distance equal to the initial radius of the sensing beam a_0. The longer negative correlation of echo signal power fluctuations is likely caused by random displacements of the sensing beam waist along the optical axis, when at long delays τ it is highly probable that fluctuations of $\bar{P}_S(t) - \langle P_S \rangle$ and $\bar{P}_S(t + \tau) - \langle P_S \rangle$ have opposite signs. The variation of C_n^2 and R does not lead to somewhat significant differences from the depicted results.

Section 1.3 presents the calculated characteristics of an echo signal for the case of the type of CO_2 ($\lambda = 10.6\ \mu m$) lidar that is used most widely among existing cw CDLs. Measurements with this lidar have limitations in the range R caused by the worsening of spatial resolution (longitudinal dimension of the sensing volume Δz increases) with the increase of focal length of the probing beam F. According to the calculations of R and Δz by (1.54) and (1.57), for a typical lidar with $a_0 = 7.5$ cm (telescope diameter of 30 cm), the conditions $F - R < 100$ m and $\Delta z < 1$ km can be true, only if the probing beam is focused to a distance no longer than roughly 1 km. The calculations of the SNR with the use of (1.51) and the algorithm described in this section for numerical simulation of laser beam ($\lambda = 10.6\ \mu m$) propagation in the turbulent atmosphere at $F = 1$ km and different C_n^2 have shown that the decrease of SNR (compared to the case of $C_n^2 = 0$) is roughly 10% at $C_n^2 = 10^{-13}$ m$^{-2/3}$ and 50% at $C_n^2 = 10^{-12}$ m$^{-2/3}$. Thus, for the cw CO_2 lidar at $C_n^2 \leq 10^{-13}$ m$^{-2/3}$, the influence of optical (refractive) turbulence on SNR can be neglected. We should emphasize, however, that pulsations of the refractive index can significantly affect the values of R and Δz (and, consequently, the statistics of lidar estimate of the radial velocity V_r), as discussed in Chapter 2 (Section 2.5).

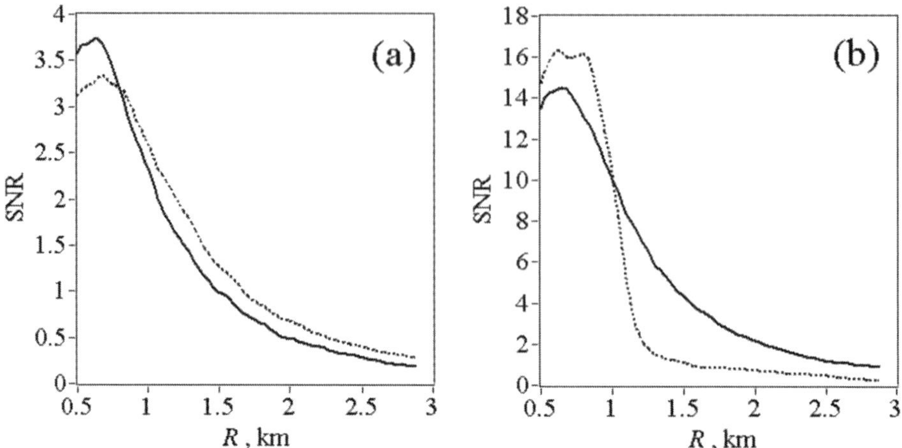

Figure 1.12 Signal-to-noise ratio versus range R at elevation angle $\varphi = 2°$ (solid curves) and $\varphi = 28°$ (dashed curves) measured on (a) August 27, 2003, and (b) August 28, 2003, at Tarbes airfield. (© 2005 American Meteorological Society. Used with permission. From [77].)

In the case of pulsed CDLs, sensing range R limitations are connected with the level of echo signal power with respect to the noise level. Using the data of Figure 1.8, for $\langle P_S(R) \rangle / P_0 \equiv \mathrm{SNR}(R)/P_0$ we can obtain the dependence of the SNR on the range R. For the same parameters, which were used to estimate SNR by (1.97) [$\lambda = 2$ μm, $\beta_\pi = 1.3 \cdot 10^{-7}$(m sr)$^{-1}$, $\eta = 0.8$, $E_P = 2$ mJ, $B_F = 50$ MHz, $T_A = 1$, and $a_0 = 2.5$ cm], from (1.117) we find $P_0 = 6.2$. Then, according to the data in Figure 1.8, at $C_n^2 = 10^{-13}$ m$^{-2/3}$ (bold curve 2) SNR varies from 1 to 15 in the range 0.5 km $\le R \le 3$ km. Figure 1.12 shows the results of the field experiment on measurement of SNR with a 2-μm pulsed CDL with the above parameters and the probing beam focused to a distance of 1.5 km [77]. The information about β_π and C_n^2 during these measurements is unavailable. Nevertheless, we can see from Figure 1.12(b) that the dependence of SNR on R at $\varphi = 2°$ corresponds quite well to that shown by bold curve 2 in Figure 1.8. However, the SNR(R) measured a day earlier at $\varphi = 2°$ (Figure 1.12(a)) is roughly four times smaller. Such a wide difference may be connected with the variability of the backscattering coefficient β_π, which is proportional to the concentration of aerosol particles. Actually, for example, after aerosol washout by rain the concentration of particles may decrease by an order of magnitude and then return to the background level over time [53, 54].

Along with fluctuations of the probing beam intensity caused by optical turbulence (fluctuations of the refractive index), there are turbulent variations of the concentration of aerosol particles (backscatter coefficient β_π) [84]. That is why the characteristics σ_T^2 and $B_T(\tau)$ cannot be determined correctly from the field experiment data.

1.5   Conclusions

1. For continuous-wave CDLs, in which the probing beam is focused to short distances (less than 50m for the CO_2 lidar), a very small sensing volume V_{eff} is formed, which contains on average less than one efficiently scattering aerosol

particle $N_{eff} < 1$. In measurements taken by pulsed CDLs for a practical V_{eff}, the number $N_{eff} \gg 1$ (for 2-μm lidar, $N_{eff} > 10^4$).

2. When the cw CDL forms a sensing volume with a very low number of efficiently scattering particles $N_{eff} < 0.1$, the relative variance of the echo signal power σ_{PS}^2 can exceed by tens and hundreds the value of σ_{PS}^2 in the case of $N_{eff} > 10$.

3. The temporal correlation function of the echo signal power of cw CDLs has a two-scale character at the focusing of the probing beam to distances from 30m to 150m. The smallest temporal scale is determined by the time, for which the distance between scattering particles transferred by the turbulent wind flow changes, on average, by a value of about the wavelength. Another scale is determined by the time of transfer of individual large particles by the cross wind over the cross section of the probing beam in the waist zone. In the case of the CO_2 lidar, the correlation time of echo signal power fluctuations can vary from 0.5 μs to 5 ms.

4. When the probing beam of a cw CDL is focused to a distance $F < 100$m, one-dimensional probability density functions of the amplitude and power of the echo signal differ widely from the Rayleigh and exponential distributions following from the Gaussian statistics of the echo signal. Deviations from the Gaussian statistics increase with a decrease of F.

5. Due to the effect of backscattering amplification, the mean power of the echo signal of a pulsed CDL can exceed by several times (nearly eight times for the 2-μm lidar) the mean power of the echo signal at the bistatic sensing. Correspondingly, estimates of the echo signal power in the bistatic approximation can be significantly understated.

6. Due to turbulent fluctuations of the refractive index of air, the lidar echo signal statistics for a pulsed CDL differ from Gaussian. However, calculations and experimental results have shown that the variance of the echo signal power can maximally exceed by 1.5 times that for the normal probability distribution of the echo signal.

7. The correlation time for fluctuations in the echo signal power of a pulsed CDL caused by turbulent fluctuations of the refractive index is determined by the time, for which the refractive index turbulent inhomogeneities are transferred to the distance equal to the initial radius of the probing beam.

References

[1]   Goulard, R. (Ed.), *Combustion Measurements: Modern Techniques and Instrumentation*, Academic Press, New York, 1976, p. 483.

[2]   Klimkin, V., Papyrin, A., and Soloukhin, R., *Optical Methods for Recording of Fast Processes*, Novosibirsk, 1980.

[3]   Dubnishcheev, Yu, N., and Rinkevichyus, B.R., *Methods of Laser Doppler Anemometry*, Nauka, Moscow, 1982, p. 304.

[4]   Chanin, M.L., et al., "A Doppler lidar for measuring winds in the middle atmosphere," *Geophys. Re. Lett.* Vol. 16, 1989, pp. 1273–1276.

[5]   Rees, D., and McDermid, I.S., "Doppler lidar atmospheric wind sensor: Reevaluation of a 355-nm incoherent Doppler lidar," *Applied Optics*, Vol. 29, 1990, pp. 4133–4157.

[6]   Rees, D., et al., "The Doppler wind and temperature system of Alomar lidar," *J. Atmos. Terr. Phys.*, Vol. 58, 1996, pp. 1827–1842.

[7]   McGill, M.J., Skinner, W.R., and Irgang, T.D., "Analysis techniques for the recovery of winds and backscatter coefficients from a multiple channel incoherent Doppler lidar," *Applied Optics*, Vol. 36, 1997, pp. 1253–1268.

[8]   Korb, C.L., et al., "Theory of the double-edge technique for Doppler lidar wind measurements," *Applied Optics*, Vol. 37, 1998, pp. 3097–3104.

[9]   McGill, M.J., et al., "Modeling the performance of direct-detection Doppler lidar systems including cloud and solar background variability," *Applied Optics*, Vol. 38, 1999, pp. 6388–6397.

[10] Protopopov, V.V., and Ustinov, N.D., *Laser Heterodyning*, Nauka, Moscow, 1985, p. 288.

[11] Frehlich, R.G., "Conditions for optimal performance of monostatic coherent laser radar," *Optics Letter*, Vol. 15, 1990, pp. 643–645.

[12] Frehlich, R.G., and Kavaya, M.J., "Coherent laser radar performance for general atmospheric turbulence," *Applied Optics*, Vol. 30, 1991, pp. 5325–5337.

[13] Rye, B.J., and Frehlich, R.G., "Optimal truncation and optical efficiency of an apertured coherent lidar focused on an incoherent backscatter target," *Applied Optics*, Vol. 31, No. 15, 1992, pp. 2891–2899.

[14] Frehlich, R.G., "Effects of refractive turbulence on coherent laser radar," *Applied Optics*, Vol. 32, No. 12, 1993, pp. 2122–2139.

[15] Frehlich, R.G., "Optimal local oscillator field for a Monostatic coherent laser radar with a circular aperture," *Applied Optics*, Vol. 32, No. 24, 1993, pp. 4569–4577.

[16] Frehlich, R.G., "Heterodyne efficiency for a coherent laser radar with diffuse or aerosol targets," *Journal of Modern Optics*, Vol. 41, No. 11, 1994, pp. 2115–2129.

[17] Zhao, Y., Post, M.J., and Hardesty, R.M., "Receiving efficiency of monostatic pulsed coherent lidars. 1: Theory," *Applied Optics*, Vol. 29, No. 28, 1990, pp. 4111–4119.

[18] Lawrence, T.R., et al., "A laser velocimeter for remote wind sensing," *Review of Scientific Instruments*, Vol. 43, No. 3, 1972, pp. 512–518.

[19] Sonnenschein, C.M., and Horrigan, F.A., "Signal-to-noise relationship for coaxial systems that heterodyne backscatter from the atmosphere," *Applied Optics*, Vol. 10, No. 7, 1971, pp. 1600–1604.

[20] Clifford, S.F., and Wandzura S., "Monostatic heterodyne lidar performance: the effect of the turbulent atmosphere," *Applied Optics*, Vol. 20, No. 3, 1981, pp. 514–516.

[21] Hawley, J.G., et al., "Coherent launch-site atmospheric wind sounder: Theory and experiment," *Applied Optics*, Vol. 32, 1993, pp. 4557–4567.

[22] Targ, R., et al., "Coherent lidar airborne windshear sensor: Performance evaluation," *Applied Optics*, Vol. 30, 1991, pp. 2013–2026.

[23] Hardesty, R.M., et al., "Characteristics of coherent lidar returns from calibration targets and aerosols," *Applied Optics*, Vol. 20, No. 21, 1981, pp. 3763–3769.

[24] Harris, M., et al., "Single-particle laser Doppler anemometry at 1.55 μm," *Applied Optics*, Vol. 40, No. 6, 2001, pp. 969–973.

[25] Frehlich, R.G., and Yadlowsky, M.J., "Performance of mean-frequency estimators for Doppler radar and lidar," *Journal of Atmospheric and Oceanic Technology*, Vol. 11, No. 5, 1994, pp. 1217–1230.

[26] Hardesty, R.M., "Performance of a discrete spectral peak frequency estimator for Doppler wind velocity measurements," *IEEE Trans. on Geoscience and Remote Sensing*, Vol. GE-24, No. 5, 1986, pp. 777–783.

[27] Banakh, V.A., et al., "Measurements of turbulent energy dissipation rate with a cw Doppler lidar in the atmospheric boundary layer," *Journal of Atmospheric and Oceanic Technology*, Vol. 16, No. 8, 1999, pp. 1044–1061.

[28] Banakh, V.A., Werner, Ch., and Smalikho, I.N., "The effect of aerosol microstructure on the error in estimating wind velocity with a Doppler lidar," *Atmos. Oceanic Opt.*, Vol. 13, No. 8, 2000, pp. 685–691.

[29] Banakh, V.A., Werner, Ch., and Smalikho, I.N., "Effect of turbulent fluctuations of refractive index on the time spectrum of wind velocity measured by Doppler lidar," *Atmos. Oceanic Opt.*, Vol. 13, No. 9, 2000, pp. 741–746.

[30] Banakh, V.A., et al., "Multi-aperture coherent reception in the turbulent atmosphere," *Atmos. Oceanic Opt.*, Vol. 13, No. 10, 2000, pp. 850–854.

[31] Banakh, V.A., Smalikho, I.N., and Werner, Ch., "Effect of aerosol particle microstructure on statistics of cw Doppler lidar signal," *Applied Optics*, Vol. 39, No. 30, 2000, pp. 5393–5402.

[32] Banakh, V.A., Smalikho, I.N., and Werner, Ch., "Numerical simulation of effect of refractive turbulence on the statistics of a coherent lidar return in the atmosphere," *Applied Optics*, Vol. 39, No. 30, 2000, pp. 5403–5414.

[33] Smalikho, I.N., "Pulse coherent lidar echo-signal power fluctuations caused by atmospheric turbulence," *Atmos. Oceanic Opt.*, Vol. 25, No. 1, 2012, pp. 82–88.

[34] Banakh, V.A., Werner, Ch., and Smalikho, I.N., "Effect of refractive turbulence on Doppler lidar operation in atmosphere. Numerical simulation," *Proc. 10th Coherent Laser Radar Conference*, Mount Hood, OR, USA, 28 June–2 July 1999, pp. 82–85.

[35] Banakh, V.A., Werner, Ch., and Smalikho, I.N., "Effect of aerosol particle microstructure on accuracy of cw Doppler lidar estimate of wind velocity," *Proc. 10th Coherent Laser Radar Conference*, Mount Hood, OR, USA, 28 June–2 July 1999, pp. 132–135.

[36] Banakh, V.A., Smalikho, I.N., and Werner, Ch., "Aerosol particle microstructure dependence of accuracy of cw Doppler estimate of wind velocity," *SPIE Proc. Atmospheric Propagation, Adaptive Systems, and Laser Radar Technology for Remote Sensing*, Barcelona, Spain, 25–28 September 2000, Vol. 4167, pp. 270–280.

[37] Born, M., and E. Wolf, *Principles of Optics*, 6th ed., Pergamon, New York, 1986.

[38] Shifrin, K.R., *Light Scattering in Turbid Medium*, Gostekhizdat, Moscow, 1951, p. 288.

[39] Hulst, van de H.C., *Light Scattering by Small Particles*, Dover Publications, New York, 1981.

[40] Zuev, V.E., and Kabanov, M.V., *Transfer of Optical Signals in the Earth's Atmosphere (under Conditions of Noise)*, Sov. Radio, Moscow, 1977, p. 368.

[41] Zuev, V.E., *Propagation of Laser Radiation in the Atmosphere*, Radio i Svyaz, Moscow, 1981, p. 288.

[42] Ishimaru, A., *Wave Propagation and Scattering in Random Media, Volume 1: Single Scattering and Transport Theory*, Academic Press, New York, 1978, p. 572.

[43] Ishimaru, A., *Wave Propagation and Scattering in Random Media, Volume 1I: Multiple Scattering, Turbulence, Rough Surfaces and Remote Sensing*, Academic Press, New York, 1978, p. 572.

[44] Zuev, V.E., and Kabanov, M.V., *Modern Problems of Atmospheric Optics. Part 4. Optics of Atmospheric Aerosol*, Gidrometeoizdat, Leningrad, 1987, p. 256.

[45] Van Vliet, K.M., "Noise limitations in solid state photodetectors," *Applied Optics*, Vol. 6, No. 7, 1967, pp. 1145–1169.

[46] Crosignani, B., Di Porto, P., and Bertolotti, M., *Statistical Properties of Scattered Light*, Academic Press, New York, 1975.

[47] Tatarskii, V.I., *Wave Propagation in a Turbulent Medium*, McGraw-Hill, New York, 1961.

[48] Gurvich, A.S., et al., *Laser Radiation in the Turbulent Atmosphere*, Nauka, Moscow, 1976, p. 280.

[49] Mironov, V.L., *Laser Beam Propagation in the Turbulent Atmosphere*, Nauka, Novosibirsk, 1981, p. 248.

[50] Banakh, V.A., and Mironov, V.L., *Lidar Propagation of Laser Radiation in the Turbulent Atmosphere*, Nauka, Novosibirsk, 1986, p. 174.

[51] Lukin, V.P., *Atmospheric Adaptive Optics*, Nauka, Novosibirsk, 1986, p. 248.

[52] Zuev, V.E., Banakh, V.A., and Pokasov, V.V., *Modern Problems of Atmospheric Optics. Part 5. Optics of the Turbulent Atmosphere*, Gidrometeoizdat, Leningrad, 1988, p. 272.

[53] Krekov, G.M., and Rakhimiv, R.F., *Optical-Lidar Model of Continental Aerosol*, Nauka, Novosibirsk, 1982, p. 197.

[54] Zuev, V.E., and Krekov, G.M., *Optical Models of the Atmosphere*, Gidrometeoizdat, Leningrad, 1986, p. 256.

[55] Köpp, F., Schwiesow, R.L., and Werner, Ch., "Remote measurements of boundary layer wind profiles using a cw Doppler lidar," *Journal of Climate Applied Meteorology*, 1984, Vol. 23, No. 1, pp. 148–158.

[56] Bendal, J.S., and Piersol, A.C., *Random Data: Analysis and Measurement Procedures*, Wiley, New York, 1971.

[57] Frehlich, R.G., "Estimation of velocity error for Doppler lidar measurements," *Journal of Atmospheric and Oceanic Technology*, Vol. 18, No. 10, 2001, pp. 1628–1639.

[58] Monin, A.S., and Yaglom, A.M., *Statistical Fluid Mechanics, Volume II: Mechanics of Turbulence*, MIT Press, Cambridge, MA, 1971.

[59] Storm, R., *Probability Theory, Mathematical Statistics, Statistical Quality Control* [Russian translation], Mir, Moscow, 1970, p. 368.

[60] Hall, F.F., et al., "Wind measurement accuracy of the NOAA pulsed infrared Doppler lidar," *Applied Optics*, Vol. 23, No. 15, 1984, pp. 2503–2506.

[61] Kane, T.J., et al., "Coherent laser radar at 1.06 μm using Nd:YAG lasers," *Optics Letters*, Vol. 12, 1987, pp. 232–241.

[62] Kavaya, M.J., et al., "Remote wind profiling with a solid-state Nd:YAG coherent lidar systems," *Optics Letters*, Vol. 14, 1989, pp. 776–778.

[63] Henderson, S.W., et al., "Eye-safe coherent laser radar system at 2 μm using Tm. Ho: YAG lasers," *Optics Letters*, Vol. 16, 1991, pp. 773–775.

[64] Henderson, S.W., et al., "Coherent laser radar at 2 μm using solid-state lasers," *IEEE Trans. Geosci. Remote Sens.*, Vol. 31, No. 1, 1993, pp. 4–15.

[65] Gordienko, V.M., et al., "Coherent CO_2 lidars for measuring wind velocity and atmospheric turbulence," *Optical Engineering*, Vol. 33, No. 10, 1994, pp. 3206–3213.

[66] Huffaker, R.M., and Hardesty, R.M., "Remote sensing of atmospheric wind velocities using solid-state and CO_2 coherent laser systems," *Proc. IEEE*, Vol. 84, 1996, pp. 181–204.

[67] Drobinski, Ph., Dabas, A., and Flamant, P. H., "Remote measurements of turbulent wind spectra by heterodyne Doppler lidar technique," *Journal of Applied Meteorology*, Vol. 39, 2000, pp. 2434–2451.

[68] Werner, Ch., et al., "Wind infrared Doppler lidar instrument," *Optical Engineering*, Vol. 40, No. 1, 2001, pp. 115–125.

[69] Reitebuch, O., et al., "Experimental validation of wind profiling performed by the airborne 10 μm heterodyne Doppler lidar WIND," *Journal of Atmospheric and Oceanic Technology*, Vol. 18, No. 8, 2001, pp. 1331–1344.

[70] Reitebuch, O., et al., "Determination of airflow across the alpine ridge by a combination of airborne Doppler lidar, routine radiosounding and numerical simulation," *Quart. J. Roy. Meteorol. Soc.*, Vol. 129, 2003, pp. 715–728.

[71] Ando, T., et al., "Development of low cost all coherent Doppler LIDAR (CDL) system," *Proc. 13th Coherent Laser Radar Conference*, Kamakura, Japan, 2005, pp. 170–173.

[72] Hannon, S.M., Pelk, J.V., and Henderson, S.W., "Recent wind and aerosol measurements using WindTracer," *Proc. 13th Coherent Laser Radar Conference*, Kamakura, Japan, 2005, pp. 84–87.

[73] Ando, T, Kameyama, S., and Hirano, Y., "All-fiber coherent Doppler LIDAR technologies at Mitsubishi Electric Corporation," *Proc. 14th International Symposium for the Advancement of Boundary Layer Remote Sensing*, IOP Conf. Series, *Earth and Environmental Science*, Vol. 1, 2008, doi:10.1088/1755-1307/1/1/012011.

[74] Davis, J.C., et al., "Doppler lidar measurements of boundary layer winds and sensible heat flux," *Proc. 14th International Symposium for the Advancement of Boundary Layer Remote Sensing*, IOP Conf. Series, *Earth and Environmental Science*, Vol. 1, 2008, doi:10.1088/1755-1307/1/1/012029.

[75] Hannon, S.M., "Wind resource assessment using long range pulsed Doppler lidar," *Proc. 15th Coherent Laser Radar Conference*, Toulouse, France, 2009, pp. 59–62.

[76] Parmentier, R., et al., "Windcube™ pulsed lidar compact wind profiler: Overview on more than two years of comparison with calibrated sensors at different location," *Proc. 15th Coherent Laser Radar Conference*, Toulouse, France, 2009, pp. 267–270.

[77] Smalikho, I.N., Köpp, F., and Rahm, S., "Measurement of atmospheric turbulence by 2-μm Doppler lidar," *Journal of Atmospheric and Oceanic Technology*, Vol. 22, No. 11, 2005, pp. 1733–1747.

[78] Banakh, V.A., et al., "Measurements of wind velocity and direction with coherent Doppler lidar in conditions of a weak echo signal," *Atmos. Oceanic Opt.*, Vol. 23, No. 5, 2010, pp. 381–388.

[79] Ray B.J. "Refractive-turbulent contribution to incoherent backscatter heterodyne lidar returns," *J. Opt. Soc. Am.*, Vol. 71. No. 8, 1981, pp. 687–691.

[80] Frehlich, R., "Simulation of laser propagation in a turbulent atmosphere," *Applied Optics*, Vol. 39, No. 3, 2000, pp. 393–397.

[81] Kandidov, V.P., "Monte Carlo method in nonlinear statistical optics," *Physics-Uspekhi*, Vol. 39, No. 12, 1996, pp. 1243.

[82] Banakh, V.A., Smalikho, I.N., and Falits, A.V., "Efficiency of the use of the subharmonic method in computer simulation of laser beam propagation in a turbulent atmosphere," *Atmos. Oceanic Opt.*, Vol. 25, No. 2, 2012, pp. 106–109.

[83] Byzova, N.L., Ivanov, V.N., and Garger, E.K., *Turbulence in Atmospheric Boundary Layer*, Gidrometeoizdat, Leningrad, 1989, p. 263.

[84] Balin, Yu, S., Razenkov, I.A., and Rostov, A.P., "Lidar studies of fluctuations of aerosol concentration in the ground atmospheric layer," *Atmos. Oceanic Opt.*, Vol. 7, No. 7, 1994, pp. 513–516.

Statistics of Lidar Estimates of the Radial Velocity and Doppler Spectrum Width

2.1 Introduction

The CDL echo signal carries information about the radial velocity of aerosol particles entrained by a wind flow. The wind flow in the atmosphere is always turbulent. Therefore, the velocities of scattering particles differ. A question arises: Which velocity does the Doppler frequency shift estimated from the lidar data correspond to? Since the correlation time of the echo signal power is on the order of a microsecond, a stable estimation of the spectral density of the echo signal power can be obtained from lidar measurement data over a period of time of around tens or a few milliseconds. For this time period, the spatial distribution of the radial velocity remains nearly unchanged. With neglected noise and instrumental spectral broadening, the estimate of the spectral density is a weighted distribution of radial velocities of the particles being in the sensing volume. Since the sensing volume is finite, this distribution is not symmetric about the spectral maximum, and the Doppler frequency corresponding to the spectral maximum by no means always corresponds to the radial velocity of wind at the center of the sensing volume (or at the point of maximum intensity of the probing beam for cw CDLs). With allowance for the Doppler relation [see (1.17) in Chapter 1], the first spectral moment is the estimate of the mean value of radial velocities of particles being in the sensing volume, that is, the estimate of the radial wind velocity averaged over the sensing volume at a fixed instant of time. The second central spectral moment (or, in other words, the squared width of the Doppler spectrum) is an estimate of the "instantaneous" variance of radial velocities of scattering particles.

The information about wind turbulence parameters is usually derived from measurements of statistical characteristics of wind velocity such as its variance, structure function, and spectrum [1–9]. In contrast to point measurers (for example, cup or acoustic anemometers), the analysis of velocity characteristics measured by a coherent Doppler lidar requires the spatial averaging of the radial velocity over the sensing volume to be taken into account. For this purpose, it is necessary to know the weighting function of the spatial averaging of velocity. In [10, 11] for the case of cw CDLs, an equation for the weighting function has been derived. This equation has been used in [12, 13] to obtain the relationship between the mathematical expectation of the second spectral moment and the dissipation rate of the turbulent energy. However, the conditions of applicability of the weighting function in this form were not determined. The point is that the influence of turbulent pulsations of the refractive index of air on the formation of the sensing volume, which may be significant for cw CDLs, was neglected in [10, 11]. The possibilities for gaining information about the turbulent energy dissipation rate from the temporal structure function and the time spectrum of wind velocity measured by a cw CDL are studied in [14–24].

The radial velocity averaged over the lidar sensing volume is estimated with some error due to echo signal fluctuations and noise. The error in lidar estimates of the radial wind velocity calculated within the framework of the theory developed in [25, 26], which assumed the Gaussian statistics of the echo signal, in the case of cw CDLs is significantly understated (by an order of magnitude at a small sensing volume) compared to the data of field experiments [16, 20]. Section 1.3 demonstrated that the one–dimensional probability density of the echo signal of a cw CDL can differ from the normal distribution due to the influence of the microstructure of aerosol particles. The influence of this factor on the error of estimation of the radial velocity averaged over the sensing volume was considered in [27].

In contrast to cw lidar, investigation of the spatial structure of turbulence with a pulsed CDL does not require application of the Taylor hypothesis of frozen turbulence [5, 8, 28]. The structure function of the radial velocity and the width of the Doppler spectrum estimated from the raw pulsed CDL data are studied in [29–32] and R. Frehlich's papers [33–35]. The most important results of these studies, obtained independently by us and by Frehlich, are published, respectively, in [29] and [33]. The results are not contradictory and form the basis for the development of methods that provide information about wind turbulence from estimates of the spatial structure function of the radial velocity measured by a lidar.

Chapter 2 is devoted to the study of the statistical characteristics of the radial velocity and the width of the Doppler spectrum for both cw and pulsed CDLs and summarizes the results of a number of studies [14–17, 19–24, 27, 29–32, 36–40].

2.2 Estimation of Spectral Moments

2.2.1 Weighting Functions of Averaging over the Sensing Volume

First, we consider the case where the signal-to-noise ratio is large (i.e., SNR $\gg$ 1) when the term $Z_N(t)$ in (1.28) is negligibly small. From L of successively measured arrays of the complex normalized echo signal $Z_S(t,l)$, where $l = 1, 2, 3, ..., L$ is a serial number of an array, we can estimate the spectral density of the echo signal power as follows [41]:

$$\hat{S}_s(f) = \frac{1}{2LT_W} \sum_{l=1}^{L} \left| \int_{-\infty}^{\infty} dt\, W(t) Z_s(t,l) \exp(-2\pi jft) \right|^2 \tag{2.1}$$

where $W(t)$ is the function of the time window symmetric about the point $t = 0$ ($|W(0)| = 1$), and $T_W = \int_{-\infty}^{+\infty} dt\, |W(t)|^2$ is the effective width of the window, which determines the frequency resolution $\Delta f = 1/T_W$ in the measured spectrum. The estimates of the zero $\hat{P}_s$, first $\hat{f}_r$, and second (central) $\hat{\sigma}_f^2$ spectral moments are calculated with these equations [26]:

$$\hat{P}_S = \int_{-\infty}^{+\infty} df\, \hat{S}_S(f) \tag{2.2}$$

$$\hat{f}_r = \hat{P}_S^{-1} \int_{-\infty}^{+\infty} df\, \hat{S}_S(f) \tag{2.3}$$

$$\hat{\sigma}_f^2 = \hat{P}_S^{-1} \int_{-\infty}^{+\infty} df\, (f - \hat{f}_r)^2\, \hat{S}_S(f) \tag{2.4}$$

Here, $\hat{P}_S$ is an estimate of the normalized power of the echo signal (SNR), for which, on substitution of (2.1) into (2.2) and integration with respect to f, we have

$$\hat{P}_S(f) = \frac{1}{2LT_W} \sum_{l=1}^{L} \int_{-\infty}^{+\infty} dt\, |W(t)|^2 |Z_S(t,l)|^2 \tag{2.5}$$

For stationary conditions, this estimate is unbiased, that is, $\langle \hat{P}_S \rangle = \langle P_S \rangle$. The values of $\hat{f}_r$ and $\hat{\sigma}_f$ with allowance for the Doppler relation [see (1.17)] are proportional to the estimates of the radial velocity $\hat{V}_r$ and the width of the Doppler spectrum $\hat{\sigma}_S$ (width of the echo signal power spectrum in velocity units), that is,

$$\hat{f}_r = (2/\lambda)\hat{V}_r \tag{2.6}$$

and

$$\hat{\sigma}_f = (2/\lambda)\hat{\sigma}_S \tag{2.7}$$

In the case of cw CDL, the measurement of one spectrum lasts $\Delta t \sim LT_W$. At $T_W = 50$ μs and $L = 1{,}000$, $\Delta t \sim 50$ ms. Section 1.3 showed that the correlation scale of echo signal power fluctuations does not exceed 5 ms. Consequently, for a measurement time of 50 ms, fluctuations of the estimate of the spectral density of the echo signal power are significantly averaged. On the other hand, for this time the radial wind velocity $V_r(z)$ remains nearly unchanged within the sensing volume. Therefore, for analysis of the spectral density of the echo signal power and spectral moments, we use the conditional averaging, that is, in (2.1) and (2.5) we replace the operator $L^{-1}\sum_{l=1}^{L}(\dots)$ with the operator of averaging $\langle\dots\rangle_m$ over an ensemble of realizations of microstructural parameters: number of particles in the elementary volume $dN_s/(dz d^2\rho)$, backscattering cross sections $|\alpha_i|^2$, and coordinates $\{z_i, \rho_i\}$ of particles. In this case, in place of $\hat{S}_S(f)$, $\hat{P}_S$, $\hat{f}_r$, $\hat{\sigma}_f$, $\hat{V}_r$, and $\hat{\sigma}_S$, we use the variables designated as, respectively, $\overline{S}_S(f)$, $\overline{P}_S$, $\overline{f}_r$, $\overline{\sigma}_f$, $\overline{V}_r$, and $\overline{\sigma}_S$. Thus,

$$\overline{S}_S(f) = \frac{1}{2T_W}\left\langle \left| \int_{-\infty}^{\infty} dt\, W(t) Z_S(t,l) \exp(-2\pi j f t) \right| \right\rangle_m \tag{2.8}$$

$$\overline{P}_S = \frac{1}{2T_W}\left\langle \int_{-\infty}^{+\infty} dt\,|W(t)|^2\,|Z_S(t,l)|^2 \right\rangle_m \tag{2.9}$$

$$\overline{f}_r = \overline{P}_S^{-1}\int_{-\infty}^{+\infty} df\,f\,\overline{S}_S(f) \tag{2.10}$$

$$\overline{\sigma}_f^2 = \overline{P}_S^{-1}\int_{-\infty}^{+\infty} df\,(f-\overline{f}_r)^2\,\overline{S}_S(f) \tag{2.11}$$

$$\overline{V}_r = (\lambda/2)\overline{f}_r \tag{2.12}$$

$$\overline{\sigma}_S = (\lambda/2)\overline{\sigma}_f \tag{2.13}$$

After the substitution of (1.49) into (2.8) and (2.9), averaging, and transformations, for cw CDL we obtain

$$\overline{S}_S(f) = \overline{P}_S\int_0^{\infty} dz\,Q_s(z)S_W(f - 2V_r(z)/\lambda) \tag{2.14}$$

where

$$\overline{P}_S = \frac{\eta\lambda^2 P_P}{h\nu B_F\pi^2 a_0^4}\int_0^{\infty} dz\,\beta_\pi(z)\int_{-\infty}^{+\infty} d^2\rho\,I_{PN}^2(z,\rho,t) \tag{2.15}$$

$$Q_s(z) = \frac{\beta_\pi(z)\int_{-\infty}^{+\infty} d^2\rho\,I_{PN}^2(z,\rho,t)}{\int_0^{\infty} dz'\,\beta_\pi(z')\int_{-\infty}^{+\infty} d^2\rho\,I_{PN}^2(z',\rho,t)} \tag{2.16}$$

is the function characterizing the spatial resolution along the axis of the probing beam (weighting function of averaging over the sensing volume) and

$$S_W(f) = T_W^{-1}\left|\int_{-\infty}^{+\infty} dt\,W(t)\exp(-2\pi j ft)\right|^2 \tag{2.17}$$

is the spectrum of the time window function. From (2.10) through (2.14) on the integration with respect to f, we have

$$\overline{V}_r = \int_0^{\infty} dz\,Q_s(z)V_r(z) \tag{2.18}$$

and

$$\bar{\sigma}_S^2 = \bar{\sigma}_V^2 + (\lambda/2)^2 \sigma_{fl}^2 \tag{2.19}$$

where

$$\bar{\sigma}_V^2 = \int\limits_0^\infty dz\, Q_s(z)[V_r(z) - \bar{V}_r]^2 \tag{2.20}$$

$$\sigma_{fl}^2 = \int\limits_{-\infty}^{+\infty} df\, f^2\, S_W(f) \tag{2.21}$$

is the instrumental broadening of the echo signal power spectrum. As follows from (2.18) and (2.20), $\bar{V}_r$ and $\bar{\sigma}_V^2$ are, respectively, the radial velocity averaged over the sensing volume along the beam axis (low-frequency spatial filtering of velocity) and broadening of the Doppler spectrum due to inhomogeneity of the radial velocity inside the sensing volume (high-frequency spatial filtering of velocity).

In the case of pulsed CDLs, l in (2.1) is the serial number for the launch of the probing pulse into the atmosphere (i.e., the serial number of the shot). The measurement time of the array of $Z_S(t,l)$ for estimation of the spectrum of (2.1) is equal to $\Delta t = LT_P$, where T_P^{-1} is the pulse repetition frequency. Let $T_P^{-1} = 500$ Hz and $L = 25$. Then, as in the earlier case for cw CDLs, the time $\Delta t = 50$ ms. With this measurement time, it is also worth using conditional averaging here to obtain the equations for estimation of the spectral density of the echo signal and spectral moments.

After the substitution of (1.91) into (2.8) and (2.9) and averaging, for the case of the pulsed lidar, we obtain

$$\bar{S}_S(f) = \frac{2\bar{P}_S}{cT_W\sqrt{\pi}\sigma_P}$$

$$\int\limits_{+\infty}^{+\infty} dz' \left| \int\limits_{+\infty}^{+\infty} dt\, W(t)\exp\left[-\frac{(t - 2z'/c)^2}{2\sigma_P^2} - 2\pi j\left(f - \frac{2V_r(R + z')}{\lambda}\right)t \right]\right|^2 \tag{2.22}$$

where $\bar{P}_S$ is described by (1.104). In deriving (2.14) and (2.22), we have taken into account that changes of $\bar{P}_S$ for the time T_W can be neglected.

For the Gaussian temporal window

$$W(t) = \exp\left(-\frac{t^2}{2\sigma_W^2}\right) \tag{2.23}$$

the effective window width is determined as $T_W = \sqrt{\pi}\sigma_W$. Substituting (2.23) into (2.22) and integrating with respect to t, we obtain [32]

$$\overline{S}_S(f) = \overline{P}_S \int\limits_{+\infty}^{+\infty} dz' Q_s(z') \frac{1}{\sqrt{2\pi}\sigma_{fl}} \exp\left\{ -\frac{[f - 2V_r(R + z')/\lambda]^2}{2\sigma_{fl}^2} \right\} \tag{2.24}$$

where

$$Q_s(z') = \Delta z^{-1} \exp[-\pi(z'/\Delta z)^2] \tag{2.25}$$

is the function characterizing the spatial resolution along the axis of the probing beam,

$$\Delta z = \sqrt{\pi}\sqrt{\sigma_P^2 + \sigma_W^2} \cdot c/2 \tag{2.26}$$

is the effective longitudinal size of the sensing volume (or the spatial resolution of the measured velocity) defined as

$$\Delta z = \int\limits_{+\infty}^{+\infty} dz' Q_s(z')/Q_s(0) = Q_s^{-1}(0) \tag{2.27}$$

and

$$\sigma_{fl} = \frac{1}{2\pi\sqrt{2}} \cdot \frac{\sqrt{\sigma_P^2 + \sigma_W^2}}{\sigma_P \sigma_W} \tag{2.28}$$

is the width of the spectrum in absence of turbulence, when $V_r(z)=$ const (instrumental broadening). It follows from the last equation that under the condition $\sigma_W^2 \gg \sigma_P^2$ the value of σ_{fl} is determined by the duration of the probing pulse ($\sigma_{fl} \sim \sigma_P^{-1}$). On the other hand, according to (2.26), under this condition the spatial resolution may be poor. The condition $\sigma_W = \sigma_P$ is quite optimal from the viewpoint of the spatial and frequency resolution [30].

From (2.10) through (2.13) and (2.24) through (2.28) after integration with respect to f for the case of a pulsed CDL and Gaussian temporal window, we have [30]

$$\overline{V}_R(R) = \int\limits_{+\infty}^{+\infty} dz' Q_s(z') V_r(R + z') \tag{2.29}$$

$$\overline{\sigma}_S^2(R) = \overline{\sigma}_V^2(R) + (\lambda/2)^2 \sigma_{fl}^2 \tag{2.30}$$

where

$$\overline{\sigma}_V^2(R) = \int\limits_{+\infty}^{+\infty} dz' Q_s(z')[V_r(R + z') - \overline{V}_r(R)]^2 \tag{2.31}$$

Here, as in the case of a cw CDL, $\overline{V}_r(R)$ and $\overline{\sigma}_V^2(R)$ are, respectively, the radial velocity averaged over the sensing volume along the beam axis and the broadening of the Doppler spectrum due to inhomogeneities of the radial velocity inside the sensing volume.

In the case of a rectangular window:

$$W(t) = \begin{cases} 1, |t| \le T_W/2 \\ 0, |t| > T_W/2 \end{cases} \tag{2.32}$$

from (2.32), (2.10), and (2.12) after calculations, we obtain [29] the equation for $\overline{V}_r(R)$ in the form of (2.29), where

$$Q_s(z') = \frac{1}{cT_W}\left[\mathrm{erf}\left(\frac{2z'/c + T_W/2}{\sigma_P}\right) - \mathrm{erf}\left(\frac{2z'/c - T_W/2}{\sigma_P}\right)\right] \tag{2.33}$$

and $\mathrm{erf}(x) = (2/\sqrt{\pi})\int_0^x d\xi \exp(-\xi^2)$ is the standard error function.

From (2.33) for the longitudinal size of the sensing volume Δz, according to (2.27), we have

$$\Delta z = (cT_W/2)/\mathrm{erf}[T_W/(2\sigma_P)] \tag{2.34}$$

Due to fluctuations of the lidar signal and the finiteness of the measurement time Δt, the estimated radial velocity $\hat{V}_r$ differs from that averaged over the sensing volume $\overline{V}_r$ by V_e (which is referred to in the following discussion as the random error of estimation of the radial velocity); that is,

$$\hat{V}_r = \overline{V}_r + V_e \tag{2.35}$$

where, according to (2.3), (2.6), (2.10), and (2.12),

$$V_e = (\lambda/2)\int_{+\infty}^{+\infty} df\, f[\hat{P}_S^{-1}\hat{S}_S(f) - \overline{P}_S^{-1}\overline{S}_S(f)] \tag{2.36}$$

By the same reasoning, the estimate of the width of the Doppler spectrum $\hat{\sigma}_S^2$ differs from $\overline{\sigma}_S^2$ and, with allowance for (2.19) and (2.30), can be written in the form [30]

$$\hat{\sigma}_S^2 = \overline{\sigma}_V^2 + (\lambda/2)^2 \sigma_{fl}^2 + E_\sigma \tag{2.37}$$

where E_σ is the random error of estimation of the squared width of the Doppler spectrum (difference $\hat{\sigma}_S^2 - \overline{\sigma}_S^2$).

In the considered case of SNR $\gg 1$, the estimate of the radial velocity is unbiased ($\langle V_e \rangle = 0$) [42]. The variance of the random error of estimation of the radial velocity $\sigma_e^2 = \langle [V_e - \langle V_e \rangle]^2 \rangle$ is determined by the statistical properties of the echo signal and the measurement duration Δt. It is obvious that $\bar{V}_r$ and V_e are statistically independent. If in the case of cw CDLs Δt far exceeds the correlation time of fluctuations of the echo signal power τ_C, then $V_e(t)$ and $V_e(t + \Delta t)$ are statistically independent. For pulsed CDL, according to the estimation for τ_C presented in Section 1.4, the condition $\tau_C \ll \Delta t = T_P$ is true even at $L = 1$ (τ_C of about a microsecond, and T_P is not shorter than a millisecond), and for the time between laser shots the coordinates of scattering particles transported by the wind flow in the sensing volume change cardinally (in a random way). Thus, $\langle V_e(t_i) V_e(t_k) \rangle$ can be represented in the form [15, 16, 29, 35, 43, 44]

$$\langle V_e(t_i) V_e(t_k) \rangle = \sigma_e^2 \delta_{i-k} \tag{2.38}$$

where $t_i = t_0 + i\Delta t$, $t_k = t_0 + k\Delta t$, $i = 1, 2, 3, \ldots$, $k = 1, 2, 3, \ldots$, and δ_i is the Kronecker delta ($\delta_0 = 1$, $\delta_{i \neq 0} = 0$). It follows from (2.38) that the error $V_e(t_i)$ is the white noise and its single-sided spectral density S_e ($f \in [0, 1/(2\Delta t)]$) is defined as [20]

$$S_e = 2\sigma_e^2 \Delta t \tag{2.39}$$

These properties of the error of estimation of the radial wind velocity allow the variance of the error σ_e^2 to be determined easily from the temporal structure function or the time spectrum of fluctuations of the radial velocity measured by a lidar [15, 16, 29, 35, 43, 44].

2.2.2 Estimation Algorithms

At low SNR, the noise component of the lidar signal can exert a decisive influence on the accuracy of estimation of the spectral moments. As a rule, low SNR takes place in the case of pulsed CDL at long sensing paths. From the raw lidar data, we obtain a discrete series of samples of the normalized complex signal:

$$Z(mT_s, l) = Z_S(mT_s, l) + Z_N(mT_s, l) \tag{2.40}$$

where, according to (1.28), $Z_S(mT_s, l)$ is the echo signal, $Z_N(mT_s, l)$ is noise, $l = 1, 2, 3, \ldots, L$ is the serial number of a laser shot, $T_s = 1/B_F$, and $m = 0, 1, 2, \ldots$. The noise statistic is Gaussian and $\langle Z_N \rangle = \langle Z_N^2 \rangle = 0$,

$$\langle Z_N(mT_s, l) Z_N^*(nT_s, l') \rangle = 2\delta_{m-n} \delta_{l-l'} \tag{2.41}$$

Using the fast Fourier transform (FFT), we can estimate the normalized (dimensionless) power spectrum of the measured photocurrent as

$$\hat{S}(f_k) = \frac{T_s}{2LT_W} \sum_{l=1}^{L} \left| \sum_{m=0}^{M-1} W(mT_s)Z(mT_s,l)\exp(-2\pi jmk/M) \right|^2 \qquad (2.42)$$

where the frequencies are $f_k = f_0 + k\delta f - f_I$, $\delta f = 1/(MT_s)$ is the width of a single frequency bin, and $k = 0, 1, 2, \ldots, M - 1$. In the case of the rectangular window [see (2.32)] and $T_W = MT_s$, the width of a single frequency bin δf coincides with the frequency resolution $\Delta f = 1/T_W$. If the lidar signal is filtered in the frequency passband $f_k \in [f_I - B_F/2, f_I + B_F/2]$ (that is, $f_0 = f_I - B_F/2$), then the following velocities correspond to the frequencies f_k:

$$V_k = (\lambda/2)[k\delta f - B_F/2] \equiv (k - M/2)\delta V \qquad (2.43)$$

where $\delta V = B_V/M$ is the width of a single bin and $B_V = (\lambda/2)B_F$ is the receiver bandwidth (both expressed in meters per second).

At extremely low SNR or when the echo signal is absent, in (2.42) $Z(mT_S,l)$ can be replaced with $Z_N(mT_S,l)$. As a result, we obtain the estimate of the normalized noise spectrum $\hat{S}_N(f_k)$. In this case, according to (2.41), the noise is white, that is, $S_N = \langle \hat{S}_N(f_k) \rangle = 1$. However, in the actual experiment, the spectrum $\hat{S}_N(f_k)$ can differ from the spectrum of white noise (S_N is a function of frequency f_k). Therefore, in order for the average noise component of the measured spectrum S_N to be equal to unity as for the white noise, we use the following normalization [45]:

$$\tilde{S}(f_k) = \hat{S}(f_k)/S_N(f_k) \qquad (2.44)$$

The obtained spectrum $\tilde{S}(f_k)$ can be written in the form

$$\tilde{S}(f_k) = \tilde{S}_S(f_k) + \tilde{S}_S(f_k) \qquad (2.45)$$

where $\tilde{S}_S(f_k)$ is the signal component and $\tilde{S}_N(f_k)$ is the noise component of the spectrum. In this case, $\langle \tilde{S}_S(f_k) \rangle$ is the normalized spectrum of the echo signal power and $\langle \tilde{S}_N(f_k) \rangle = 1$. To determine the mean noise level

$$\overline{S}_N = M^{-1} \sum_{k=0}^{M-1} \tilde{S}_N(f_k) \qquad (2.46)$$

from the measured spectrum $\tilde{S}(f_k)$, we can use the approaches proposed in [30, 46]. In view of the finiteness of L and M, $\overline{S}_N$ is a random value with the unit mean $\langle \overline{S}_N \rangle = 1$.

The simplest way to estimate the radial velocity is to estimate the Doppler frequency from the spectral peak, that is, $\hat{f}_r = f_0 + \delta f k_{max} - f_I$, where k_{max} is the number of the spectral channel, at which $\hat{S}(f_k)$ takes the maximal value. The estimates of the zero $\hat{P}_S$ (that is, normalized echo signal power being the signal-to-noise ratio $\hat{SNR}$), first $\hat{f}_r$, and second $\hat{\sigma}_f^2$ spectral moments can be obtained as [26, 30]

$$\hat{P}_S = \frac{1}{M} \sum_{k=k_1}^{k_2-1} [\tilde{S}(f_k)/\overline{S}_N - 1] \tag{2.47}$$

$$\hat{f}_r = \hat{P}_S^{-1} \frac{1}{M} \sum_{k=k_1}^{k_2-1} f_k [\tilde{S}(f_k)/\overline{S}_N - 1] \tag{2.48}$$

$$\hat{\sigma}_f^2 = \hat{P}_S^{-1} \frac{1}{M} \sum_{k=k_1}^{k_2-1} (f_k - \hat{f}_r)^2 [\tilde{S}(f_k)/\overline{S}_N - 1] \tag{2.49}$$

where k_1 and k_2 are the numbers of the spectral channels determined up to the first negative value of $\tilde{S}(f_k)/\overline{S}_N - 1$ to the right and to left from the index $k_{\max}$.

Because the number L is finite, the estimate of the spectrum $\hat{S}(f_k)$ is a partially averaged random value and can have many random local peaks caused by fluctuations of both the noise and the signal components of this estimate. At the very low signal-to-noise ratio, when the signal component is hidden in noise, the index $k_{\max}$ can be associated with high probability with a noise peak lying beyond the range $[\overline{f}_r - \sigma_f, \overline{f}_r + \sigma_f]$. Following the terminology used in [42], these estimates of the Doppler frequency $\hat{f}_r$ are called bad. Correspondingly, if $\hat{f}_r$ falls within the range $[\overline{f}_r - \sigma_f, \overline{f}_r + \sigma_f]$, then the estimate is called good. An increase of the number L leads to a decrease in fluctuations $\hat{S}(f_k)$ and to a growth of the probability that the index $k_{\max}$ is associated with the signal peak rather than with the noise one.

To increase the accuracy of the estimate $\hat{f}_r$, we can use Levin's method of estimation from the maximum of the logarithmic likelihood function in the frequency domain [66]. If the echo signal obeys the Gaussian statistics, then the probability density of the estimate $\tilde{S}_k = \tilde{S}(f_k)/\overline{S}_N$ has the gamma distribution [47]:

$$p(\tilde{S}_k) = (L^L/\Gamma(L))(\tilde{S}_k/S_k)^{L-1} \exp(-L\hat{S}_k/S_k)/S_k \tag{2.50}$$

where $S_k = \langle \hat{S}(f_k) \rangle$, and $\Gamma(x)$ is the gamma function. Assuming statistical independence for the estimates $\hat{S}_k$ and $\hat{S}_{k' \neq k}$, the logarithmic likelihood function can be presented in the following form:

$$\Phi(f_r) = \ln\left[\prod_{k=0}^{M-1} p(\hat{S}_k) \right] = M \ln[L^L/\Gamma(L)] +$$

$$\sum_{k=0}^{M-1} [(L-1)\ln(\hat{S}_k) - L\ln(S_k) - L\hat{S}_k/S_k] \tag{2.51}$$

The Doppler frequency $\hat{f}_r$ is estimated from the maximum of this function:

$$\max\{\Phi(f_r)\} = \Phi(\hat{f}_r) \tag{2.52}$$

This approach is used for pulsed CDLs. Strictly speaking, the estimates $\hat{S}_k$ and $\hat{S}_{k'\neq k}$ are independent when the contribution of turbulent variations of the radial wind velocity to the broadening of the Doppler spectrum is negligibly small and the rectangular window with width $T_W = MT_s$ satisfying the condition $T_W\sigma_{f1} \gg 1$ is used. Here $\sigma_{f1} = (2\sqrt{2\pi}\sigma_p)^{-1}$ is the width of the Doppler spectrum determined by the duration of the probing pulse σ_P. In this case, S_k can be replaced with the model spectrum [47, 48]

$$S_M(f_k;f_r) = \frac{\text{SNR}}{\sqrt{2\pi}\sigma_{f1}T_s}\exp\left[-\frac{(f_r - f_k)^2}{2\sigma_{f1}^2}\right] + 1 \tag{2.53}$$

At low SNR, this method is more accurate than the estimation by (2.48).

As applied to radars [26, 49], it was shown that at low SNR the estimates $\hat{f}_r$ and $\hat{\sigma}_f^2$ can be strongly biased due to noise, that is, $\langle\hat{f}_r\rangle \neq f_r$ and $\langle\hat{\sigma}_f^2\rangle \neq \sigma_f^2$, where f_r and σ_f are the true values of the Doppler frequency and the width of the Doppler spectrum. The same is true for CDL ($\langle V_e\rangle \neq 0$ and $\langle E_\sigma\rangle \neq 0$). For these estimates to be unbiased, it is necessary either to increase the order of spectral accumulation L or to apply other specialized procedures for processing of experimental data requiring the knowledge of statistical properties of the lidar echo signal.

2.3 Statistical Characteristics of Estimates of the Radial Velocity and the Doppler Spectrum Width for Continuous-Wave CDLs

The wind turbulence is assumed stationary, homogeneous, and isotropic. Equation (2.20) in this case can be written as

$$\bar{\sigma}_V^2 = \int_0^\infty dz\, Q_s(z)[V_r'(z)]^2(\bar{V}_r')^2 \tag{2.54}$$

where $V_r' = V_r - \langle V_r\rangle$, $\bar{V}_r' = \bar{V}_r - \langle\bar{V}_r\rangle$, and $\langle\bar{V}_r\rangle = \langle V_r\rangle$ is the mean radial velocity. In the general case, the function $Q_s(z)$ in (2.54) is random due to turbulent variations of the backscattering coefficient $\beta_\pi(z)$ (concentration of aerosol particles) and the intensity of the probing beam $I_{PN}^2(z,\rho,t)$ [see (2.16)]. Turbulent fluctuations of the concentration of the background aerosol are about 10% and their spatial scales usually far exceed the longitudinal size of the lidar sensing volume [50]. Therefore, we can consider $\beta_\pi(z) = \text{const}$ in (2.16). As was shown in [21], under certain conditions the turbulent fluctuations of the refractive index can influence markedly the intensity of the CDL probing beam and, consequently, the form of the function $Q_s(z)$, even for a lidar wavelength of $\lambda = 10.6$ μm. For lidars with a shorter wavelength, this effect is more significant. If the lidar measurements are conducted under conditions close to the conditions of neutral temperature stratification, then the values of C_n^2 are small and we can use (1.53) for the intensity $I_{PN}(z,\rho,t)$ in (2.16). As a result, for $Q_s(z)$ we have [10–12, 19]

$$Q_s(z) = \frac{1}{\Delta z} \cdot \frac{(F/L_d)^2 / [1 + (F/L_d)^2]}{(1 - z/F)^2 + (z/L_d)^2} \tag{2.55}$$

where Δz is described by (1.57).

2.3.1 Variance of the Lidar Estimate of Radial Velocity and Mathematical Expectation of the Squared Width of the Doppler Spectrum

Let the mathematical expectation of turbulent broadening of the Doppler spectrum be denoted as σ_t^2; that is, $\sigma_t^2 = \langle \overline{\sigma}_V^2 \rangle$. Assuming that $Q_s(z)$ is a deterministic function [for example, it is described by (2.55)] and proposing the homogeneity of turbulence, from (2.18) and (2.54) we obtain

$$\sigma_t^2 = \sigma_V^2 - \sigma_{\overline{V}}^2 \tag{2.56}$$

where for the variance of the radial wind velocity averaged over the sensing volume $\sigma_{\overline{V}}^2 = \langle (\overline{V}_r')^2 \rangle$ we have from (2.18)

$$\sigma_{\overline{V}}^2 = \sigma_V^2 \int\limits_0^\infty dz_1 Q_s(z_1) \int\limits_0^\infty dz_2 Q_s(z_2)\, K_V(z_1 - z_2) \tag{2.57}$$

The correlation coefficient of wind velocity fluctuations $K_V(z)$ can be written as

$$K_V(z) = \sigma_V^{-2} \int\limits_{-\infty}^{+\infty} d\kappa_z S_V(\kappa_z) \exp(2\pi j z \kappa_z) \tag{2.58}$$

where

$$S_V(\kappa_z) = \sigma_V^2 \int\limits_{-\infty}^{+\infty} dz\, K_V(z) \exp(-2\pi j z \kappa_z) \tag{2.59}$$

is the spatial spectrum of wind velocity fluctuations. From (2.56) through (2.58) we obtain [14]

$$\sigma_{\overline{V}}^2 = \int\limits_{-\infty}^{+\infty} d\kappa_z\, S_V(\kappa_z) H_S(\kappa_z) \tag{2.60}$$

and

$$\sigma_t^2 = \int\limits_{-\infty}^{+\infty} d\kappa_z\, S_V(\kappa_z)[1 - H_S(\kappa_z)] \tag{2.61}$$

where

$$H_s(\kappa_z) = \left| \int_0^\infty dz\, Q_s(z) \exp(2\pi j z \kappa_z) \right|^2 \tag{2.62}$$

is the transfer function of the low-frequency spatial filter. Upon substitution of (2.55) into (2.62), under the condition $F \ll L_d$, we obtain the approximate equation [14]

$$H_s(\kappa_z) = \exp(-4\Delta z |\kappa_z|) \tag{2.63}$$

From (1.75) and (2.58), we can find

$$S_V(0) = 2\sigma_V^2 L_V \tag{2.64}$$

If the condition $\Delta z \gg L_V$ is fulfilled, we can take

$$\int_{-\infty}^{+\infty} d\kappa_z\, S_V(\kappa_z) \exp(-4\Delta z |\kappa_z|) \approx S_V(0) \int_{-\infty}^{+\infty} d\kappa_z\, \exp(-4\Delta z |\kappa_z|) = S_V(0)/(2\Delta z) \tag{2.65}$$

As a result, from (2.60), (2.61), and (2.63), with allowance for (2.64), (2.65), and the equality $\int_{-\infty}^{+\infty} d\kappa_z\, S_V(\kappa_z) = \sigma_V^2$, we arrive at the following asymptotic equations:

$$\sigma_{\bar{V}}^2 = \sigma_V^2 \cdot L_V / \Delta z \tag{2.66}$$

and

$$\sigma_t^2 = \sigma_V^2 \cdot [1 - L_V / \Delta z] \tag{2.67}$$

from which it follows that at $\Delta z / L_V \to \infty$ the variance $\sigma_{\bar{V}}^2 \to 0$, and σ_t^2 saturates to the level σ_V^2.

At high frequencies (within the inertial interval of turbulence $l_V^{-1} \gg |\kappa_z| \gg L_V^{-1}$), the spectrum of wind velocity obeys the Kolmogorov-Obukhov law [4–6, 51–55]

$$S_V(\kappa_z) = 0.0365\, C_K \varepsilon^{2/3} |\kappa_z|^{-5/3} \tag{2.68}$$

If the condition $\Delta z \ll L_V$ is fulfilled, then we can use (2.68) in (2.61). As a result, with the use of (2.63) it becomes possible to perform the analytical integration with respect to κ_z in (2.61) and to obtain a simple equation for σ_t^2 in the form [14]

$$\sigma_t^2 = C_K (2/\pi)^{2/3} (\varepsilon \Delta z)^{2/3} \tag{2.69}$$

It follows from (2.69) that for a small sensing volume ($\Delta z \ll L_V$) the mathematical expectation of the turbulent broadening of the Doppler spectrum is determined by the dissipation rate of the turbulent energy ε.

To take into account the outer scale of wind turbulence, we use the von Karman model [4–7]:

$$S_V(\kappa_z) = 2\sigma_V^2 L_V \left[1 + (8.42 \, L_V \kappa_z)^2\right]^{-5/6} \tag{2.70}$$

From (2.68) and (2.70), we find

$$\varepsilon = \frac{1.972}{C_K^{3/2}} \cdot \frac{\sigma_V^3}{L_V} \tag{2.71}$$

It follows from (2.56) that the estimates of the radial wind velocity (first spectral moment) and the width of the Doppler spectrum (second spectral moment) contain the information about the variance of wind velocity σ_V^2. We use a numerical simulation to study the following issue: What is the variance of radial velocity estimates obtained from the maximum of the Doppler spectrum $\sigma_{\max}^2 = \langle[\hat{V}_{r\max} - \langle V_r \rangle]^2\rangle$ [where $\hat{V}_{r\max} = k_{\max}\Delta V - B_V'/2$, B_V' is the bandwidth (in velocity units), within which the spectral moments are estimated, and ΔV is the velocity resolution in the Doppler spectrum]?

Let SNR $\gg 1$ and the number L [see (2.1)] be large enough that we can neglect the errors in estimation of the radial velocity and the turbulent broadening of the Doppler spectrum (V_e and E_σ). In the case of the rectangular time window, from (2.17) and (2.32) within the selected bandwidth $B_F' = (2/\lambda)B_V'$ for the spectrum of the time window, we have

$$S_W(f) = T_W \operatorname{sinc}^2(\pi T_W f) \begin{cases} 1, |f| \le B_F'/2 \\ 0, |f| > B_F'/2 \end{cases} \tag{2.72}$$

where $\operatorname{sinc}(x) = \sin(x)/x$. Upon substitution of (2.72) into (2.21) and integration with respect to f under the condition that $T_W B_F'$ is an integer number, for the instrumental broadening $\sigma_{VI}^2 = (\lambda/2)^2 \sigma_{fI}^2$ (in velocity units), we obtain

$$\sigma_{VI}^2 = \frac{\Delta V B_V'}{2\pi^2} \tag{2.73}$$

where $\Delta V = \lambda/(2T_W)$. Based on (2.14), (2.55), and (2.72), the normalized Doppler spectrum $\bar{S}_S(V_k)$ was simulated as [27]

$$\bar{S}_S(V_k) = \bar{P}_S \delta z \sum_{i=0}^{I} Q_{rs}(i\delta z) \operatorname{sinc}^2[\pi(V_k - V_r(i\delta z))/\Delta V)] \tag{2.74}$$

where $V_k = \Delta V k'$, $k' = k - M'/2$, $k = 0, 1, 2, ..., M'$, and $M' = B'_V / \Delta V$.

With the use of model (2.70), random realizations $V_r(i\delta z)$ at $\langle V_r \rangle = 0$ were simulated with the aid of FFT in the following way [27]:

$$V_r(i\delta z) = \text{Re}\left\{ \sum_{k_z=0}^{N_z-1} \xi_{k_z} \left[\frac{1}{\delta z N_z} S_V\left(\frac{k'_z}{\delta z N_z} \right) \right]^{1/2} \exp(-2\pi j k_z i / N_z) \right\} \qquad (2.75)$$

where $k'_z = k_z$ at $k_z < N_z/2$ and $k'_z = N_z - k_z$ at $k_z \geq N_z/2$. In (2.75), the ξ_{k_z} are pseudorandom complex numbers corresponding to the white noise with the Gaussian statistics at zero mean and unit variance for the real and imaginary parts of these numbers. The length of realization $N_z \delta z$ should satisfy the condition $N_z \delta z \gg L_V$, and the step δz can be considered as an analog of l_V.

The simulation was performed for a lidar with $\lambda = 10.6$ μm, $a_0 = 7.5$ cm ($L_d = 3,334$m) at $\Delta V = 0.1$ m/s ($T_W = 53$ μs), $B'_V = 10$ m/s, $\sigma_V = 1$ m/s, $L_V = 100$m ($\varepsilon = 6.97 \cdot 10^{-3}$ m^2/s^3), $\delta z = 0.1$m, and different focal lengths of the probing beam F. In (2.74), the number I was taken equal to $[3F/\delta z]$. From every simulated realization of the spectrum $\overline{S}_S(V_k)$, in addition to the estimates of the radial velocity from the position of the spectral maximum $\hat{V}_{r\,max} = k_{max}\Delta V - B'_V/2$, the first and second spectral moments were calculated, respectively, as

$$\overline{V}_r = \sum_{k=0}^{M'} V_k \overline{S}_S(V_k) \bigg/ \sum_{k=0}^{M'} \overline{S}_S(V_k) \qquad (2.76)$$

and

$$\overline{\sigma}_S^2 = \sum_{k=0}^{M'} (V_k - \overline{V}_r)^2 \overline{S}_S(V_k) \bigg/ \sum_{k=0}^{M'} \overline{S}_S(V_k) \qquad (2.77)$$

For the calculation of each of $\sigma_{max}^2, \sigma_{\overline{V}}^2 = \langle \overline{V}_r^2 \rangle$, and $\sigma_t^2 = \langle \overline{\sigma}_S^2 \rangle - \sigma_{VI}^2$, we used 10^5 independent realizations of the modeled characteristic. According to (2.73), the instrumental broadening of the spectrum was $\sigma_{VI}^2 \approx 0.05$ (m/s)^2.

Figure 2.1 shows the calculated dependences of σ_{max}^2, $\sigma_{\overline{V}}^2$, and σ_t^2 on the focal length of the probing beam F as solid curves. According to (1.57), the longitudinal size of the sensing volume Δz varied from 2.3m ($F = 50$m), when $\Delta z \ll L_V$, to 413m ($F = 700$m), when $\Delta z \gg L_V$. The satisfiability of (2.56) follows from the data of numerical simulation. Thus, from lidar measurements of $\sigma_{\overline{V}}^2$ and σ_t^2 we can obtain the information about the variance of wind velocity σ_V^2. An analysis of the results shown in the figure as curves 2 and 3 has shown that under the condition $\Delta z < L_V$ the variance $\sigma_{\overline{V}}^2$ exceeds the turbulent broadening of the Doppler spectrum σ_t^2, and when $\Delta z > L_V$, the inequality $\sigma_{\overline{V}}^2 < \sigma_t^2$ is true.

The dashed curves in Figure 2.1 demonstrate $\sigma_{\overline{V}}^2$ and σ_t^2 calculated by, respectively, (2.60) and (2.61), where $H_s(\kappa_z)$ is calculated by the approximate equation (2.63), and the velocity spectrum $S_V(\kappa_z)$ is described by the von Karman model of

(2.70). Deviations of these curves from solid curves 2 and 3 up to $F \approx 250\text{m}$ are negligibly small, but increase with an increase of F, because approximation (2.63) becomes rougher as the ratio L_d/F decreases, and at $F = 700\text{m}$ the use of (2.63) understates $\sigma_{\overline{V}}^2$ and overstates σ_S^2 by $\sim 10\%$.

It follows from the data shown as curve 1 in Figure 2.1 that the variance $\sigma_{\max}^2$ decreases with an increase of the focal length of the probing beam F (increase of longitudinal size of the sensing volume Δz). Thus, when estimating the radial velocity from the position of the maximum in the Doppler spectrum, velocity fluctuations become partially averaged, that is, $\sigma_{\max}^2 < \sigma_V^2$, whereas $\sigma_{\max}^2 > \sigma_{\overline{V}}^2$. In contrast to $\sigma_{\overline{V}}^2$ and σ_t^2, the relationship between the variance $\sigma_{\max}^2$ and turbulent parameters of the wind flow is unknown. Therefore, measurements of $\sigma_{\max}^2$ are not informative in terms of wind turbulence parameters.

The results shown in Figure 2.1 ignore the influence of the echo signal and noise fluctuations on the spectral moments to be estimated; that is, it is assumed that we can take $V_e = 0$ and $E_\sigma = 0$ for random errors of estimation of the first and second moments. Assume that we have arrays of estimates of the radial velocity $\hat{V}_r$ and the Doppler spectrum width $\hat{\sigma}_S$ obtained from the raw lidar data with the use of (2.47) through (2.49). Note, too, that we believe in this case that the sample size is sufficient for the calculation of statistical characteristics such as the variance $\sigma_{\hat{V}}^2 = \langle [\hat{V}_r - \langle \hat{V}_r \rangle]^2 \rangle$ and the mathematical expectation $\langle \hat{\sigma}_S^2 \rangle$. Taking into account that $\overline{V}_r$ and V_e are statistically independent, from (2.35) we obtain for $\sigma_{\hat{V}}^2$

$$\sigma_{\hat{V}}^2 = \sigma_{\overline{V}}^2 + \sigma_e^2 \tag{2.78}$$

that is, the variance of the lidar estimate of the radial wind velocity is a sum of the variance of the radial velocity averaged over the sensing volume and the variance of

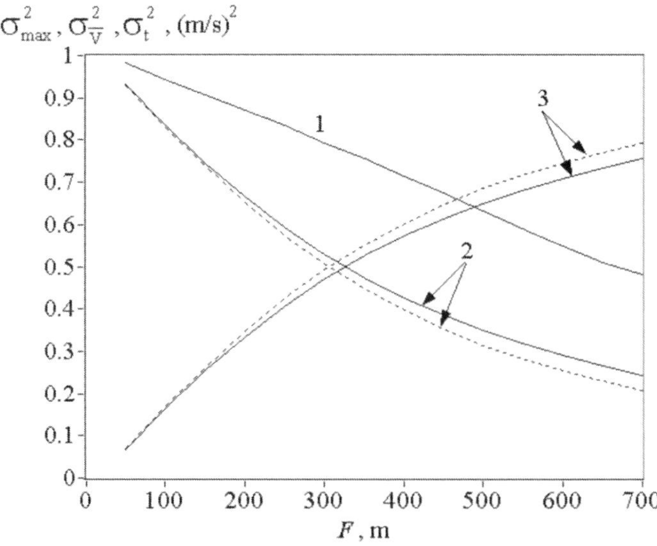

Figure 2.1 Dependence of the variances of estimates of the radial velocity $\sigma_{\max}^2$ (curve 1), $\sigma_{\overline{V}}^2$ (curves 2), and turbulent broadening of the Doppler spectrum σ_t^2 (curves 3) on the focal length of the probing beam F. Solid curves correspond to the results of numerical simulation, whereas dashed curves are for the calculation by (2.60) and (2.61) with the use of (2.63) and (2.70).

the random error of the estimate of the radial velocity. From (2.18) through (2.21) and (2.37) in the general case (with allowance for inhomogeneity of the mean wind) for $\langle \hat{\sigma}_S^2 \rangle$ we have the equation

$$\langle \hat{\sigma}_S^2 \rangle = \sigma_t^2 + \sigma_{\langle V \rangle}^2 + \sigma_{VI}^2 + \langle E_\sigma \rangle \tag{2.79}$$

from which it follows that the mean value of estimates of the squared width of the Doppler spectrum is a sum of the mathematical expectation of turbulent broadening σ_t^2, broadening $\sigma_{\langle V \rangle}^2$ caused by inhomogeneity of the mean wind, instrumental broadening σ_{VI}^2 of the Doppler spectrum, and the error $\langle E_\sigma \rangle$. According to (2.18) and (2.20),

$$\sigma_{\langle V \rangle}^2 = \int_0^\infty dz Q_s(z) \langle V_r(z) \rangle^2 - \left[\int_0^\infty dz Q_s(z) \langle V_r(z) \rangle \right]^2 \tag{2.80}$$

It is obvious that $\sigma_{\langle V \rangle}^2 = 0$ at $\langle V_r(z) \rangle = \text{const.}$

It follows from (2.56), (2.78), and (2.79) that if the condition

$$\sigma_V^2 \gg (\sigma_e^2 + \sigma_{\langle V \rangle}^2 + \sigma_{VI}^2 + \langle E_\sigma \rangle) \tag{2.81}$$

is fulfilled, then the variance of wind velocity σ_V^2 can be determined from experimental lidar data as a result of summation of measured values of $\sigma_{\hat{V}}^2$ and $\langle \hat{\sigma}_S^2 \rangle$. Otherwise, it is necessary to take into account the terms in the right-hand side of inequality (2.81). The instrumental broadening σ_{VI}^2 is known, as a rule. The variance σ_e^2 can be determined from experiment [16, 20]. The results of numerical simulation [30] show that the random error E_σ does not vanish ($\langle E_\sigma \rangle \neq 0$) as a result of averaging—even at very high SNR. Nevertheless, using the approaches for the experimental data processing proposed in [19, 30], we can reduce the value of $\langle E_\sigma \rangle$ to zero. It is more difficult to take into account the broadening of the Doppler spectrum due to the regular wind inhomogeneity within the sensing volume $\sigma_{\langle V \rangle}^2$, because for that it is necessary to know the profile of mean wind $\langle V_r(z) \rangle$. The term $\sigma_{\langle V \rangle}^2$ can be rather large in the presence of wind shear and jet flow.

2.3.2 Temporal Structure Function and Spectrum of Wind Velocity Measured with a Continuous-Wave CDL

The most widely used method for measurement of the dissipation rate of turbulent energy by wind sensors (for example, cup or acoustic anemometers) is based on the measurement of the temporal structure function or the temporal spectrum of wind velocity fluctuations and estimation of the dissipation rate ε within the inertial interval of turbulence with the use of the Taylor hypothesis of frozen turbulence. The results of investigation of the temporal structure function and the temporal spectrum of velocity measured by cw CDLs are reported in [14–16].

According to (2.35), the sequence of samples of the radial velocity measured by a lidar at time instants $t = t_0 + m\Delta t$ $(m = 0,1,2,...)$ at a fixed position of the probing beam in space can be written as

$$\hat{V}_r(t) = \bar{V}_r(t) + V_e(t) \tag{2.82}$$

where the first term is described by (2.18), and the second one is the random error in estimation of the radial velocity caused by echo signal fluctuations. Since $\bar{V}_r$ and V_e are statistically independent, the equation for the temporal structure function $D_{\hat{V}}(\tau) = \langle[\hat{V}_r'(t + \tau) - \hat{V}_r'(t)]^2\rangle$ with allowance for (2.38) at $\tau \neq 0$ can be represented in the following form:

$$D_{\hat{V}}(\tau) = D_{\bar{V}}(\tau) + 2\sigma_e^2 \tag{2.83}$$

where $D_{\bar{V}}(\tau) = \langle[\bar{V}_r'(t + \tau) - \bar{V}_r'(t)]^2\rangle$ is the temporal structure function of the radial wind velocity averaged over the sensing volume. For the structure function on the assumption of stationary and statistically homogeneous wind flow with the use of the Taylor hypothesis of frozen turbulence, the equation

$$D_{\bar{V}}(\tau) = 2[\sigma_{\bar{V}}^2 - B_{\bar{V}}(\tau)] \tag{2.84}$$

was derived in [14], where

$$B_{\bar{V}}(\tau) = \int_{-\infty}^{+\infty} d^2\kappa S_V^{(2)}(\boldsymbol{\kappa})H_s(\kappa_z)\exp(2\pi j\boldsymbol{\kappa}\langle V\rangle\tau) \tag{2.85}$$

is the time correlation function of the radial wind velocity averaged over the sensing volume, $\boldsymbol{\kappa} = \{\kappa_z, \kappa_y\}$, $S_V^{(2)}(\boldsymbol{\kappa})$ is the two-dimensional spatial spectrum of wind velocity fluctuations, $\langle V\rangle = \{\langle V_z\rangle, 0, \langle V_y\rangle\}$ is the vector of mean wind velocity, and $H_s(\kappa_z)$ is the transfer function of the low-frequency spatial filter of (2.62). In the case of isotropic turbulence and the von Karman model, the spectrum $S_V^{(2)}(\boldsymbol{\kappa})$ has the form [5]

$$S_V^{(2)}(\kappa_z,\kappa_y) =$$

$$\frac{1}{6\pi}\frac{\sigma_V^2(8.42L_V)^2}{[1 + (8.42L_V)^2(\kappa_z^2 + \kappa_y^2)]^{4/3}}\left[1 + \frac{8}{3}\cdot\frac{(8.42L_V\kappa_y)^2}{1 + (8.42L_V)^2(\kappa_z^2 + \kappa_y^2)}\right] \tag{2.86}$$

In the high-frequency spectral range under the condition $(8.42L_V)^2(\kappa_z^2 + \kappa_y^2) \gg 1$ from (2.86) with allowance for (2.71), we have

$$S_V^{(2)}(\kappa_z,\kappa_y) = \frac{1}{6\pi}\frac{C_K\varepsilon^{2/3}}{(8.42\cdot 1.972)^{2/3}}(\kappa_z^2 + \kappa_y^2)^{-4/3}\left[1 + \frac{8}{3}\cdot\frac{\kappa_y^2}{\kappa_z^2 + \kappa_y^2}\right] \tag{2.87}$$

Consider the case that the condition $|\langle V \rangle \tau| \ll L_V$ is true. Then, using (2.84), (2.85), (2.87), and (2.63), after the change of variables $\kappa_z = \kappa \sin\varphi'$ and $\kappa_y = \kappa \cos\varphi'$ and integration with respect to κ, we obtain the equation [14]

$$D_{\bar{V}}(\tau) = 1.2 C_K \varepsilon^{2/3} (\Delta z)^{2/3} \int_0^\pi d\varphi' \left[1 - \frac{8}{11} \sin^2(\varphi' - \gamma') \right] \times$$

$$\left[\mathrm{Re}\left(|\sin(\varphi' - \gamma')| + j \frac{\pi}{2\Delta z} \sin\varphi' \cdot |U\tau| \right)^{2/3} - |\sin(\varphi' - \gamma')|^{2/3} \right] \quad (2.88)$$

where $\gamma' = \arcsin(\langle V_y \rangle / U)$ is the angle between the beam axis and the wind direction, and $U = |\langle V \rangle|$ is the absolute value of the vector of mean wind velocity. Note that (2.88) is inapplicable if two conditions are fulfilled simultaneously [14]:

$$\Delta z / |\langle V_z \rangle| \gg \tau_\eta \text{ and } |\langle V_y \rangle| < \sigma_V \quad (2.89)$$

where $\tau_\eta = (v_k/\varepsilon)^{1/2} \sim 0.1\text{s}$ is the characteristic timescale of Lagrange wind velocity [5] and v_k is the kinematic viscosity of air.

If the condition $\Delta z \ll U|\tau| \ll L_V$ is true, in (2.88) we can take $\Delta z \to 0$ (point sensing volume), and $D_{\bar{V}}(\tau)$ in this case is described by the well-known equation [5]

$$D_{\bar{V}}(\tau) = D_V(\tau) = C_K [1 + (1/3) \sin^2 \gamma'](\varepsilon U |\tau|)^{2/3} \quad (2.90)$$

In another limiting case, when $|U\tau| \ll \Delta z \ll |U\tau/\sin\gamma'|$ (the case of large sensing volume when the wind direction coincides with the beam axis, $\gamma' = 0$), from (2.88) we have the asymptotic equation

$$D_{\bar{V}}(\tau) = 0.4 C_K \varepsilon^{2/3} (\Delta z)^{-4/3} (\langle V_z \rangle \tau)^2 \quad (2.91)$$

However, in this case both conditions of (2.89) can be true simultaneously.

Under the condition $\Delta z \gg |U\tau/\sin\gamma'|$ (the case of a large sensing volume in the presence of lateral wind, $\gamma' \neq 0$), from (2.88) we obtain [14]

$$D_{\bar{V}}(\tau) = 2.67 C_K \varepsilon^{2/3} \frac{|\langle V_y \rangle \tau|^{5/3}}{\Delta z} \quad (2.92)$$

If, in addition, the condition $\langle V_y \rangle^2 \gg \sigma_V^2$ is true, then (2.92) is applicable for any Δz, in particular, for $\Delta z \gg L_V$.

From (2.85), (2.87), and (2.63), for the single-sided spectral density averaged over the sensing volume $S_{\bar{V}}(f) = 2 \int_{-\infty}^{+\infty} d\tau B_{\bar{V}}(\tau) \exp(-2\pi j f \tau)$ $(f \geq 0)$ within the inertial interval, that is, when $f \gg U/L_V$, we obtain [14]

$$S_{\bar{V}}(f) = S_V(f) H_t(f) \quad (2.93)$$

where

$$S_V(f) = 0.073C_K\left(1 + \frac{1}{3}\sin^2\gamma'\right)\varepsilon^{2/3}U^{2/3}f^{-5/3} \tag{2.94}$$

is the temporal spectrum of the radial wind velocity for the sensing volume of "point" dimensions $(z = R)$ [5],

$$H_t(f) = C_2\left(1 + \frac{1}{3}\sin^2\gamma'\right)^{-1}\int_{-\infty}^{+\infty}d\xi(1 + \xi^2)^{-4/3}\left[1 - \frac{8}{11}\frac{(\cos\gamma' - \xi\sin\gamma')^2}{1 + \xi^2}\right]$$
$$\times\exp\left(-\frac{4\Delta zf}{U}\left|\cos\gamma' - \xi\sin\gamma'\right|\right) \tag{2.95}$$

is the transfer function of the temporal low-frequency filter and

$$C_2 = (55/27)[\Gamma(1/3)/\Gamma(11/16)](4\sqrt{\pi}) \approx 0.91$$

In particular cases, the integral in (2.95) can be taken analytically. Thus, at $\Delta z \to 0$ the function $H_t(f) \to 1$. For the case, when the probing beam is aligned with the mean wind $(\gamma' = 0)$, from (2.95) we have

$$H_t(f) = \exp(-4\Delta zf/U) \tag{2.96}$$

Under the condition

$$(4\Delta z\left|\sin\gamma'\right|/U)f \gg 1 \tag{2.97}$$

the major contributor to integral (2.95) is a small area about the point $\xi_m = \cot\gamma'$, whose boundaries are determined by the sharp decrease of the exponent with the distance from ξ_m. Therefore, we can take $\xi = \xi_m$ in the factor outside the exponent in (2.95) and carry out the integration. As a result, we obtain

$$H_t(f) = C_2\left(1 + \frac{1}{3}\sin^2\gamma'\right)^{-1}\frac{U\left|\sin\gamma'\right|^{5/3}}{2\Delta zf} \tag{2.98}$$

Thus, in this case, we have for the spectrum of velocity averaged over the sensing volume [14]

$$S_{\bar{V}}(f) = 0.033C_K\varepsilon^{2/3}\left|U\sin\gamma'\right|^{5/3}\frac{1}{\Delta z}f^{-8/3} \tag{2.99}$$

It follows from (2.99) that if the condition (2.97) is fulfilled, the spectrum of radial wind velocity averaged over the sensing volume, in the inertial interval of turbulence has the power frequency dependence f^v with the exponent $v = -8/3$ rather than $-5/3$ as in the case of no averaging [$\Delta z = 0$; see (2.94)]. In (2.99) the longitudinal dimension of the sensing volume Δz can exceed the outer scale of turbulence L_V.

Because every estimate of the radial wind velocity obtained from lidar data is a sum of the radial velocity averaged over the sensing volume and the error of velocity estimation [see (2.82)], the single-sided spectral density of velocity fluctuations $S_{\hat{V}}(f)(f \in [0, 1/(2\Delta t)])$ also consists of two terms [16]:

$$S_{\hat{V}}(f) = S_{\bar{V}}(f) + S_e \qquad (2.100)$$

where $S_{\bar{V}}(f)$ is the spectrum of radial velocity averaged over the sensing volume, which is described in the inertial interval of turbulence by (2.93) through (2.95), and S_e is the noise component of the spectrum [see (2.39)]. Numerous field experiments have shown [15, 16, 20] that at frequencies f close to the Nyquist frequency $f_N = 1/(2\Delta t)$, the spectrum $S_{\hat{V}}(f)$ is a white noise; that is, $S_{\hat{V}}(f) \approx S_e$, especially in the case of the large longitudinal dimension of the sensing volume Δz, when velocity fluctuations are strongly averaged. If the assumptions made in the derivation of (2.93) through (2.95) and (2.100) are true, then from the spectrum of the wind velocity measured by a lidar we can acquire information about the dissipation rate of turbulent energy ε and the noise $S_e(\sigma_e^2)$ at any dimensions Δz through the application of the corresponding procedure of fitting of theoretical calculations to experimental data [16]. With allowance for (2.83), the estimate of the dissipation rate ε can be obtained from the difference of structure functions $\Delta D(\tau) = D_{\hat{V}}(\tau) - D_{\hat{V}}(\Delta t)$ at $\tau = m\Delta t$, $m = 2, 3, ..., M$ and $M\Delta t \ll L_V/U$, since $\Delta D(\tau) = D_{\bar{V}}(\tau) - D_{\bar{V}}(\Delta t)$ is independent of σ_e^2, and $D_{\bar{V}}(\tau)$ is described by (2.88) including the parameter we seek, ε. The information about the mean wind velocity U and wind direction γ' can be obtained from the data for a conically scanning lidar (see Section 3.2).

Figure 2.2 exemplifies the differences of the structure functions $\Delta D(\tau)$ (Figure 2.2(a)) and the spectrum $S_V(f)$ (Figure 2.2(b)) calculated from simultaneous measurements of the radial wind velocity by a sonic anemometer (curves 1) and a Doppler lidar (curves 2) [15, 56]. From the data of the sonic anemometer for $\Delta D(\tau)$, the dissipation rate was estimated as $\varepsilon = 3 \cdot 10^{-3} \text{m}^2/\text{s}^3$. The results of calculation by (2.88) and (2.93) with the use of this value of ε are shown as dashed curves 1' (sonic anemometer) and 2' (lidar). This figure illustrates the influence of the spatial averaging of the radial wind velocity over the sensing volume in the case of the temporal structure function and spectrum. With regard to the averaging, the lidar and sonic anemometer data for the structure function and the spectrum are in good agreement.

From the spectrum of fluctuations of the wind velocity measured by the lidar, we can estimate the noise component of the spectrum S_e and then calculate the variance of the error of estimation of the radial wind velocity σ_e^2 with (2.42). The temporal spectrum of the radial wind velocity averaged over the sensing volume $S_{\bar{V}}(f)$ can be estimated by the subtraction of noise component S_e from the measured spectrum $S_{\hat{V}}(f)$. This procedure is applicable only for the frequency f range, in which the condition $S_{\hat{V}}(f) - S_e \gg S_e \sqrt{2/n_d}$ is true, where n_d is the number of degrees of freedom used for the averaging of the measured spectrum [41]. Figure 2.3 exemplifies the spectrum $S_{\bar{V}}(f)$ (squares, $n_d = 24$) obtained from measurements of the radial velocity at an altitude $h = 200$m at $\Delta z = 100$m ($F = 330$m, angle of site $\varphi = 30°$), the mean horizontal wind velocity $U = 15.6$ m/s, and angle $\gamma' = 75°$ [16]. For these parameters, condition (2.97) is fulfilled at frequencies $f \geq 0.2$ Hz. Consequently, the experimental spectrum $S_{\bar{V}}(f)$ in accordance with asymptotic equation (2.99) should have the power dependence of the frequency with the exponent $v = -8/3$. The solid curve is the result of least-squares fitting of the dependence of $\ln S_{\bar{V}}$ on $\ln f$ to the function $\ln \mu + v \ln f$. The resultant value of v appears to be very close to the theoretical value equal to $-8/3$. Consequently, from the equality of μ to the factor of $f^{-8/3}$ in the right-hand side of (2.99) we can estimate the turbulent energy dissipation rate ε. The dashed curve in Figure 2.3 is the Kolmogorov-Obukhov spectrum $S_V(f) \sim f^{-5/3}$ calculated by (2.94) with the use of the obtained estimate ε. It can be seen that already at frequency $f \sim 0.3$ Hz the spectral amplitude calculated from the lidar data is an order of magnitude smaller than the amplitude of the Kolmogorov-Obukhov spectrum at the same frequency due to the spatial averaging over the sensing volume ($\Delta z = 100$m $\sim L_V$).

The conducted field experiments [15, 16, 20] have shown that at small longitudinal dimensions of the sensing volume Δz and arbitrary velocity of the lateral wind or even at large $\Delta z > L_V$, if the lateral wind during the experiment was strong enough for condition (2.97) to be fulfilled, the spectra of fluctuations of the wind velocity measured by the lidar were in agreement with the theory considered in this section and, consequently, the results of these measurements carry the information about the dissipation rate ε. Lidar measurements of the velocity at the mean wind direction coinciding with the direction of propagation of the probing beam ($\gamma' \approx 0$, horizontal path, $h = 3$m) were conducted in [16]. In this case, Δz was equal to 220m ($F = 500$m) far exceeding L_V at an altitude of 3m. The analysis of the spectra obtained under these conditions shows that they do not obey theoretical regularities (2.93) through (2.95). On the assumption of the power dependence of these spectra on the frequency $S_{\bar{V}}(f) \sim f^v$, the least-squares method was applied to determine the exponent v, which appeared to be approximately equal to -2.

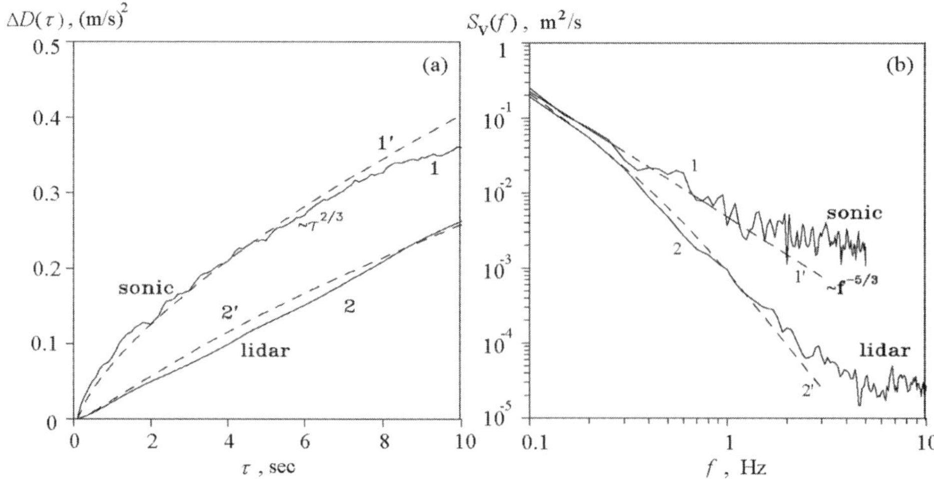

Figure 2.2 Differences of (a) structure functions $\Delta D(\tau)$ and (b) spectra $S_V(f)$ of wind velocity measured by lidar and sonic anemometer. (© 1999 American Meteorological Society. Used with permission. From [19].)

The analysis of results of other series of lidar measurements has shown that the velocity spectra obtained for different altitudes h, when $\Delta z \geq L_V$ and $\gamma \leq 15°$, also has the power dependence $S_{\bar{V}}(f) \sim f^{-2}$. A probable reason for this discrepancy between theory and experiment is the use of the Taylor hypothesis of frozen turbulence in the derivation of (2.93) through (2.95), whereas for the correct description of the spectrum of velocity measured by a lidar at the weak lateral wind and the large sensing volume, it is necessary to take into account the lateral transfer of aerosol particles falling within the sensing volume by turbulent vortices of different scales and the evolution of these vortices in time. Another reason may be the influence of turbulent pulsations of the refractive index of air on the sensing volume formed by the lidar. This issue is considered in Section 2.5.

2.4 Error in Estimation of the Radial Velocity from Continuous-Wave CDL Data

Section 2.3 showed that the temporal spectrum of lidar estimates of the radial wind velocity (first spectral moment) $S_{\hat{V}}(f)$ is a sum of the spectrum of radial wind velocity averaged over the sensing volume $S_{\bar{V}}(f)$ and the noise component of the spectrum S_e. The value of S_e can be determined from the lidar spectrum of velocity $S_{\hat{V}}(f)$ at high frequencies, where $S_{\hat{V}}(f) \approx S_e$. Then, using (2.39) we can calculate the variance of the error in the estimate of the radial wind velocity σ_e^2. In [26], the asymptotic equation for the variance of the error in the estimate of the radial wind velocity σ_{en}^2 was derived as applied to Doppler radars. In [25], this equation was proposed for the case of cw CDLs. Under the condition SNR $\gg 1$, this equation has the form

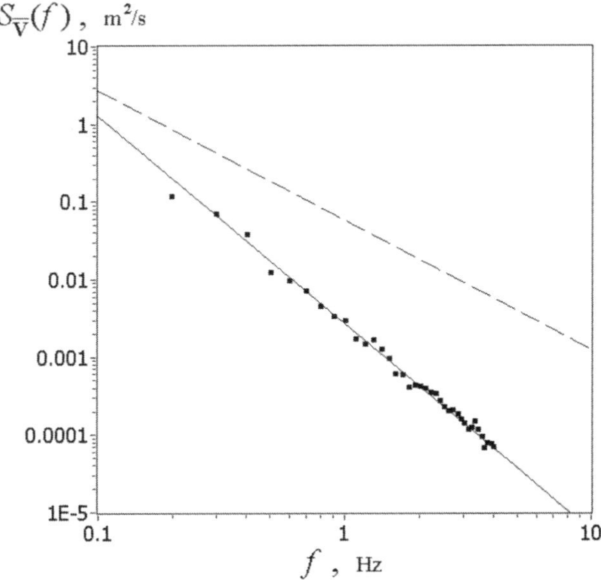

Figure 2.3 Temporal spectrum of the radial wind velocity averaged over the sensing volume for the altitude $h = 200$ and $\Delta z = 100$m at strong lateral wind (squares). The solid curve is a result of fitting of the measured spectrum to the power dependence of the frequency, and the dashed curve is a result of calculation by (2.91). (© 1999 American Meteorological Society. Used with permission. From [19].)

$$\sigma_{en}^2 = \frac{1}{8\sqrt{\pi}} \cdot \frac{\lambda}{\Delta t} \sigma_S \tag{2.101}$$

where σ_{en} is used in place of σ_e to distinguish the error estimated from the temporal spectrum of velocity $S_{\hat{V}}(f)(\sigma_e)$ and the error calculated by (2.101) with the use of measured Doppler spectrum width σ_S (σ_{en}). If $\sigma_t^2 \gg (\sigma_{\langle V \rangle}^2 + \sigma_{VI}^2)$, then $\sigma_S^2 \approx \sigma_t^2$, and σ_S^2 is described by (2.69) for $\Delta z \ll L_V$, or it coincides with the variance of the wind velocity σ_V^2 for $\Delta z \gg L_V$. From (2.101) it follows that for $\lambda = 10.6\ \mu m$, $\Delta t = 0.05s$, and $\sigma_S = 1$ m/s the error $\sigma_{en} \approx 4 \cdot 10^{-3}$ m/s.

Field experiments [20] were conducted under different turbulent conditions (different ε) and $\Delta z \ll L_V$. The error σ_e was estimated from the same raw data (Doppler spectra) by two methods: from the calculations of (1) $S_{\hat{V}}(f)$ with the use of (2.39) and (2) σ_S with the use of (2.101). The ratios σ_e/σ_{en} obtained from these experiments at different σ_S are shown in Figure 2.4, from which it follows that σ_e is greater than σ_{en} approximately by an order of magnitude.

Equation (2.101) has been derived on the assumption of the Gaussian statistics of the echo signal. In Section 1.3 it was shown that the echo signal statistics of cw CDLs differ from the Gaussian ones due to the influence of the aerosol microstructure on the statistical properties of the scattered radiation, especially at small dimensions of the sensing volume. In [20, 27], the influence of the microstructure of scattering particles on the error σ_e at different dimensions of the sensing volume was studied theoretically with the aid of numerical simulation of the cw CDL signal.

The normalized complex signal $Z(t)$ was simulated in the following way. According to (1.28), (1.13), (1.49), (1.53), and (1.59), $Z(t)$ can be represented in the form

$$Z(t) = \sqrt{\mathrm{SNR}}\, q^{-1} \sum_{i=1}^{N_s} \sigma_\pi^{1/2}(a_i) \frac{a_0^2}{a_b^2(z_i)}$$

$$\times \exp\left[-\frac{(\rho_i V_\perp t)^2}{a_b^2(z_i)} + j\Psi_i + 4\pi j \frac{V_r(z_i)t}{\lambda} \right] Z_N(t) \tag{2.102}$$

where $q = \pi^{2/3} a_0^2 \sqrt{\beta_\pi/\lambda}$, $a_b(z) = a_0[(1 - z/F)^2 + (z/L_d)^2]^{1/2}$ is the beam radius at the distance z from the lidar, $\rho = \{x, y\}$, $V_\perp = \{V_x, 0\}$, V_x is the transverse component of the wind velocity, and $t = [(l' - 1)M + m]T_s$, where $m = 0, 1, 2, \ldots, M - 1$ and $l' = 1, 2, \ldots, L$. We divide the sensing path in the vicinity of the focus F into N_z layers, each with the thickness δz satisfying the condition $\delta z \ll \Delta z \ll N_z \delta z$ and represent $Z(t)$ as

$$Z(t) = \sqrt{\mathrm{SNR}}\, q^{-1} \sum_{i=0}^{N_z-1} \frac{a_0^2}{a_b^2(\delta z i')} \sum_{k=N_t}^{N_V+N_t-1} \left[\sum_{l=1}^{n} \sigma_\pi^{1/2}(a_{ikl}) \exp(j\psi_{ikl}) \right]$$

$$\times \exp\left\{ -\left[\frac{\delta x(k' - N_V/2)}{a_b(\delta z i')} \right] + j\frac{4\pi}{\lambda} V_r(\delta z i)t \right\} + Z_N(t) \tag{2.103}$$

where $i' = F/\delta z - N_z/2 + i$, $n = [\rho_0 \delta V(\delta z i)]$ is the number of particles in the elementary volume $\delta V(\delta z i) = \delta z \cdot \delta z \cdot \sqrt{\pi/2} a_b(\delta z i')$, $\delta x = \sqrt{\pi} a_b(F)/N_x$ is the dimension of the elementary volume along the axis x, N_x is the partition number of the effective beam

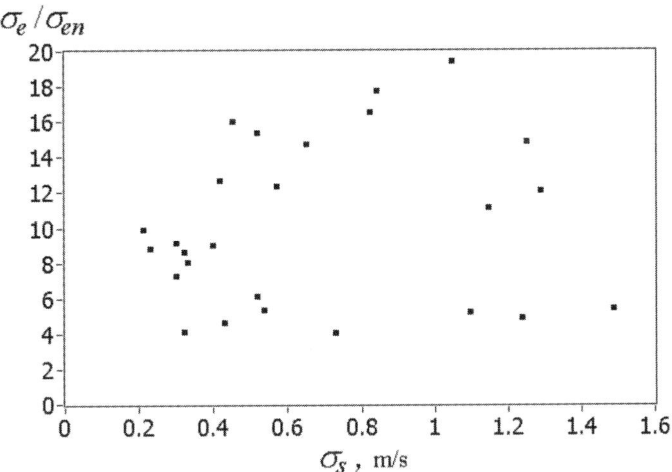

Figure 2.4   Experimental value of the ratio σ_e/σ_{en} obtained at different σ_s.

width along the axis x in the focal plane $z = F$, $N_V = [2\sqrt{\pi} a_b (\delta z i')/\delta x]$ is the number of volumes δV in the ith layer, $N_t = [V_x t/\delta x]$, and $k' = k - N_t$. It follows from (2.103) that to obtain one value of Z, it is necessary to generate $N_z N_x n$ independent values of the particle radius a_{ikl}, the same number of independent values of the phase Ψ_{ikl}, N_z values of the radial velocity V_r, and one complex number Z_N. The procedures for the simulation of random realizations for the particle radius a, phase Ψ, and velocity V_r are described in Sections 1.3 and 2.3. As in Section 1.3, to accelerate the simulation algorithm, in (2.103) we substitute $\sum\limits_{l=1}^{n} \sigma_\pi^{1/2}(a_{ikl})\exp(j\psi_{ikl})$ with $\sigma_\pi^{1/2}(a_{max})\exp(j\psi)$. The noise component of the complex signal $Z_N(t)$ was simulated on the assumption of the Gaussian distribution of the probability density of the real and imaginary parts (which are statistically independent) and the covariance function in the form $(1/2)\langle Z_N(t_m) Z_N^*(t_{m'}) \rangle = \delta_{m-m'}$.

In the numerical simulation, the following parameters were taken: $\lambda = 10.6\ \mu$m, $P_P = 4$W, $\Delta f = 1/(MT_s) = 20$ kHz, $M = 256$, $L = 100$ and $1,000$, $\sigma_V = 1.5$ m/s, $L_V = 50$m, $\langle V_r \rangle = 13$ m/s, and $V_x = 2$ m/s, and the model described by (1.60) was used for $\rho_s(a)$. M is the number of spectral channels and L is the number of laser shots used for the spectral accumulation. In addition, the simulation was carried out assuming identical dimensions for the aerosol particles; that is, $\rho_s(a) = \rho_0 \chi(a_p - a)$, where $\chi(x)$ is the Heaviside function ($\chi(x) = 0$, $x < 0$ and $\chi(x) = 1$, $x \geq 0$).

The first term in (2.103) at a fixed instant can be written as a sum $\sum\limits_{i=0}^{N_z-1} A_i \exp(j\psi_i)$, where A_i and Ψ_i are, respectively, the amplitude (in relative units) and phase of the wave scattered in the ith layer of the sensing volume. Figure 2.5 exemplifies the simulation of instantaneous distribution of A_i (where the A_i are normalized to the maximal amplitude in every realization) along the sensing path z in the form of vertical lines at different F. The normalized distributions $Q_s(z)$ calculated by (2.55), that is, functions characterizing the spatial resolution Δz, are shown as dashed curves in the plots for A_i. It follows from the figure that, in contrast to the cases with $F = 50$m and $F = 100$m, at $F = 10$m the separation between the position of the maximum amplitude A_i and the focus can exceed Δz several times. Since at $F = 10$m the formed sensing volume contains few optically active particles (see Table 1.1), it is quite possible that one large aerosol particle lying at the wing of the distribution $Q_s(z)$ far from the focal zone $F \pm \Delta z/2$ makes the predominant contribution to the lidar echo signal.

Examples of simulated realizations of the distribution of radial wind velocity fluctuations $V'_r = V_r - \langle V_r \rangle$ along the axis z are shown to the right in Figure 2.5. One can see that at $F = 10$m, within the effective sensing volume $F - \Delta z/2 < z < F + \Delta z/2$ the velocity changes very slightly even under the conditions of strong turbulence ($\sigma_V = 1.5$ m/s). At the same time, the appearance of a large particle in the wing of the distribution $Q_s(z)$ can extend significantly the range of the deviation of V'_r from $V'_r(F)$.

The cw CO_2 lidar (see Figure 1.2) has such elements as a spectral analyzer designed with the use of surface acoustic waves [57, 58] and integrator, which averages unsmoothed estimates of spectral amplitudes. For simpler calculations, we use the rectangular time window. Then the estimate of the spectrum of normalized signal amplitude has the form

$$\hat{S}_A(f_k) = L^{-1} \sum_{l=1}^{L} \left| \sum_{m=0}^{M-1} Z\big(((l-1)M + m)T_s\big)\exp(-2\pi jkm/M) \right| \tag{2.104}$$

where $f_k = f_0 + k\Delta f$, $k = 0, 1, 2, \ldots, M - 1$, M is the number of spectral channels, and $\Delta f = 1/(MT_s)$ is the frequency resolution. The estimate of the normalized signal power spectrum can be written as $\hat{S}(f_k) = (4/\pi)(2M)^{-1}\hat{S}_A^2(f_k)$. For the estimation of the radial velocity $\hat{V}_r = (\lambda/2)\hat{f}_r$, we use algorithm (2.48). In numerical simulation, the radial velocity profile $V_r(\delta zi)$ is known. Therefore, we can calculate the spectral amplitude averaged over signal fluctuations

$$\overline{S}_A(f_k) = \sqrt{(\pi/2)[M \cdot \overline{S}_S(f_k) + 1]} \tag{2.105}$$

where $\overline{S}_S(f_k)$ is calculated by (2.74) upon substitution of V_k with $(\lambda/2)f_k$ [$\Delta V = \lambda/(2MT_s)$, $\overline{P}_S \equiv \mathrm{SNR}$], and the radial velocity averaged over the sensing volume

$$V_r = \delta z \sum_{i=0}^{N_z-1} Q_s(\delta zi')V_r(\delta zi) \tag{2.106}$$

The left column of Figure 2.6 shows examples of simulations of the time dependence of the instantaneous power of lidar signal $P = (1/2)|Z|^2$ (with noise neglected; that is, at $Z_N = 0$) at different values of F. The presence of peaks with the width $\tau \sim a_b(F)/V_x$ is a consequence of the dominant influence of one large particle transported by the lateral wind through axis z on power $P(t)$. Such behavior of $P(t)$ at a small sensing volume is also observed in field experiments [59]. The analysis of simulated data allows one to determine the distance from the lidar to the dominant particle, the particle residence time in the sensing beam, and the radial component

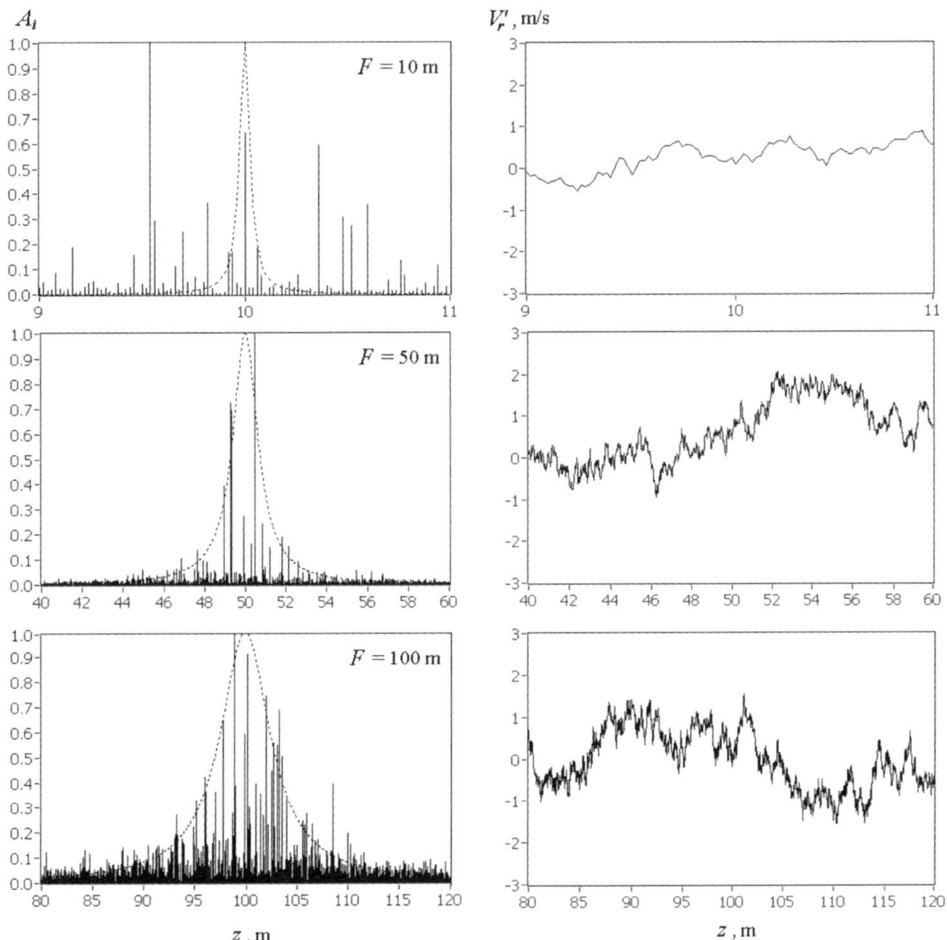

Figure 2.5 Examples of simulation of instantaneous distributions of the amplitudes A_i and fluctuations of the radial wind velocity V_r' along the axis z. (© 2000 Optical Society of America. From [27].)

of particle motion, which may be a reason for the appearance of a sharp peak in the measured spectrum.

The Doppler spectra of the amplitude $S_A(V) = \hat{S}_A(\lambda f_k/2)$, obtained using the corresponding simulations of $P(t)$ and (2.105), are shown as solid curves in the right part of Figure 2.6 along with the spectra $\overline{S}_A(2V/\lambda)$ calculated by (2.105) with the use of the corresponding realizations $V_r(\delta zi)$ shown as dashed curves. The sharp peaks of some spectra $S_A(V)$ are associated with the contribution of individual large particles to the spectrum.

According to (2.35), the values of $\hat{V}_r$ and $\overline{V}_r$ obtained in every realization of the numerical simulations were used to calculate random errors $V_e = \hat{V}_r - \overline{V}_r$. Then the bias $b_r = \langle V_e \rangle (\langle \overline{V}_r \rangle = \langle V_r \rangle)$ and the variance $\sigma_e^2 = \langle [V_e - \langle V_e \rangle]^2 \rangle$ of the random error of radial velocity estimation were determined with the use of 1,000 independent realizations of V_e for the averaging. From the data derived from numerical experiments, it follows that the estimate of the radial velocity is unbiased, that is, $b_r = 0$.

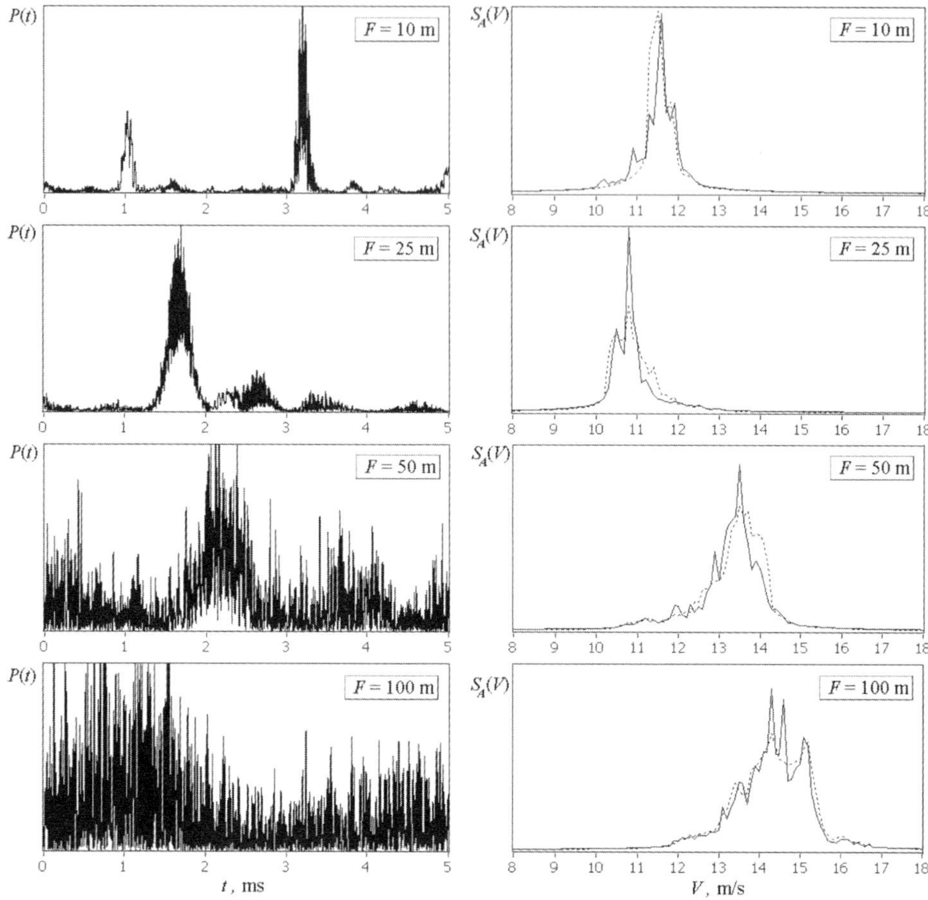

Figure 2.6   Examples of simulations of power $P(t)$ and estimates of the spectrum $S_A(V)$ of the lidar echo signal (solid curves) and calculations of $\overline{S}_A(2V/\lambda)$ (dashed curves). (© 2000 Optical Society of America. From [27].)

Figure 2.7 shows the dependence of the error σ_e on the focal length F obtained in [27]. Curve 1 corresponds to the calculations obtained using numerical simulation on the assumption that all particles are of identical size [that is, $\sigma_\pi(a_{ikl}) = $ const in (2.103)] and the noise can be neglected, that is, $Z_N = 0$. The dashed curve shows the results of calculation by (2.101) with the use of (2.61) and (2.69). One can see (compare curve 1 and dashed curve) that these two approaches give close results. With an increase of F (or Δz), the error σ_e increases, and when the condition $\Delta z \gg L_V$ is fulfilled, it should saturate to the level described by (2.101) at $\sigma_S = \sigma_V$. Curve 2 in the figure shows σ_e calculated from the model data with allowance for the aerosol microstructure. The noise was ignored in the simulation ($Z_N = 0$). It can be seen that, with the microstructural factor taken into account, the error of velocity estimation increases by an order of magnitude (if we compare curves 1 and 2). In addition, in contrast to curve 1, curve 2 shows a decrease in σ_e with an increase in F. The inclusion of noise in the calculations (SNR is estimated in Section 1.3) has practically no effect on the value of σ_e.

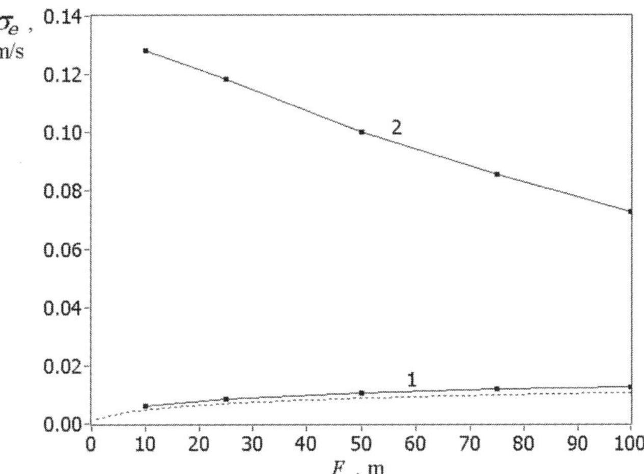

Figure 2.7 Error of lidar estimation of the radial wind velocity as a function of the focal length of the probing beam. Solid curve 1: result of numerical simulation on the assumption of equal scattering amplitudes of aerosol particles; dashed curve 1: calculation by (2.101); curve 2: result of numerical simulation with the use of the aerosol particle size distribution function in the form (1.68). Integral time $T = 50$ ms. (© 2000 Optical Society of America. From [27].)

The reason for an increase in σ_e with a decrease in the focal length F is as follows. With a decrease in F, the sensing volume V_{eff} decreases and, as shown in Section 1.3, the number of efficiently scattering particles N_{eff} within this volume decreases. Consequently, the probability increases that the major contribution to the measured spectrum of echo signal power comes from one or several large scattering particles, which are beyond the range $[F - \Delta z/2, F + \Delta z/2]$ (see, for example, Figure 2.5 at $F = 10$m) and move due to the wind turbulence with the velocity V_{r1} markedly different from the radial wind velocity averaged over the sensing volume $\overline{V}_r$. Numerical experiments at $F = 10$m demonstrate that the deviations $\left| V_{r1} - \overline{V}_r \right|$ can be equal to several tens of percent of the rms deviation of the wind velocity σ_V (on average, 10% to 15%). With an increase of the sensing volume, the number of efficiently scattering particles increases, and the averaging over velocities of scattering particles becomes better.

Thus, as in the field experiment (see Figure 2.4), the error σ_e obtained from the data of numerical simulation with allowance for the microstructure of scattering particles exceeds the error of estimation of the radial velocity calculated by (2.101), on average, by an order of magnitude.

To find the dependence of the variance σ_e^2 on the focal length of the probing beam (on the longitudinal dimension of the sensing volume Δz), the field experiments, whose results are reported in [20], were conducted on different days at different wind turbulence intensities. In all of the experiments, the time for measurement of one Doppler spectrum was $\Delta t = 50$ ms. The minimal sensing range (focal length F) was 50m ($\Delta z \approx 2.3$m). Aerosol microstructure parameters were not measured, but rough estimates show that the aerosol concentration could vary widely for different measurements. The SNR remained rather high because we could neglect the influence system noise on the error σ_e.

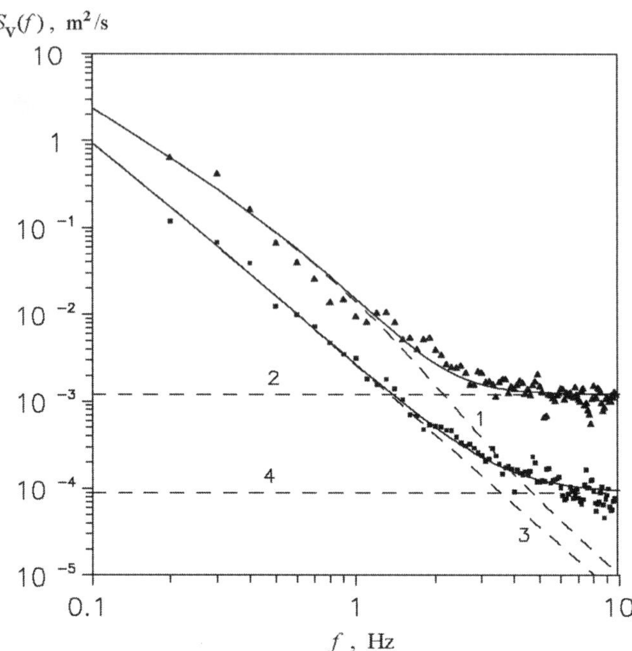

Figure 2.8 Temporal spectra of wind velocity measured by lidar at $F = 50$m ($\Delta z = 2.3$m, triangles) and 330m ($\Delta z = 100$m, squares). Dashed lines 2 and 4: noise spectra S_e at $F = 50$m and 330m, respectively; solid curves: calculation by (2.93) through (2.95) and (2.100); dashed curves 1 and 3: calculation by the same equations but at $S_e = 0$. (© 1999 American Meteorological Society. Used with permission. From [19].)

Figure 2.8 depicts the spectra of wind velocity measured by lidar with an interval of 15 min at $F = 50$m ($\Delta z = 2.3$m) and $F = 330$m ($\Delta z = 100$m). Dashed lines 2 and 4 show the noise levels of the spectra S_e calculated from measurements of $S_{\hat{V}}(f)$ at frequencies $f \geq 6$ Hz, at which the condition $S_e \gg S_{\hat{V}}(f)$ is fulfilled (see solid curves and dashed curves 1 and 3). It can be seen that S_e and [with allowance for (2.39)] σ_e^2 at $F = 50$m are approximately an order of magnitude greater than at $F = 330$m. This result does not contradict the result of numerical simulation shown in Figure 2.7 as curve 2, when the error σ_e decreases with an increase of F.

Figure 2.9 shows the dependence of the variance σ_e^2 on F. The symbols connected by dashed lines are for σ_e^2 values calculated from experimental data, whereas the solid curve is for σ_e^2 values calculated from numerical simulation data. In contrast to the field experiment, in which the dependences of σ_e^2 on F were obtained under different conditions of dynamic turbulence, mean wind velocity, and parameters of aerosol microstructure, the solid curve was calculated at the single fixed value $\sigma_V = 1$ m/s (transverse component of the mean wind velocity $V_\perp = 4$ m/s). With an increase in F (and, consequently, an increase in $V_{eff} \sim F^4$), the number of particles in the sensing volume increases drastically, leading to a significant increase in the computational time. That is why we succeeded in obtaining the model dependence only up to $F = 150$m. It follows from the figure that the variance σ_e^2 decreases with an increase of the sensing volume. On average, the behavior of σ_e^2 estimated from the data of field experiments is analogous to the behavior of σ_e^2 determined from the numerical simulation.

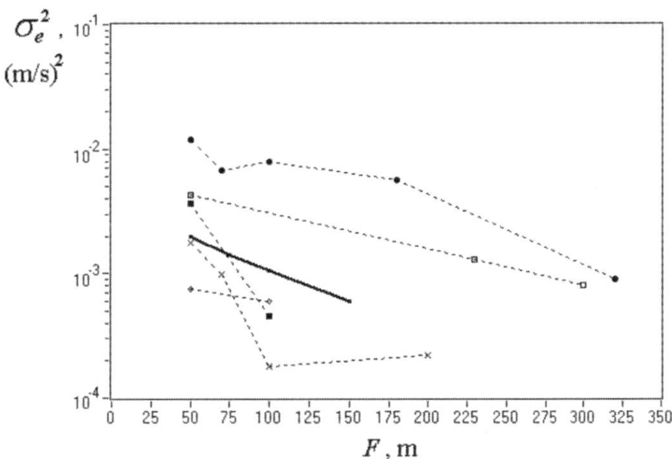

Figure 2.9 Dependence of σ_e^2 on F. Symbols connected by dashed lines: field experiment; solid curve: numerical experiment.

2.5 Influence of Turbulent Fluctuations of the Refractive Index on the Temporal Spectrum of Wind Velocity Measured by Continuous-Wave CDL

It was noted in Section 2.3 that at large Δz and weak lateral wind (or when $\gamma' \approx 0$) experimental data differ from theoretical calculations of the temporal spectra (structure functions) of wind velocity measured by lidar. In addition to the violation of conditions of applicability of the Taylor hypothesis of frozen turbulence, a possible reason for the discrepancy between theoretical and experimental results may be the influence of turbulent fluctuations of the refractive index of air on the temporal spectrum of wind velocity measured by lidar. Actually, in the derivation of (2.88) and (2.93) through (2.95), we used (2.55); that is, we assumed that the function $Q_s(z)$ characterizing the spatial resolution of the measured velocity is not affected by the atmospheric turbulence and remains unchanged during the measurement. This approximation is well founded, if we take into account that for the wavelength $\lambda = 10.6$ µm, relatively short sensing paths (up to 1 km), and the structure characteristic of the refractive index in the surface atmospheric layer $C_n^2 \leq 10^{-12}\,\mathrm{m}^{-2/3}$, the influence of turbulent fluctuations of the refractive index on the SNR can be neglected. Nevertheless, fluctuations of the refractive index can, although insignificantly, change the function $Q_s(z)$, which can ultimately affect the temporal spectrum of wind velocity measured by lidar. Let us consider the influence of the refractive turbulence on the spectrum in more detail.

Based on (2.16) and (2.18) with regard to averaging of the probing beam intensity over the time of measurement of one Doppler spectrum Δt, we can represent the radial wind velocity averaged over the sensing volume at the moment $t = t_0 + m\Delta t$, $m = 0, 1, 2, \ldots$, in the form [21]

$$\bar{V}_r(t) = \int\limits_0^\infty dz\, Q_s(z,t) V_r(z,t) \tag{2.107}$$

where the spatial resolution function $Q_s(z,t)$ is expressed as

$$Q_s(z,t) = \frac{\int\limits_{t-\Delta t}^{t} dt' \int\limits_{-\infty}^{+\infty} d^2\rho I_{PN}^2(z,\rho,t')}{\int\limits_{0}^{\infty} dz \int\limits_{t-\Delta t}^{t} dt' \int\limits_{-\infty}^{+\infty} d^2\rho I_{PN}^2(z,\rho,t')} \qquad (2.108)$$

To take into account random distortions of the probing beam intensity $I_{PN}(z,\rho,t)$ caused by refractive turbulence, we use numerical simulation (see Section 1.4).

The two-dimensional function $Q_s(z,t)$ was simulated without averaging over the time Δt, that is, $\Delta t \to 0$ in (2.108). Figure 2.10 exemplifies the simulation of random realizations of $Q_s(z,t)$ (solid curves) for the parameters $F = 500\text{m}$, $C_n^2 = 10^{-13}\text{ m}^{-2/3}$ (Figure 2.10(a)) and $C_n^2 = 10^{-12}\text{ m}^{-2/3}$ (Figure 2.10(b)) [21]. The dashed curve shows the function $Q_s(z)$ at $C_n^2 = 0$ calculated by (2.55). It can be seen that at $C_n^2 = 10^{-13}\text{ m}^{-2/3}$ the fluctuations of the refractive index lead mostly to only random displacements of the center of the sensing volume determined by the point of maximum $z = z_{\max}$ of the function $Q_s(z,t)$, leaving the shape of the resolution function nearly unchanged. At $C_n^2 = 10^{-12}\text{ m}^{-2/3}$, in addition to random displacements of the sensing volume, its longitudinal dimension Δz increases significantly.

The obtained random realizations of $Q_s(z,t)$ were used to calculate such statistical characteristics as the variance $\sigma_z^2 = \langle z'^2_{\max} \rangle$, coefficient of temporal correlation $K_z(\tau) = \langle z'_{\max}(t + \tau)z'_{\max}(t) \rangle / \sigma_z^2$ of displacements of the sensing volume center, where $z'_{\max} = z_{\max} - \langle z_{\max} \rangle$, and the mean longitudinal dimension of the sensing volume determined as [21] $\Delta z = 1/\langle Q_s(z_{\max}) \rangle$ ($\langle Q_s(z) \rangle = q_s(z) \Big/ \int\limits_0^{\infty} dz' q_s(z')$ and $q_s(z) = \int\limits_{-\infty}^{+\infty} d^2\rho < I_{PN}^2(z,\rho) >$). It was assumed in the simulation that turbulent inhomogeneities of the refractive index are transported by the mean wind flow in the plane perpendicular to the probing beam optical axis with their shape unchanged (according to the hypothesis of frozen turbulence).

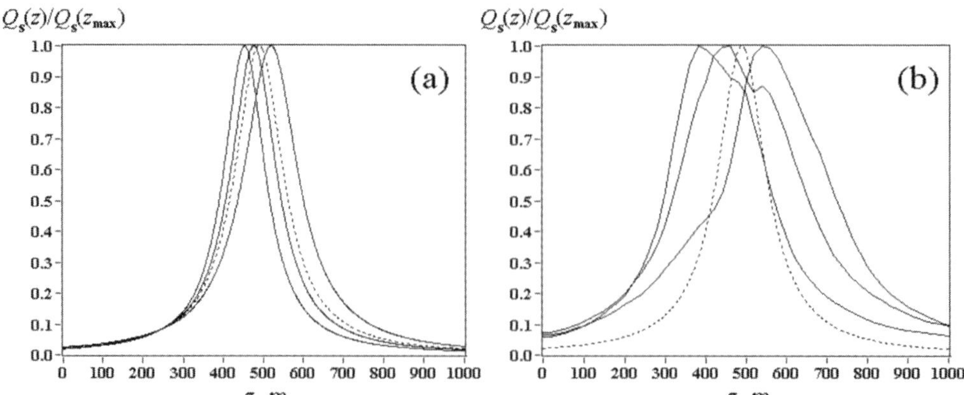

Figure 2.10 Examples of simulation of random realizations of the function $Q_s(z,t)$ at $C_n^2 = 10^{-13}\text{ m}^{-2/3}$ (a) and $C_n^2 = 10^{-12}\text{ m}^{-2/3}$ (b) (solid curves); calculation by (2.55) (dashed curve).

In Figures 2.11(a) and 2.11(b), dots show the rms deviation σ_z calculated from the data of numerical simulation as a function of C_n^2 and the focal length F [21]. It turns out that $\sigma_z \sim \sqrt{C_n^2}$ at fixed F and $\sigma_z \sim F^2$ at fixed C_n^2. Solid lines demonstrate the results of calculation by the empirical equation $\sigma_z = \alpha_1 F^2 \sqrt{C_n^2}$ obtained from the data of numerical simulation, where $\alpha_1 = 148$ m$^{-2/3}$. It follows from the depicted data that the rms deviation of displacements σ_z and the longitudinal dimension of the sensing volume $\Delta z = \Delta z_0$ at $C_n^2 = 0$ [see (1.58) at $F \ll L_d$] are proportional to F^2. The ratio $\sigma_z / \Delta z_0 = 2\alpha_0^2 \lambda^{-1} \alpha_1 \sqrt{C_n^2}$, with allowance that $\lambda = 10.6$ μm and $a_0 = 7.5$ cm, is equal to 0.016 at $C_n^2 = 10^{-14}$ m$^{-2/3}$ and 0.16 at $C_n^2 = 10^{-12}$ m$^{-2/3}$.

The dots in Figure 2.11(c) show the longitudinal dimension of the sensing volume Δz as a function of C_n^2 at the focal length $F = 500$m. The solid horizontal line corresponds to the value $\Delta z = \Delta z_0$ in the absence of turbulence (see Table 1.1). It can be seen that the value of Δz differs only slightly from Δz_0 up to $C_n^2 \approx 10^{-13}$ m$^{-2/3}$, and at $C_n^2 = 10^{-12}$ m$^{-2/3}$ it increases nearly 1.5 times.

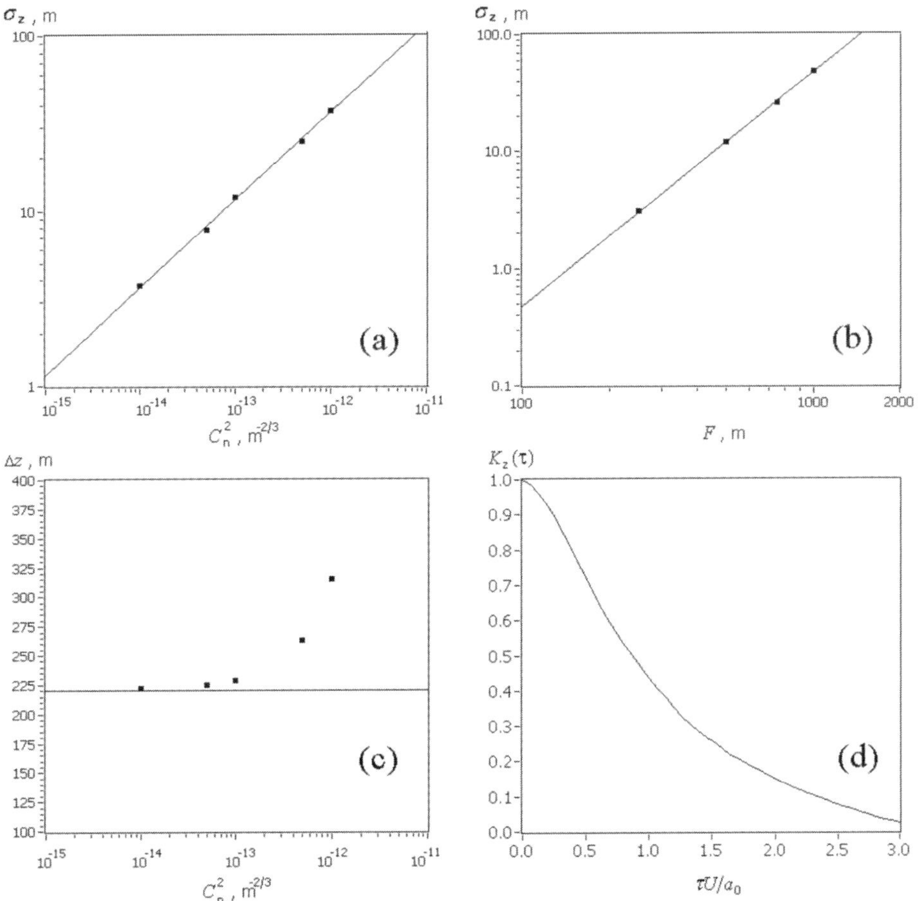

Figure 2.11   (a) Dependence of the rms deviation of displacements of the sensing volume σ_z on C_n^2 at $F = 500$m; (b) rms deviation of displacements of the sensing volume σ_z as a function of F at $C_n^2 = 10^{-13}$ m$^{-2/3}$; (c) dependence of the mean value of the longitudinal dimension of the sensing volume Δz on C_n^2 at $F = 500$m; (d) coefficient of temporal correlation of displacements of the sensing volume caused by refractive turbulence.

The coefficient of temporal correlation of displacements of the sensing volume caused by turbulent fluctuations of the refractive index at $F = 500$m and $C_n^2 = 10^{-13}$ m$^{-2/3}$ is shown in Figure 2.11(d) [21]. The analogous dependences for $K_z(\tau)$ were also obtained for other values of F and C_n^2. The analysis of the results of $K_z(\tau)$ calculations demonstrates that the correlation time of displacements of the sensing volume τ_z is determined by the time of transport of turbulent inhomogeneities of the refractive index with the mean wind velocity U to the distance equal to the initial radius of the probing beam a_0 ($\tau_z \approx a_0/U$). For example, at $a_0 = 7.5$ cm and $U = 4$ m/s, the correlation time is $\tau_z \approx 0.02$s. It is known [8] that in the atmospheric boundary layer the correlation time of wind velocity fluctuations is $\tau_V \sim 10$s to 20s. Consequently, τ_V exceeds τ_z by roughly three orders of magnitude. This means that to simulate $\overline{V}_r(t)$ based on (2.107) and (2.108), we needs an array of $\int_{-\infty}^{+\infty} d^2\rho\, I_{PN}^2(z,\rho,t)$ values approximately 10^3 times greater than the array of $V_r(z,t)$ values. Such a great volume of calculations seriously restricts the possibility of implementing the algorithm of direct simulation of $\overline{V}_r(t)$ and forces us to use approximate methods.

As can be seen from Figure 2.10(a), the shape of the instantaneous ($\Delta t \to 0$) distributions $Q_s(z,t)$ at $C_n^2 = 10^{-13}$ m$^{-2/3}$ differs only slightly from the shape of $Q_s(z)$ calculated at $C_n^2 = 0$ (see the dashed curve). Turbulent inhomogeneities of the refractive index lead only to random deviations of the focusing range of the probing beam $\tilde{F}(t)$ from the focal length F. This allows us to approximate the function $Q_s(z,t)$ based on the representation of the instantaneous intensity $I_{PN}(z,\rho,t)$ in (2.108), by analogy with (1.53), in the form [21]

$$I_{PN}^2(z,\rho,t) = \exp[-\rho^2 / (\tilde{g}(z,t)a_0^2)]/(\tilde{g}(z,t)) \tag{2.109}$$

where $\tilde{g}(z,t) = (1 - z/\tilde{F}(t))^2 + (z/L_d)^2$ and $L_d = 2\pi a_0^2/\lambda$. Substituting (2.109) into (2.108) and integrating with respect to ρ and z, we obtain for

$$Q_s(z,t) = \frac{\int_{t-\Delta t}^{t} dt'\, \tilde{g}^{-1}(z,t')}{L_d\left\{\dfrac{\pi}{2}\Delta t + \int_{t-\Delta t}^{t} dt'\, \arctan[L_d/\tilde{F}(t')]\right\}} \tag{2.110}$$

From (2.110) at $\Delta t \to 0$, we can easily find the relation between $\tilde{F}(t)$ and $z_{\max}(t)$:

$$\tilde{F}(t) = \frac{1}{2} L_d \left[\frac{L_d}{z_{\max}(t)} - \sqrt{\frac{L_d^2}{z_{\max}^2(t)} - 4} \right] \tag{2.111}$$

The comparison of the functions $Q_s(z,t)$ simulated directly [through the substitution of random intensity distribution $I_{PN}(z,\rho,t)$ into (2.108)] and with the use of the corresponding values of $z_{\max}(t)$ in (2.111) and (2.110) shows that the difference between them is small up to $C_n^2 = 10^{-13}$ m$^{-2/3}$. In the case of large C_n^2 (for example, $C_n^2 = 10^{-12}$ m$^{-2/3}$), the use of approximations (2.110) and (2.111) becomes impossible due to the significant increase of the longitudinal dimension of the sensing volume (see Figure 2.11(c)).

Thus, the problem of determination of $Q_s(z,t)$ reduces to the simulation of the time dependence of displacement of the sensing volume $z_{max}(t)$, where $t \in [0,T]$. In this case, the time interval T and the sampling frequency $1/\Delta t'$ ($\Delta t' \neq \Delta t$) should satisfy the conditions $T \gg \tau_V$ and $\Delta t' \ll \tau_z$. From two simulated discrete series $z_{max}(m\Delta t')$, where $m = 1, 2, 3, ..., 512$ with $\Delta t' = \delta y/U$ and $\Delta t' = a_0/U$ ($a_0/\delta y = 15$, $a_0 = 7.5$ cm), two estimates for the spectrum of displacements of the sensing volume

$$S_z(f) = \sigma_z^2 \int_{-\infty}^{+\infty} d\tau K_z(\tau) \exp(-2\pi j f \tau)$$ were obtained in [21]. The moving smoothing was

performed with the use of a rectangular spectral window including nine spectral frequencies. Figure 2.12 depicts the obtained spectra as solid curves 1 ($\Delta t' = \delta y/U$) and 2 ($\Delta t' = a_0/U$). Based on these data, the asymptotic equation was found for the spectrum $S_z(f)$ in the form [21]

$$S_z(f) = \sigma_z^2 \frac{2\tau_z}{[1 + (4.48\tau_z f)^2]^{4/3}} \tag{2.112}$$

where $\tau_z = \int_0^{\infty} d\tau K_z(\tau) = a_0/U$ is the integral correlation scale of displacements of the

sensing volume. The calculation by (2.112) is shown as a dashed curve in Figure 2.12.

The analysis of simulated results has shown that the probability density function of displacements of the sensing volume $p(z_{max})$ is close to the normal distribution. Thus, using spectrum (2.112) and calculated values of σ_z^2 (see Figure 2.11), we can

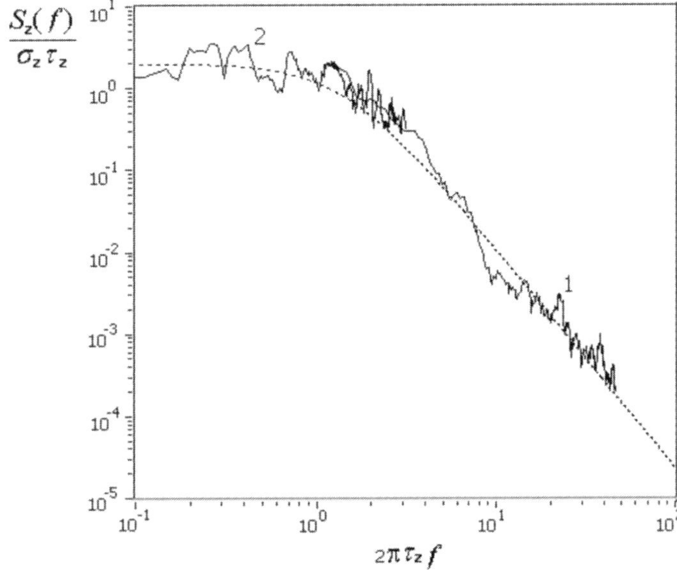

Figure 2.12 Temporal spectrum of displacements of the sensing volume. Curves 1 and 2: estimates of the spectrum obtained from data of numerical simulation with sampling frequency $1/\Delta t = U/\delta z$) and $1/\Delta t = U/a_0$, respectively; dashed curve: calculation by (2.112).

simulate random realizations of $z_{\mathrm{max}}(t)$ in the spectral range, omitting the procedure of simulation of the probing beam propagation in the turbulent atmosphere.

To calculate the spectrum of velocity measured by lidar $S_{\overline{V}}(f)$, when determining $\overline{V}_r(t)$, it is necessary, according to (2.107), to simulate not only $Q_s(z,t)$, but also the radial velocity $V_r(z,t)$. Random realizations of $V_r(z,t)$ were simulated in [21] by the algorithm described in Section 2.3 with the use of the von Karman model of (2.86), two-dimensional FFT, and the Taylor hypothesis of frozen turbulence. The results shown in Figure 2.13 were obtained at $\sigma_V = 1$ m/s and $L_V = 50$m.

The data of numerical simulation of $\overline{V}_r(t)$ were used to calculate the spectra $S_{\overline{V}}(f)$. Figure 2.13(a) depicts these spectra at $F = 500$m and different values of the mean velocity of lateral wind U (curves 1, 2, and 3 correspond to $U = 1$ m/s, while curves 4, 5, and 6 were obtained at $U = 4$ m/s) and $C_n^2 = 0$ (curves 1 and 4), $C_n^2 = 10^{-14}$ m$^{-2/3}$ (curves 2 and 5), and $C_n^2 = 10^{-13}$ m$^{-2/3}$ (curves 3 and 6). In the absence of fluctuations of the refractive index ($C_n^2 = 0$), the spectra $S_{\overline{V}}(f)$ calculated with the use of numerical simulation data in the high-frequency range ($f \geq 0.1$ Hz) are in satisfactory agreement with calculations by asymptotic equation (2.99) with regard for (2.71). The results of calculation by this equation are shown as dashed curves in Figure 2.13. The averaging of the radial wind velocity over the sensing volume acts as a low-frequency filter of fluctuations of the measured velocity, and the spectrum $S_{\overline{V}}(f)$ is proportional to $f^{-8/3}$ in place of the dependence $S_{\overline{V}}(f) \sim f^{-5/3}$ corresponding to point measurements.

Random displacements of the sensing volume $z_{\mathrm{max}} - \langle z_{\mathrm{max}} \rangle$ caused by turbulent fluctuations of the refractive index lead to an additional spread in velocity estimates. Since the correlation time τ_z is comparable to the time for measurement of one Doppler spectrum Δt, random displacements of the sensing volume should lead ultimately to an increase of fluctuations of the measured velocity at high frequencies as compared to the case of $C_n^2 = 0$. It can be seen from Figure 2.13(a) that the

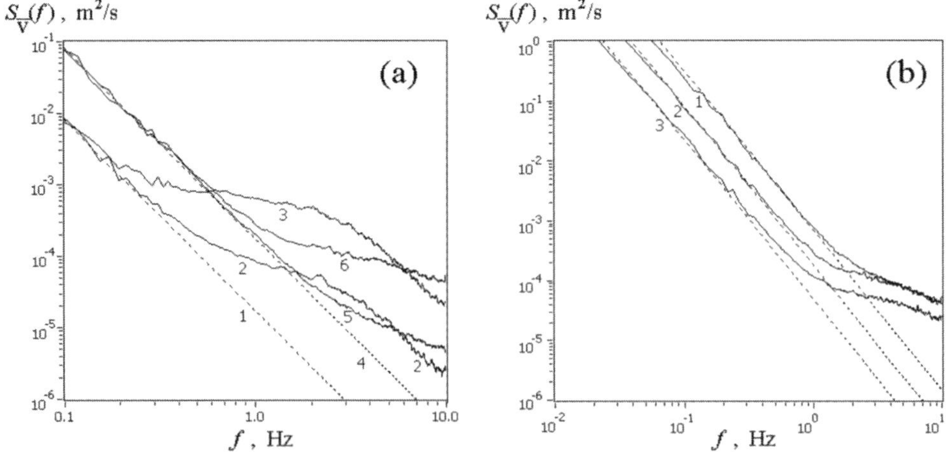

Figure 2.13 (a) Spectra of wind velocity estimated from cw CDL data at $F = 500$m and different values of the mean wind velocity $U = 1$ m/c (curves 1–3) and 4 m/s (curves 4–6); $C_n^2 = 0$ (curves 1 and 4), $C_n^2 = 10^{-14}$ m$^{-2/3}$ (curves 2 and 5); $C_n^2 = 10^{-13}$ m$^{-2/3}$ (curves 3 and 6). (b) Spectra of wind velocity estimated from CDL data at $F = 250$m (1), 500m (2), and 1 km (3); $C_n^2 = 10^{-13}$ m$^{-2/3}$; $U = 4$ m/s.

increase of C_n^2 causes the significant increase in the energy of fluctuations in the high-frequency range of the spectrum $S_{\overline{V}}(f)$.

In Figure 2.13(b), solid curves show the spectra $S_{\overline{V}}(f)$ at $C_n^2 = 10^{-13}\,\mathrm{m}^{-2/3}$, $U = 4$ m/s, and different focal lengths F. One can see that the greater F is (and, correspondingly, the greater the longitudinal dimension of the sensing volume Δz), the more pronounced the effect of the increase in the spectral density at high frequencies. At $F = 250$m, power dependence (2.99) for the spectrum $S_{\overline{V}}(f)$ keeps true up to the frequency $f_h \approx 1$ Hz (compare solid and dashed curves 1). At $F = 1$ km, the frequency interval, in which (2.99) is true becomes much narrower: $f_h \approx 0.3$ Hz (see curves 3). This fact should be taken into account when determining the dissipation rate of the kinetic energy of turbulence ε from lidar data by (2.99). The narrowing of the frequency interval, within which ε is estimated, leads to an increase in the error of ε estimation (see Section 3.2). It is clear that measurements of ε at $F = 250$m are more accurate than at $F = 1$ km.

The influence of turbulent fluctuations of the refractive index on the temporal spectrum of the wind velocity measured by Doppler lidar depends on the magnitude of lateral wind. The higher U, the stronger the averaging of random displacements of the sensing volume $z_{\max} - \langle z_{\max} \rangle$ for the time Δt and the less significant the effect of refractive turbulence.

Figure 2.13(a) demonstrates that at a weak lateral wind of $U = 1$ m/s (if, of course, at this wind the hypothesis of frozen turbulence is acceptable), it is impossible to estimate the dissipation rate of turbulent energy even at relatively moderate (for the atmospheric surface layer) turbulence: $C_n^2 = 10^{-14}\,\mathrm{m}^{-2/3}$. At a strong lateral wind, when the condition $\Delta t \gg a_0/U$ is true, fluctuations of $Q_s(z)$ caused by turbulent fluctuations of the refractive index of air are averaged to a significant extent. In this case, in (2.108) we can replace the integration with respect to time t' with the averaging over an ensemble of realizations. Numerical experiments have shown that at this averaging the longitudinal dimension of the sensing volume Δz increases, and for $R \le 500$m and $C_n^2 \le 10^{-13}\,\mathrm{m}^{-2/3}$ the value of Δz exceeds the longitudinal dimension of the sensing volume at $C_n^2 = 0$ (that is, Δz_0) by no more than 5%. Thus, at $\Delta t = 50$ ms and under the condition $\Delta t \gg a_0/U$, even if $R = 500$m and $C_n^2 = 10^{-13}\,\mathrm{m}^{-2/3}$, from the spectrum of wind velocity measured by lidar we can acquire the information about the dissipation rate of the turbulent energy through the use of (2.93) through (2.95).

2.6   Statistics for Radial Velocity Estimates and the Width of the Doppler Spectrum for Pulsed CDLs

Based on cw CDL data measured at a fixed position of the probing beam in space (no beam scanning), the spatial structure of wind turbulence can be studied only through calculation of the temporal structure function $D_{\hat{V}}(\tau)$ or the temporal spectrum $S_{\hat{V}}(f)$ of wind velocity fluctuations and the width of the Doppler spectrum σ_S. In this case, for the analysis of $D_{\hat{V}}(\tau)$ and $S_{\hat{V}}(f)$, the Taylor hypothesis of frozen turbulence should be used. In the case of pulsed CDLs, however, it is possible to estimate the radial velocity $\hat{V}_r(R, t_i)$, where $t_i = t_0 + i\Delta t$, $i = 1, 2, 3, \ldots$ and $\Delta t = LT_P$, from the raw data of one laser shot ($L = 1$) or a series of shots ($L > 1$) at different distances R from the lidar to the center of the sensing volume. This fact allows us to calculate the spatial longitudinal structure function of wind velocity fluctuations $D_{\hat{V}}(r) = \langle [\hat{V}_r'(R+r) - \hat{V}_r'(R)]^2 \rangle$, where $\hat{V}_r' = \hat{V}_r - \langle \hat{V}_r \rangle$ directly.

According to (2.35), the lidar estimate of the radial wind velocity $\hat{V}_r(R, t_i)$ can be represented as a sum of the velocity $\overline{V}_r(R, t_i)$ averaged over the sensing volume and the error of estimation of this velocity $V_e(R, t_i)$. Assuming the homogeneity and stationarity of turbulence and taking into account the statistical independence of $\overline{V}_r(R, t_i)$ and $V_e(R, t_i)$, we obtain for the longitudinal structure function $D_{\hat{V}}(r)$ [29]

$$D_{\hat{V}}(r) = D_{\overline{V}}(r) + D_e(r) \tag{2.113}$$

where

$$D_{\overline{V}}(r) = \langle [\overline{V}_r'(R+r) - \overline{V}_r'(R)]^2 \rangle \tag{2.114}$$

is the structure function of the radial wind velocity averaged over the sensing volume,

$$D_e(r) = \langle [V_e'(R+r) - V_e'(R)]^2 \rangle \tag{2.115}$$

is the structure function of the error of velocity estimation, and $V' = V - \langle V \rangle$ are fluctuations of the value of V.

We assume that the estimates $\hat{V}_r(R, t_i)$ are obtained from measured Doppler spectra with use of (2.48) (that is, first spectral moment). Then, in (2.114) for $\overline{V}_r(R, t_i)$ we can use (2.29) and write the structure function $D_{\overline{V}}(r)$ in the following form [29, 30]:

$$D_{\overline{V}}(r) = 4 \int_0^\infty d\kappa_z S_V(\kappa_z) H_p(\kappa_z)[1 - \cos(2\pi r \kappa_z)] \tag{2.116}$$

where

$$H_p(\kappa_z) = \left| \int_{-\infty}^{+\infty} dz' Q_s(z') \exp(-2\pi j \kappa_z z') \right|^2 \tag{2.117}$$

is the transfer function of the low-frequency spatial filter in the case of a pulsed CDL. Assuming that the probing pulse is Gaussian [see (1.90)], upon substituting (2.33) and (2.25) into (2.117) and integrating with respect to the variable z', we obtain, respectively, [29]

$$H_p(\kappa_z) = \exp\{-2(\pi \Delta p \kappa_z)^2\} \left[\frac{\sin(\pi \Delta R \kappa_z)}{\pi \Delta R \kappa_z} \right]^2 \tag{2.118}$$

for the case of rectangular window and

$$H_p(\kappa_z) = \exp\{-2\pi(\Delta z \kappa_z)^2\} \tag{2.119}$$

for the case of the Gaussian window [30]. In (2.118) and (2.119), $\Delta p = \sigma_P c/2$, $\Delta R = cT_W/2$, and $\Delta z = \sqrt{\pi}\sqrt{\sigma_P^2 + \sigma_W^2} \cdot c/2$.

For 2-μm CDLs, the longitudinal dimension of the sensing volume Δz, at which the spatial and frequency resolution is optimal, equals ~40m ($\sigma_P = 120$ ns) and ~90m ($\sigma_P = 240$ ns) depending on the pulse duration σ_P. For the atmosphere, these values of Δz can be both smaller and greater than the integral (outer) scale of turbulence L_V. If the dimension Δz is comparable with or larger than L_V, the inertial interval, in which the local structure of turbulence is determined only by the dissipation rate ε, can be far smaller than the longitudinal dimension of the sensing volume. Consequently, in this case, ε cannot be estimated with acceptable accuracy from the structure function of wind velocity measured by lidar with neglect of the outer scale of turbulence. To take into account the scale L_V, we use the von Karman model of (2.116) for the spectrum $S_V(\kappa_z)$. Then, from (2.116), (2.70), and (2.71), we obtain [30]

$$D_{\bar{V}}(r) = \varepsilon^{2/3} G_s(r, L_V) \tag{2.120}$$

where

$$G_s(r, L_V) = 5.088 C_K L_V^{5/3} \int\limits_0^\infty d\kappa_z \frac{[1 - \cos(2\pi r \kappa_z)] H_p(\kappa_z)}{[1 + (8.42 L_V k_z)^2]^{5/6}} \tag{2.121}$$

With an increase of r, the structure function $D_{\bar{V}}(r)$ saturates to the level $2\sigma_{\bar{V}}^2$, where $\sigma_{\bar{V}}^2 = \langle(\bar{V}_r')^2\rangle$ is the variance of radial velocity averaged over the sensing volume. According to (2.116), it is described by the equation

$$\sigma_{\bar{V}}^2 = 2\int\limits_0^\infty d\kappa_z S_V(\kappa_z) H_p(\kappa_z) \tag{2.122}$$

From (2.29) through (2.31), for the mean square estimate of the Doppler spectrum width $\langle\hat{\sigma}_S^2\rangle$ we obtain the equation analogous to (2.79), in which the instrumental broadening of the spectrum $\sigma_{VI}^2 = (\lambda/2)^2 \sigma_{fl}^2$ in the case of the Gaussian window is given by (2.28), and the equation for the turbulent broadening of the Doppler spectrum σ_t^2 at homogeneous turbulence has the form [30]

$$\sigma_t^2 = 2\int\limits_0^\infty d\kappa_z S_V(\kappa_z)[1 - H_p(\kappa_z)] \tag{2.123}$$

It follows from (2.122) and (2.123) that, as in the case for a cw CDL, measurements of $\sigma_{\bar{V}}^2$ and σ_t^2 carry information about the variance of wind velocity σ_V^2 ($\sigma_V^2 = \sigma_{\bar{V}}^2 + \sigma_t^2$). The instrumental broadening of the Doppler spectrum σ_{VI}^2 in the case of pulsed CDLs is much higher than that of cw CDLs, and, as a rule, at weak and moderate wind turbulence $\sigma_{VI} > \sigma_t$ [30, 60].

Parameters $\sigma_{\bar{V}}^2$ and σ_t^2 calculated from pulsed CDL measurement data can also be used for determining the dissipation rate of turbulent energy ε if the spatial spectrum of wind velocity fluctuations $S_V(\kappa_z)$ is known. With the use of the von Karman model, from (2.123), (2.69), (2.71), and (2.119) we have [30]

$$\sigma_t^2 = \varepsilon^{2/3} G_W(\Delta z, L_V) \tag{2.124}$$

where

$$G_W(\Delta z, L_V) = 0.2485 C_K \Delta z^{2/3} \int_0^\infty \frac{d\xi\{1 - \exp[\xi^2/(2\pi)]\}}{[\xi^2 + (0.746\Delta z/L_V)^2]^{5/6}} \tag{2.125}$$

and $\Delta z = \sqrt{\pi}\sqrt{\sigma_P^2 + \sigma_W^2} \cdot c/2$.

Under the condition $\Delta z \ll L_V$, we can take $\Delta z/L_V = 0$ in (2.125) and, on integration with respect to ξ, obtain a simple equation for σ_t^2:

$$\sigma_t^2 = 0.274 C_K (\varepsilon \Delta z)^{2/3} \tag{2.126}$$

Here, to estimate the dissipation rate, it is sufficient to determine only the turbulent broadening of the Doppler spectrum σ_t^2.

According to (2.113), the lidar estimate of the structure function $D_{\hat{V}}(r)$ is a sum of the functions $D_{\bar{V}}(r)$ and $D_e(r)$. Taking into account that the radial wind velocity averaged over the sensing volume remains nearly unchanged (even with the use of scanning by the probing beam), that is, $\bar{V}_r(R, t_{i+1}) \approx \bar{V}_r(R, t_i)$, for the rather short time interval Δt between measurements of radial velocities (for example, 50 ms), from experimental data we can find

$$\Delta V_e(R) = \hat{V}_r(R, t_{i+1}) - \hat{V}_r(R, t_i) \tag{2.127}$$

which is the difference of uncorrelated random errors $V_e(R, t_{i+1}) - V_e(R, t_i)$. Then, from the array of these differences, we can calculate the structure function $D_e(r)$ as

$$D_e(r) = \frac{1}{2}\langle [\Delta V_e(R + r) - \Delta V_e(R)]^2 \rangle \tag{2.128}$$

which can be represented in the form

$$D_e(r) = 2\sigma_e^2[1 - K_e(r)] \tag{2.129}$$

where $\sigma_e^2 = \langle V_e^2 \rangle - \langle V_e \rangle^2$ is variance and $K_e(r) = \langle V_e(R + r)V_e(R) \rangle / \sigma_e^2$ is the correlation coefficient of random errors in estimation of the radial velocity.

The variance σ_e^2 depends on the SNR, the width of the temporal window σ_W, the probing pulse duration σ_P, the number of pulses used for accumulation of spectra L, and the method of estimation of the radial wind velocity. In addition, wind turbulence can influence the value of σ_e^2. To study statistical properties of the random error $V_e(R)$, we use numerical simulation of the signal of a pulsed CDL.

Reference [37] shows that the longitudinal (along the probing beam axis) scale of correlation of echo signal power fluctuations $\bar{P}_S(R) - \langle P_S(R) \rangle$ caused by turbulent fluctuations of the refractive index is comparable to the sensing path length and, consequently, far exceeds Δz. Therefore, fluctuations of the refractive index have no effect on the estimate of the radial velocity $\hat{V}_r(R)$ and are neglected in the numerical simulation of random realizations of the pulsed CDL signal. The algorithm is based on (1.28), (1.91), and (1.94) with allowance for the fact that the one-dimensional probability density function of the echo signal of a pulsed CDL has a Gaussian distribution. The sensing path is divided into thin layers each with thickness δz, and samples of the normalized complex signal $Z(mT_s)$ are simulated as [29]

$$Z(mT_s) = Z_S(mT_s) + Z_N(mT_s) \tag{2.130}$$

where $m = 0, 1, 2, 3, \ldots$ is the sample number,

$$Z_S(mT_s) = \sqrt{\frac{\text{SNR}\delta z}{\sqrt{\pi}\Delta p}} \sum_{i=0}^{N_P-1} \xi'(Km + i)\exp\left\{-\frac{1}{2}\left[\frac{\delta z}{\Delta p}\left(i - \frac{N_P}{2}\right)\right]^2\right\}$$
$$\times \exp\left\{j\frac{4\pi}{\lambda}T_s m V_r[\delta z(Km + i)]\right\} \tag{2.131}$$

N_P is the partition number of the section $\delta z N_P$, $\xi'(i)$ is the Gaussian white noise with zero mean and unit variance for the real and imaginary parts, $K \geq 1$ is an integer number defined as $K = (ct_s/2)/\delta z$, and $\Delta p = \sigma_P c/2$. The condition $\delta z \ll \Delta p \ll \delta z N_P/2$ should be fulfilled in the simulation. The noise component of the lidar signal $Z_N(mT_s)$ is simulated as in (2.103). The radial velocity is simulated with the use of (2.75).

From simulated realizations, with the use of (2.42), some temporal window, and (2.47) and (2.48), we obtain an estimate of the radial velocity $\hat{V}_r(R)$. Since the simulated realization of the velocity $V_r(z)$ is known, we can use (2.29) to calculate the radial velocity averaged over the sensing volume $\bar{V}_r(R)$. Then the random error of estimation is determined as $V_e(R) = \hat{V}_r(R) - \bar{V}_r(R)$. The estimate bias $b_r = \langle V_e \rangle$, variance σ_e^2, and the correlation coefficient $K_e(r)$ of the error in estimation of the radial velocity are calculated with the number of $V_e(R)$ realizations necessary to conduct the averaging. The numerical simulation allows us also to calculate the probability density $p(V_e)$.

For the case of a very high signal-to-noise ratio (SNR $\gg$ 1), homogeneous wind $V_r(z) = $ const, and the Gaussian window of (2.23), simple analytical equations

$$\sigma_e^2 = \frac{(\lambda/2)^2}{(4\pi)^2 L\sigma_P\sqrt{\sigma_P^2 + \sigma_W^2}} \tag{2.132}$$

and

$$K_e(r) = \exp[-(\pi/2)(r/\Delta z)^2] \tag{2.133}$$

were derived in [32] based on (2.131) and (2.1) through (2.3) with regard to $\hat{P}_S =$

SNR. It follows from (2.133) that the correlation scale $L_e = \int\limits_0^\infty dr K_e(r)$ is determined by the longitudinal dimension of the sensing volume Δz [see (2.26)]; that is, $L_e = \Delta z/\sqrt{2}$. An analysis of (2.132) shows that a decrease of the probing pulse duration σ_P leads to an unlimited increase of the error σ_e, even when the noise component of the lidar signal can be neglected. With a decrease in the width of the time window σ_W, when σ_P is fixed, the variance σ_e^2 saturates to a certain level according to (2.132). At the same time, with a decrease of σ_W, along with an improvement of the spatial resolution (the value of Δz decreases to the limit determined by the probing pulse duration), the Doppler spectrum broadens [see (2.28)], the spectral maximum decreases, and the noise component of the spectrum begins to exert the decisive effect on the accuracy of estimation of the radial velocity (even at SNR $\gg$ 1); that is, the error of estimation increases due to noise. The practice shows [30] that the equality $\sigma_W = \sigma_P$ is optimal from the viewpoint of spatial resolution and the accuracy of velocity estimation.

The data of the field and numerical experiments shown in Figure 2.14 were obtained at the following parameters: $\lambda = 2$ μm, $\sigma_W = \sigma_P = 0.2488$ μs, and $L = 25$. In this case, according to (2.26) and (2.28), $\Delta z = 94$m and $\sigma_{VI} = (\lambda/2)\sigma_{fI} = 0.64$ m/s.

Figure 2.14(a) shows the result of measurements of the error σ_e at different SNR [30, 32] as dots. The depicted data of the field experiment were obtained for different altitudes at different levels of turbulence. According to the results of lidar measurements [30], ε varied from ~$5 \cdot 10^{-5}$ to ~$5 \cdot 10^{-3}$ m^2/s^3, and L_V took values from ~50m to ~200m. Curves 1, 2, and 3 demonstrate the results of calculation of the error σ_e from the data of numerical simulation, where the following values of parameters were taken: $L_V = 100$ m, $\varepsilon = 0$ (1), $\varepsilon = 10^{-3}$ m^2/s^3 (2), and $\varepsilon = 10^{-2}$ m^2/s^3 (3). One can see that the field and numerical experiments give close results. The horizontal line shows the result of calculation by (2.132) ($\sigma_e = 0.054$ m/s). Expectedly, the calculated value of error is smaller than the data of numerical and field experiments, because (2.132) ignores the influence of lidar signal noise and wind turbulence. Nevertheless, already at SNR = 4 the data of numerical simulation (curve 1) and field experiment at weak turbulence become close to the value of σ_e calculated by (2.132). The comparison of curves 1, 2, and 3 allows us to judge the degree of influence of the wind turbulence on the error of estimation of the radial velocity. Thus, the ratio of the σ_e values shown by curves 3 and 1 is about 1.5. As can be seen from the figure, at SNR < 1 the decrease in the signal-to-noise ratio leads to a sharp increase in σ_e, although for the considered values of SNR > 0.2 the probability of a bad estimate (connected with a noise peak in the Doppler spectrum) of the radial velocity is negligibly low and the estimates $\hat{V}_r$ obtained at this SNR are unbiased, that is, $b_r = 0$.

We can see from Figure 2.14(b) that the values of the correlation coefficient $K_e(r)$ calculated by (2.133) nearly coincide with the data of the field experiment at the high SNR. A decrease of the SNR level causes a slight decrease in the correlation scale L_e, which is also observed in the numerical experiment [29].

The SNR influence on the accuracy of lidar measurements of the radial velocity is studied, in particular, in [42,43, 45, 47, 48, 61–64]. Figure 2.15 shows the probability density functions of the estimate of radial velocity $p(\hat{V}_r)$ at different SNR. These functions were obtained through the simulation of 10^5 independent realizations of the signal with the use of (2.130) and (2.131) at $\lambda = 2$ μm, $\sigma_P = 120$ ns ($\Delta p = 18$ m),

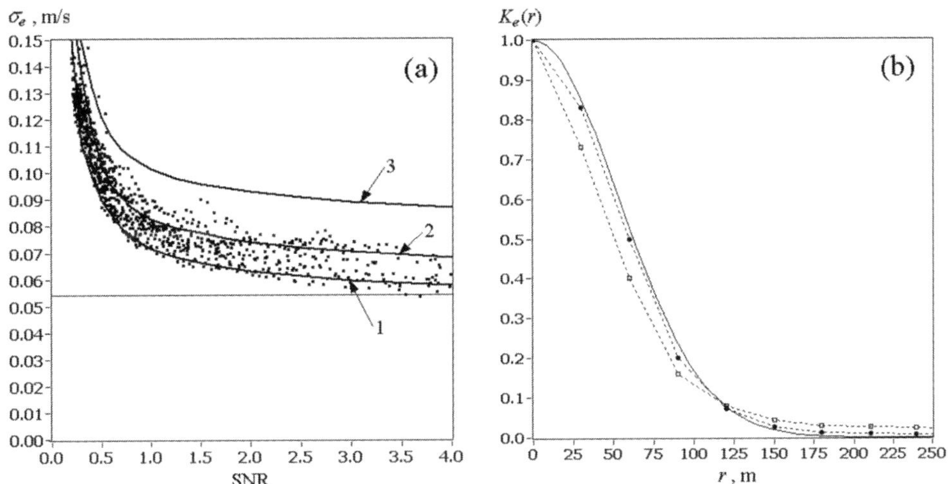

Figure 2.14 Error σ_e of measurement of the radial velocity averaged over the sensing volume as a function of the (a) SNR and (b) correlation coefficient $K_e(r)$. Data of field experiment (lidar measurements in Tarbes on August 27, 2003 [32]) are shown by dots (a), circles (SNR = 3) and squares (SNR = 0.3) (b). Curves 1 ($\varepsilon = 0$), 2 ($\varepsilon = 10^{-3}$ m^2/s^3, $L_V = 100$ m), and 3 ($\varepsilon = 10^{-2}$ m^2/s^3, $L_V = 100$ m) are for the data of numerical experiment. The horizontal line (a) demonstrates the result of calculation by (2.132). The solid curve (b) shows the result of calculation by (2.133).

$T_s = 20$ ns (3 m, $B_F = 50$ MHz), and $V_r = 0$. The estimates of the spectra (at $L = 1$) were obtained with the use of $M = 16$ samples of the signal $Z(mT_s)$ with addition of 112 zeros (rectangular time window) for each of them. Thus, every spectrum consisted of 128 values with the frequency bin $\delta f = B_F/128 = 0.39$ MHz ($\delta V = 0.39$ m/s).

These spectra were used to estimate the radial velocity and to calculate the probability density function $p(\hat{V}_r)$ after smoothing over two neighboring points. For the cases of SNR = 0.1 and SNR = 0.01, the accumulation (averaging) of spectra was used before the estimation of the velocity ($L = 1$ and 10 for SNR = 0.1 and $L = 1$, 10, 100, and 1,000 for SNR = 0.01). L is the number of laser shots used for spectral accumulation. The figure shows the values of the probability P of obtaining the estimate in the range from -1 m/s to $+1$ m/s. It can be seen that in the case of very low SNR and $L = 1$ the function $p(\hat{V}_r)$ has a nearly uniform distribution. The use of the methods of velocity estimation from maximum likelihood allows some improvement in the accuracy to be achieved [65], but in the case of a very weak signal the lidar measurement of velocity is impossible without spectral accumulation at a large number of laser shots. If at low SNR the velocity is estimated from every laser shot and the results are averaged, then this estimate is biased (at any wind velocity the estimate is close to zero).

The probability density distributions shown in Figure 2.15 can be approximately (exactly at large values of L) described by the Gauss distribution at a uniform pedestal. With guidance from [42, 47, 48], we have succeeded in constructing a model for the probability density function $p(\hat{V}_r)$ in the form

$$p(\hat{V}_r) = \frac{1 - b_e}{\sqrt{2\pi}\sigma_g} \exp\left[-\frac{(\hat{V}_r - \langle V_r \rangle)^2}{2\sigma_g^2} \right] + \frac{b_e}{B_V} \tag{2.134}$$

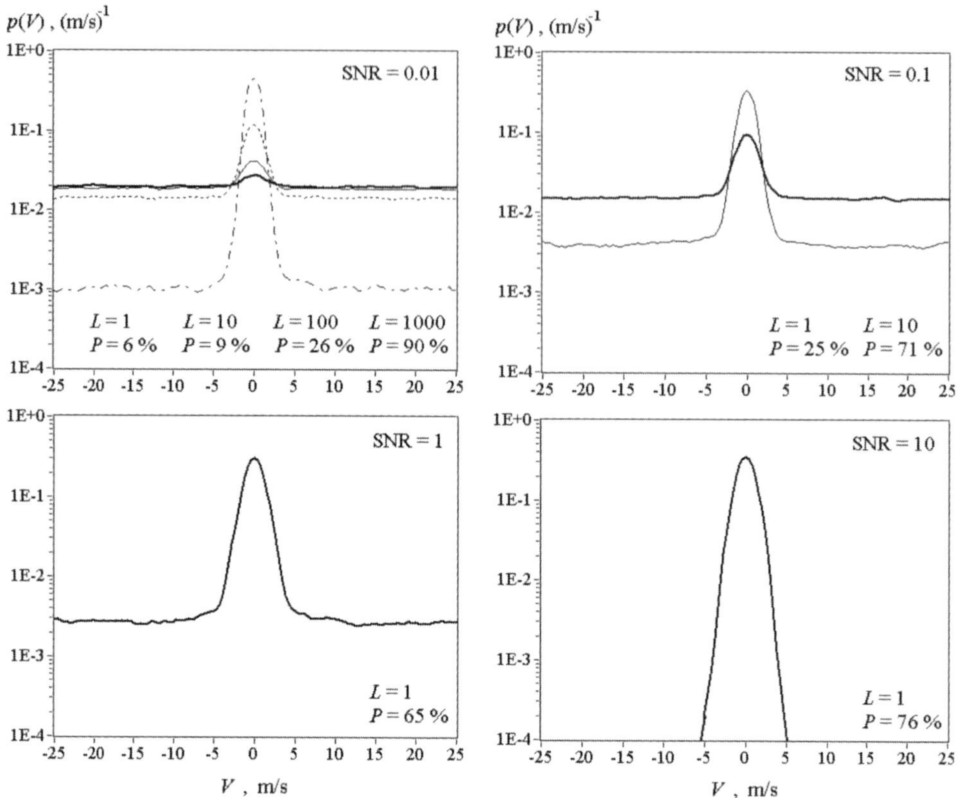

Figure 2.15 Probability density function of the lidar estimate of radial velocity at different SNR and $L = 1$ (bold solid curves), 10 (thin solid curves), 100 (dashed curves), and 1000 (dot-dash curves).

where b_e is the fraction of bad estimates, σ_g^2 is the variance of good estimates of the radial velocity, and $B_V = (\lambda/2)B_F$. Then, for the bias $\mathrm{Bias}[\hat{V}_r] = \langle(\hat{V}_r - \langle V_r\rangle)\rangle$ and the variance $\mathrm{Var}[\hat{V}_r] = \langle(\hat{V}_r - \langle V_r\rangle)^2\rangle$ from (2.134) under the condition $\langle V_r\rangle^2 + \sigma_g^2 \ll (B_V/2)^2$ we have

$$\mathrm{Bias}[\hat{V}_r] = b_e\langle V_r\rangle \tag{2.135}$$

$$\mathrm{Var}[\hat{V}_r] = (1 - b_e)\sigma_g^2 + b_e B_V^2/12 \tag{2.136}$$

In the limiting case of $b_e = 1$, the bias and variance of the estimate are determined, respectively, by $\langle V_r\rangle$ and $B_V^2/12$. In another limiting case of $b_e = 0$, the estimate is unbiased, and its variance coincides with σ_g^2. On the other hand, for the variance of the unbiased estimate we can write the equation

$$\mathrm{Var}[\hat{V}_r] \equiv \mathrm{Var}[\overline{V}_r + V_e] = \sigma_{\overline{V}}^2 + \sigma_e^2 \tag{2.137}$$

Taking into account that at the rather high order of spectral accumulation $L \gg 1$ the variance of the error for an unbiased estimate of the radial velocity σ_e^2 can be very small (see, for example, Figure 2.14) compared to $\sigma_{\tilde{V}}^2$ at a not very weak turbulence, we can take the variance σ_g^2 equal to $\sigma_{\tilde{V}}^2$.

Reference [42] show for the case of homogeneous wind ($\sigma_V = 0$) that with a decrease of SNR the variance of good estimates σ_g^2 increases and at SNR $\ll 1$ (when $b_e \approx 1$) becomes close to the squared width of the Doppler spectrum $(\lambda/2)^2 \sigma_{f1}^2$, where $\sigma_{f1} = (2\sqrt{2}\pi\sigma_p)^{-1}$. For this limiting case, according to (1.99), at $\sigma_V \neq 0$ the value of σ_g^2 should be determined as $\sigma_V^2 + (\lambda/2)^2 \sigma_{f1}^2$. Section 3.3 of this monograph discusses the methods for estimation of the wind speed and direction from scanning CDL data developed based on the model of (2.134).

2.7   Conclusions

1. The radial velocity estimated from the Doppler spectrum is a sum of the radial velocity averaged over the sensing volume and the random error of velocity estimation, which has properties of white noise. For the correct consideration of spatial averaging of the velocity measured by lidar, it is necessary to know the weighting functions of averaging over the sensing volume, whose form for the case of estimation (from cw and pulsed CDL data) of the radial velocity from the first spectral moment is determined in Sections 2.2 and 2.3.

2. The variance of the radial velocity estimated at the maximum point of the measured Doppler spectrum is lower than the variance of the wind velocity, but higher than the variance of the first spectral moment. In the case of homogeneous wind turbulence and A SNR far higher than unity, the sum of the turbulent broadening of the Doppler spectrum and the variance of the first spectral moment, regardless of the CDL type and the form of the weighting function of averaging over the sensing volume, is a sum of the wind velocity variance, instrumental broadening of the Doppler spectrum, and variance of random error of the lidar estimate of the radial velocity.

3. The width of the lidar Doppler spectrum carries the information about the dissipation rate of turbulent energy, if the longitudinal dimension of the sensing volume is much smaller than the outer scale of turbulence. Otherwise, it is necessary to know the form of the spectrum of turbulent fluctuations of wind velocity in the entire range of spatial frequencies and, in addition to the Doppler spectrum width, it is necessary to know the variance of the radial velocity averaged over the sensing volume. Theoretical equations for the mathematical expectation of the squared width of the Doppler spectrum obtained in Sections 2.3 and 2.6 with the use of the spectral von Karman model form the basis for the development of methods of lidar measurements of wind turbulence parameters at any dimensions of the sensing volume.

4. The temporal spectrum of the wind velocity measured by the coherent lidar at the frequencies corresponding to the inertial interval of turbulence differs widely from the Kolmogorov-Obukhov spectrum due to velocity averaging over the sensing volume. At the strong lateral wind and large longitudinal dimension of the sensing volume, which may exceed the outer scale of turbulence, at frequencies satisfying condition (2.94), the lidar spectrum of velocity has the

power dependence on frequency with the exponent of $-8/3$, rather than $-5/3$, as in the case of a point measurer. Equation (2.99) allows one to estimate the dissipation rate of the turbulent energy from the temporal spectrum of the radial velocity measured by lidar.

5. In the case of cw CDLs, turbulent fluctuations of the refractive index of air cause random displacements of the sensing volume along the optical axis with the scale of temporal correlation approximately equal to the ratio of the initial radius of the sensing beam to the lateral wind velocity. These displacements of the sensing volume can, under certain conditions (high value of C_n^2 and F and weak lateral wind), lead to a significant (manifold) increase in the energy of fluctuations in the high-frequency range of the wind velocity spectrum corresponding to the inertial interval of turbulence, which provides no way to obtain a correct estimate of the dissipation rate of the turbulent energy from the spectrum of velocity measured by lidar by (2.93) through (2.95).

6. As the sensing volume formed by a cw CDL decreases, the contribution from the probing beam scattering at individual large particles (whose number decreases as well) to the measured Doppler spectrum increases. As a result, the variance of random deviations of the estimated radial wind velocity from the radial velocity averaged over the sensing volume in the turbulent wind flow increases and can exceed 10-fold and higher the variance of the error of velocity estimation calculated on the assumption of the Gaussian statistics of the lidar echo signal, which is true in the case of the very large mean number N_{eff} of efficiently scattering particles in the sensing volume ($N_{eff} > 100$).

7. The dependences of the error of estimation of the radial velocity on the SNR obtained in Section 2.6 with the use of numerical simulation are in satisfactory agreement with the data of the field experiment. The coefficient of longitudinal spatial correlation of the random error of radial velocity estimated from the pulsed CDL data is determined by the longitudinal dimension of the sensing volume and at SNR ≥ 1 is described by (2.131).

References

[1]   Gurvich, A.S., "Spectra of pulsations of the vertical component of wind velocity and their relation with micrometeorological conditions," *Tr. IFA AN SSSR*, No. 4, 1962, p. 101.

[2]   Gurvich, A.S., "Frequency spectra and probability distribution functions of the vertical wind component," *Izvestiya AN SSSR (Ser. Geofiz.)*, No. 7, 1960, p. 1042.

[3]   Zubkovskii, R.L., "Experimental investigation of spectra of pulsations of the vertical wind component in the free atmosphere," *Izvestiya AN SSSR (Ser. Geofiz.)*, No. 8, 1963, p. 1285.

[4]   Lumley, J.L., and Panofsky, H.A., *The Structure of Atmospheric Turbulence*, Interscience Publishers, New York, 1964.

[5]   Monin, A.S., and Yaglom, A.M., *Statistical Fluid Mechanics, Volume II: Mechanics of Turbulence*, MIT Press, Cambridge, MA, 1971.

[6]   Tatarskii, V.I., *Wave Propagation in a Turbulent Medium*, McGraw-Hill, New York, 1961.

[7]   Vinnichenko, N.K., et al., *Turbulence in the Free Atmosphere*, Gidrometeoizdat, Leningrad, 1976, p. 288.

[8]   Byzova, N.L., Ivanov, V.N., and Garger, E.K., *Turbulence in Atmospheric Boundary Layer*, Gidrometeoizdat, Leningrad, 1989, p. 263.

[9] Volkovitskaya, Z.I., and Ivanov, V.P., "Dissipation of turbulence energy in the atmospheric boundary layer," *Izv. AN SSSR. Fiz. Atmos. Okeana*, Vol. 6, No. 5, 1970, pp. 435–444.

[10] Lawrence, T.R., et al., "A laser velocimeter for remote wind sensing," *Review of Scientific Instruments*, Vol. 43, No. 3, 1972, pp. 512–518.

[11] Sonnenschein, C.M., and Horrigan, F.A., "Signal-to-noise relationship for coaxial systems that heterodyne backscatter from the atmosphere," *Applied Optics*, Vol. 10, No. 7, 1971, pp. 1600–1604.

[12] Gordienko, V.M., et al., "Coherent CO_2 lidars for measuring wind velocity and atmospheric turbulence," *Optical Engineering*, Vol. 33, No. 10, 1994, pp. 3206–3213.

[13] Byzova, N.L., et al., "Joint measurements of wind velocity by Doppler lidar and high-tower anemometers," *Meteorol. Gidrol.*, No. 3, 1991, pp. 114–117.

[14] Smalikho, I.N., "On measurement of the dissipation rate of the turbulent energy with a cw Doppler lidar," *Atmos. Oceanic Opt.*, Vol. 8, No. 10, 1995, pp. 788–793.

[15] Banakh, V.A., et al., "Turbulence measurements with a cw Doppler lidar in the atmospheric boundary layer," *Atmos. Oceanic Opt.*, Vol. 8, No. 12, 1995, pp. 955–959.

[16] Banakh, V.A., et al., "Fluctuation spectra of wind velocity measured with a Doppler lidar," *Atmos. Oceanic Opt.*, Vol. 10, No. 3, 1997, pp. 202–208.

[17] Banakh, V.A., and Smalikho, I.N., "Determination of the turbulent energy dissipation rate from lidar sensing data," *Atmos. Oceanic Opt.*, Vol. 10, No. 4–5, 1997, pp. 295–302.

[18] Smalikho, I.N., "Accuracy of the turbulent energy dissipation rate estimation from the temporal spectrum of wind velocity fluctuations," *Atmos. Oceanic Opt.*, Vol. 10, No. 8, 1997, pp. 559–563.

[19] Banakh, V.A., et al., "Measurements of turbulent energy dissipation rate with a cw Doppler lidar in the atmospheric boundary layer," *Journal of Atmospheric and Oceanic Technology*, Vol. 16, Nol.8, 1999, pp. 1044–1061.

[20] Banakh, V.A., Werner, Ch., and Smalikho, I.N., "The effect of aerosol microstructure on the error in estimating wind velocity with a Doppler lidar," *Atmos. Oceanic Opt.*, Vol. 13, No. 8, 2000, pp. 685–691.

[21] Banakh, V.A., Werner, Ch., and Smalikho, I.N., "Effect of turbulent fluctuations of refractive index on the time spectrum of wind velocity measured by Doppler lidar," *Atmos. Oceanic Opt.*, Vol. 13, No. 9, 2000, pp. 741–746.

[22] Banakh, V.A., et al., Turbulent energy dissipation rate measurements by coherent lidar," *SPIE Proc. Lidar Techniques for Remote Sensing II*, Paris, France, 25–26 September 1995, Vol. 2581, pp. 243–253.

[23] Banakh, V.A., et al., "Estimations of turbulent energy dissipation rate from Doppler lidar data," *Proc. 8th Coherent Laser Radar Conference*, Keystone, CO, USA, July 1995, pp. 116–118.

[24] Banakh, V.A., Werner, Ch., and Smalikho, I.N., "Effect of aerosol particle microstructure on accuracy of cw Doppler lidar estimate of wind velocity," *Proc. 10th Coherent Laser Radar Conference*, Mount Hood, OR, USA, 28 June–2 July 1999, pp. 132–135.

[25] Keeler, R.J., et al., "An airborne air motion sensing system. Pat I: Concept and preliminary experiment," *Journal of Atmospheric and Oceanic Technology*, Vol. 4, No. 3, 1987, pp. 113–127.

[26] Zrnic, D.S., "Estimation of spectral moments for weather echoes," *IEEE Trans. on Geoscience Electronics*, Vol. GE-17, No. 4, 1979, pp. 113–128.

[27] Banakh, V.A., Smalikho, I.N., and Werner, Ch., "Effect of aerosol particle microstructure on statistics of cw Doppler lidar signal," *Applied Optics*, Vol. 39, No. 30, 2000, pp. 5393–5402.

[28] Gurvich, A.S. "Influence of temporal evolution of turbulent inhomogeneities on frequency spectra," *Izv. AN SSSR. ser. Fiz. Atmos. Okeana*, Vol. 16, No. 4, 1980, pp. 345–354.

[29] Banakh, V.A., and Smalikho, I.N., "Estimation of the turbulence energy dissipation rate from the pulsed Doppler lidar data," *Atmos. Oceanic Opt.*, Vol. 10, No. 12, 1997, pp. 957–965.

[30] Smalikho, I.N., Köpp, F., and Rahm, S., "Measurement of atmospheric turbulence by 2-μm Doppler lidar," *Journal of Atmospheric and Oceanic Technology*, Vol. 22, No. 11. 2005, pp. 1733–1747.

[31] Smalikho, I.N., Köpp, F., and Rahm, S., "Measurement of atmospheric turbulence by 2-μ Doppler lidar," *Proc. Workshop Aircraft Vortices and Atmospheric Turbulence*, Institute of Atmospheric Physics, DLR, Oberpfaffenhofen, Germany, September 2004, pp. 20–31.

[32] Smalikho, I.N., Köpp, F., and Rahm, S., "Measurement of atmospheric turbulence by 2-μm Doppler lidar," DLR report No. 200, August 2004, Oberpfaffenhofen, p. 37.

[33] Frehlich, R.G., "Effect of wind turbulence on coherent Doppler lidar performance," *Journal of Atmospheric and Oceanic Technology*, Vol. 14, No. 2, 1997, pp. 54–75.

[34] Frehlich, R.G., Hannon, S.M., and Henderson, S.W., "Coherent Doppler lidar measurements of wind field statistics," *Boundary-Layer Meteorology*, Vol. 86, No. 1, 1998, pp. 223–256.

[35] Frehlich, R.G., and Cornman, L.B., "Estimating spatial velocity statistics with coherent Doppler lidar," *Journal of Atmospheric and Oceanic Technology*, Vol. 19, No. 3, 2002, pp. 355–366.

[36] Smalikho, I.N., "On random errors of measuring the wind velocity with a cw coherent lidar," *Atmos. Oceanic Opt.*, Vol. 7, No. 10, 1994, pp. 744–748.

[37] Banakh, V.A., Smalikho, I.N., and Werner, Ch., "Numerical simulation of effect of refractive turbulence on the statistics of a coherent lidar return in the atmosphere," *Applied Optics*, Vol. 39, No. 30, 2000, pp. 5403–5414.

[38] Smalikho, I.N. "On accuracy of estimates of the radial wind velocity component from Doppler lidar measurements," Dep. In VINITI, No. 1848B94, July 18, 1994.

[39] Banakh, V.A., et al., "Turbulent characteristics of wind velocity measured by Doppler lidar," *Proc. 9th Coherent Laser Radar Conference*, Linköping, Sweden, 23–27 June 1997. pp. 166–169.

[40] Banakh, V.A., Smalikho, I.N., and Werner, Ch., "Aerosol particle microstructure dependence of accuracy of cw Doppler estimate of wind velocity," *SPIE Proc. Atmospheric Propagation, Adaptive Systems, and Laser Radar Technology for Remote Sensing*, Barcelona, Spain, 25–28 September 2000, Vol. 4167, pp. 270–280.

[41] Bendat, J.S., and Piersol, A.C., *Random Data: Analysis and Measurement Procedures*, Wiley, New York, 1971.

[42] Frehlich, R.G., and Yadlowsky, M.J., "Performance of mean-frequency estimators for Doppler radar and lidar," *Journal of Atmospheric and Oceanic Technology*, Vol. 11, No. 5, 1994, pp. 1217–1230.

[43] Frehlich, R.G., Hannon, S.M., and Henderson, S.W., "Performance of a 2-μm coherent Doppler lidar for wind measurements," *Journal of Atmospheric and Oceanic Technology*, Vol. 11, No. 6, 1994, pp. 1517–1528.

[44] Frehlich, R.G., "Estimation of velocity error for Doppler lidar measurements," *Journal of Atmospheric and Oceanic Technology*, Vol. 18, No. 10, 2001, pp. 1628–1639.

[45] Frehlich, R.G., Hannon, S.M., and Henderson, S.W., "Coherent Doppler lidar measurements of winds in the weak signal regime," *Applied Optics*, Vol. 36, No. 15, 1997, pp. 3491–3499.

[46] Hildebrand, P. H., and Sekhon, R.S., "Objective determination of the noise level in Doppler spectra," *Journal of Applied Meteorology*, Vol. 13, No. 10, 1974, pp. 808–811.

[47] Ray, B.J., and Hardesty, R.M., "Discrete spectral peak estimation in incoherent backscatter heterodyne lidar. I: Spectral accumulation and Cramer-Rao lower bound," *IEEE Trans. on Geoscience and Remote Sensing*, Vol. 31, No. 1, 1993, pp. 16–27.

[48] Hardesty, R.M., "Performance of a discrete spectral peak frequency estimator for Doppler wind velocity measurements," *IEEE Trans. on Geoscience and Remote Sensing*, Vol. GE-24, No. 5, 1986, pp. 777–783.

[49] Maharatra, P. R., and Zrnic, D.S., "Practical algorithms for mean velocity estimation in pulse Doppler weather radars using a small numbers of samples," *IEEE Trans. on Geoscience Electronics*, Vol. GE-21, No. 4, 1983, pp. 491–501.

[50] Balin, Yu.S., Razenkov, I.A., and Rostov, A.P., "Lidar studies of fluctuations of aerosol concentration in the ground atmospheric layer," *Atmos. Oceanic Opt.*, Vol. 7, No. 7, 1994, pp. 513–516.

[51] Kolmogorov, A.N., "Local structure of turbulence in incompressible viscous fluid at very large Reynolds numbers," *Doklady AN SSSR*, Vol. 30, No. 4, 1941, pp. 299–303.

[52] Kolmogorov, A.N., "Scattering of energy at locally isotropic turbulence," *Doklady AN SSSR*, Vol. 32, No. 1, 1941, pp. 19–21.

[53] Obukhov, A.M., "On the distribution of energy in the spectrum of turbulent flow," *Izvestiya AN SSSR. Ser. Geogr. i Geofiz.*, Nos.4–5. 1941, pp. 453–463.

[54] Obukhov, A.M. "Statistical description of continuous fields," *Trudy Geofiz. Inst. AN SSSR*, No. 24(151), 1954, pp. 3–42.

[55] Obukhov, A.M., and Yaglom, A.N., "Microstructure of turbulent flow," *Prikl. Matematika I Mekhanika*, Vol. 15, Issue 1, 1951, pp. 3–26.

[56] Patrushev, G.Ya., Rostov, A.P., and Ivanov, A.P., "Automated ultrasonic anemometer-thermometer for measuring the turbulent characteristics in the ground atmospheric layer," *Atmos. Oceanic Opt.*, Vol. 7, No. 11–12, 1994, pp. 890–891.

[57] Alldritt, M., et al., "The processing of digital signals by a surface acoustic wave spectrum analyzer," *Journal of Physics E: Scientific Instruments*, Vol. 11, 1978, pp. 1–4.

[58] Maines, J.D., and Paige, E.G.S., "Surface-acoustic-wave devices for signal processing applications," *Proc. IEEE*, Vol. 64, No. 5, 1976, pp. 639–652.

[59] Hardesty, R.M., et al., "Characteristics of coherent lidar returns from calibration targets and aerosols," *Applied Optics*, Vol. 20, No. 21, 1981, pp. 3763–3769.

[60] Frehlich, R.G., and Conman, L.B., "Coherent Doppler lidar signal spectrum with wind turbulence," *Applied Optics*, Vol. 38, No. 36, 1999, pp. 7456–7466.

[61] Frehlich, R.G., "Comparison of 2- and 10-μm coherent Doppler lidar performance," *Journal of Atmospheric and Oceanic Technology*, Vol. 12, No. 2, 1995, pp. 415–420.

[62] Frehlich, R.G., "Simulation of coherent Doppler lidar performance in the weak-signal regime," *Journal of Atmospheric and Oceanic Technology*, Vol. 13, No. 6, 1996, pp. 646–658.

[63] Ray, B.J., and Hardesty, R.M., "Discrete spectral peak estimation in incoherent backscatter heterodyne lidar. II: Correlogram accumulation," *IEEE Trans. on Geoscience and Remote Sensing*, Vol. 31, No. 1, 1993, pp. 28–35.

[64] Ray, B.J., and Hardesty, R.M., "Detecting techniques for validating Doppler estimates in heterodyne lidar," *Applied Optics*, Vol. 36, No. 9, 1997, pp. 1940–1951.

[65] Van Trees, H.L., *Detection, Estimation, and Modulation Theory*, Wiley, New York, 1968.

[66] Levin, J.M., "Power spectrum parameter estimation," *IEEE Trans. on Information and Theory*, Vol. IT-11, 1965, pp. 100–107.

CHAPTER 3

Measuring the Wind Velocity and Direction with Coherent Doppler Lidars

3.1 Introduction

As it follows from (1.29) for the complex signal $Z_S(t)$ and (1.17) for the frequency shift f_{ri}, the raw CDL data, measured under a fixed direction of propagation for the probing beam, contain information about the radial velocities $V_r(z_i)$ of aerosol particles in the sensing volume. If these particles move with the wind, the velocity $V_r(z_i)$ is the projection of the wind velocity vector on the probing beam axis at distance z_i from the lidar. This projection is also known as line-of-sight velocity.

To determine three components of the wind velocity vector (wind speed and direction), it is necessary to conduct lidar measurements at (at least) three different directions of propagation of the probing beam. For this purpose, it was proposed [1] to change successively the azimuth angle θ of the probing beam axis at the fixed elevation angle φ during measurements; that is, to have the probing beam use conical scanning around the vertical axis of the Cartesian coordinate system with the constant angular speed ω_0 (see Figure 3.1). With this scanning geometry, in the case of statistically homogeneous wind along the horizontal, the mean value of the radial wind velocity has a sine wave dependence on the azimuth angle. Due to the wind turbulence and random errors (connected with echo signal fluctuations and noise), the obtained dependence of the lidar estimate of the radial velocity on the azimuth angle differs from the sine wave one. Therefore, to estimate the three components of the wind velocity vector from the array of radial velocities measured by a scanning lidar, it is necessary to apply some procedure that minimizes the error of fitting the sine wave dependence of the radial velocity on the azimuth angle to the measured data. The procedure of sine wave fitting to lidar estimates of the radial velocity based on the least-squares technique has gained widespread acceptance [1–11].

Estimation of the mean wind velocity and direction from the data measured by, for example, a cup or sonic anemometer (i.e., a point sensor) is performed by way of time averaging of measured data. For stationary conditions, the accuracy of the estimate of the mean wind is determined by the intensity of wind turbulence and by the ratio of the correlation time of the measured value to the duration of the measurement [12]. The lower this ratio (which, with allowance for the Taylor hypothesis of frozen turbulence, is the ratio of the outer scale of turbulence to the transfer distance of the turbulent eddies by wind flow during a specific measurement time), the more accurate the estimate of the mean wind. The 10-min average is usually used in meteorological practice.

As a rule, measurements with a coherent Doppler lidar involve conical scanning by the probing beam at a tangential speed (product of the angular speed ω_0 by the

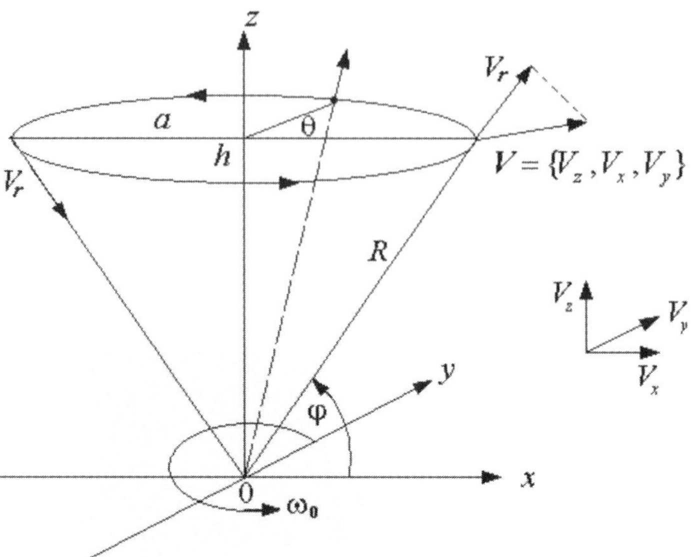

Figure 3.1 Geometry of lidar measurement for conical scanning by the probing beam. (© 1995 Optical Society of America. From [11].)

radius a of the base of the scanning cone), which significantly, by an order of magnitude, exceeds the velocity of the wind flow. If in this case the probability of bad estimates (see Section 2.6) of the radial velocity is negligibly low, then the procedure of sine wave fitting is equivalent to the averaging of turbulent fluctuations of the measured velocity over the circle of the scanning cone base [13]. In addition, radial velocities are averaged over the sensing volume along the probing beam axis (see Section 2.2). Thus, along with the time averaging (connected with the transport of turbulent vortices by a wind flow), the spatial averaging also takes place in measurements made by a coherent Doppler lidar.

If the contribution of spatial averaging is significant compared to that of the time averaging, then, obviously, the measurement of the mean wind velocity vector with a lidar takes less time than does a measurement with a point sensor. At the same time, we should keep in mind that at elevation angles φ close to 90° the estimation of horizontal components of the wind velocity vector from lidar data is impossible regardless of the duration of the measurements. Consequently, the measurement geometry, along with the temporal and spatial averaging, also affects the accuracy of lidar estimates of the mean velocity and direction of wind.

The options for measuring wind velocity and direction using a cw CDL and conical scanning are studied in [5–8, 14]. However, issues regarding the influence of wind turbulence and the duration of measurements on the accuracy of lidar estimation of the mean wind velocity were not discussed in these papers. These issues are considered in [11, 15].

The restriction on the measurement range of cw CDLs is related to worsening of the spatial resolution of the radial velocity as the focal length of the sensing beam increases. The maximal range (and, consequently, the radius of the scanning cone base) does not exceed 1 km [5]. For a pulsed CDL, the longitudinal dimension of the sensing volume is independent of the measurement range. Therefore, depending on the echo signal level, one can conduct measurements at the radius of the scanning cone base far larger than 1 km and can also obtain an estimate of the mean velocity with the required accuracy for one full scan ($0° \leq \theta \leq 360°$).

In measurements by a ground-based cw CO_2 CDL in the atmospheric boundary layer, as a rule, the SNR is high and the probability of bad estimates of the radial velocity is negligibly low. In the case of a pulsed CDL, at large ranges, the SNR can be so low that the probability of bad estimates of the radial velocity is, in contrast, close to unity. Consequently, for a pulsed lidar the problem of measurement of the mean wind velocity at the large radius of the scanning cone base lies with the low SNR rather than the wind turbulence. The options for measuring wind velocity and direction under conditions of a weak lidar echo signal have been studied in [16–22].

This chapter considers the methods for estimating wind velocity and direction from measurements of the radial velocity by continuous-wave and pulsed coherent Doppler lidars. The influence of wind turbulence, the geometry and duration of measurements, and the SNR on the accuracy of estimation of the mean wind velocity and direction from lidar data is analyzed. The methods for estimating the wind velocity vector from lidar data under conditions of a weak echo signal are described. The numerical simulation is used to study the possibility of measuring wind with a coherent Doppler lidar from aboard a satellite in space. The chapter is based on results published in [10, 11, 15–39].

3.2   Measurement of Mean Wind Velocity and Direction with a Continuous-Wave CDL

The radial velocities $\hat{V}_{ri} \equiv \hat{V}_r(R, \theta)$ for different azimuth angles θ_i ($i = 1, 2, 3, ..., n$) are estimated from an array of Doppler spectra measured by a coherent Doppler lidar by using a probing beam that does conical scanning. Estimates of three components of the wind velocity vector $\{\hat{V}_z, \hat{V}_x, \hat{V}_y\}$ can be obtained by the least-squares method through minimization of the functional [1, 9]

$$\rho(V_z, V_x, V_y) = \sum_{i=1}^{n} (V_{ri} - \hat{V}_{ri})^2 \qquad (3.1)$$

where, according to the measurement geometry shown in Figure 3.1, the radial velocity at the horizontally homogeneous wind flow can be represented in the form

$$V_{ri} = V_z \sin\varphi + V_x \cos\varphi \cos\theta_i + V_y \cos\varphi \sin\theta_i \equiv S_i \cdot V \qquad (3.2)$$

In (3.2), $S_i = \{\sin\varphi, \cos\varphi\cos\theta_i, \cos\varphi\sin\theta_i\}$ and $V = \{V_z, V_x, V_y\}$. This procedure is usually referred to as *sine wave fitting*.

The minimum point of the functional is determined through solution of the system of linear equations $\nabla\rho(V) = 0$, where $\nabla = \{\partial/\partial V_z, \partial/\partial V_x, \partial/\partial V_y\}$. The solution of this system in the formalized form can be written as follows:

$$\hat{V} = \left(\sum_{i=1}^{n} S_i S_i^T \right)^{-1} \cdot \sum_{i-1}^{n} \hat{V}_{ri} S_i \qquad (3.3)$$

where the • symbol denotes multiplication of a matrix by a vector. If the data array is obtained at the integer number N of full scans (revolutions around the vertical axis Z) with the same intervals $\Delta\theta = |\theta_{i+1} - \theta_i|$, the matrix takes the form

$$\left(\sum_{i=1}^{n} S_i S_i^T \right)^{-1} = \begin{pmatrix} \dfrac{1}{n\sin^2\varphi} & 0 & 0 \\ 0 & \dfrac{2}{n\cos^2\varphi} & 0 \\ 0 & 0 & \dfrac{2}{n\cos^2\varphi} \end{pmatrix} \qquad (3.4)$$

Then, under the condition $\Delta\theta \ll \pi/2$, estimates of the mean velocity components can be represented as

$$\hat{V}_z = \frac{1}{\sin\varphi}\frac{1}{T}\int_0^T dt\,\hat{V}_r(t) \qquad (3.5)$$

$$\hat{V}_x = \frac{2}{\cos\varphi}\frac{1}{T}\int_0^T dt\,\hat{V}_r(t)\cos(\omega_0 t) \qquad (3.6)$$

$$\hat{V}_y = \frac{2}{\cos\varphi}\frac{1}{T}\int_0^T dt\,\hat{V}_r(t)\sin(\omega_0 t) \qquad (3.7)$$

where $T = 2\pi N/\omega_0$ is the duration of measurement. The estimates of the mean wind velocity $\hat{V}_{xy}$ and direction $\hat{\theta}_V$ are determined as $\hat{V}_{xy} = \left|\hat{V}_{xy}\right|$ and $\hat{\theta}_V = \arg(\hat{V}_{xy})$, where $\hat{V}_{xy} = \hat{V}_x + j\hat{V}_y$.

We assume that the measurement time of one Doppler spectrum, from which the radial velocity is estimated, is 50 ms. According to (2.35), the estimate $\hat{V}_r(t)$ can be represented as a sum of the radial velocity averaged over the sensing volume (along the optical axis) $\overline{V}_r(t)$ and the random error $V_e(t)$. As follows from experiments (see Figure 2.9), the error in the estimation of the radial velocity σ_e from data of a cw CO_2 lidar does not exceed 0.1 m/s. The random error $V_e(t)$ has the properties of white noise. In this case, it can be easily shown based on (3.5) through (3.7) that the estimation error of components of the wind velocity vector $\hat{V}$ caused by fluctuations of $V_e(t)$ is proportional to $\sigma_e/\sqrt{n}$. Thus, for example, at $n = 100$ these errors are ~0.01 m/s; that is, they are negligibly small. Therefore, in (3.5) through (3.7) we can replace $\hat{V}_r(t)$ with $\overline{V}_r(t)$.

It is well known that the velocity of airflow in the atmosphere varies randomly in space $r = \{z, x, y\}$ and time t. The difference between the wind velocity being fully averaged over the flow and being measured for a finite time interval can be interpreted as a random error of determination of the mean velocity. As mentioned in the introduction to this chapter, in meteorological practice, in measurements at a fixed spatial point (by, for example, a cup or sonic anemometer) the data obtained for the time interval of ~10 min are used to estimate the mean wind. For this time, turbulent fluctuations of the measured velocity are averaged to a significant extent, because their correlation time is from several seconds to several tens of seconds [12, 40, 41]. In the case of scanning by a Doppler lidar, not only temporal, but also spatial averaging takes place, because the information about wind is retrieved from the sensing volume (with the longitudinal dimension Δz), which quickly (relative to the wind velocity) moves along the boundary of the scanning cone base. It is obvious that if the characteristic scales of wind flow inhomogeneities are much smaller than Δz or the diameter of the scanning cone base $2R\cos\varphi$, then turbulent fluctuations of wind velocity are spatially averaged, even if the scanning time is much shorter than 10 min. A question then arises: How long should the Doppler lidar measurement be (i.e., how many full conical scans are necessary) to estimate the mean wind velocity with the required accuracy (for example, with relative error less than 10%)?

We assume that the atmospheric turbulence is stationary and horizontally homogeneous and that the mean wind velocity vector at height h is directed along the x axis of the Cartesian coordinate system:

$$\langle V \rangle = \{0, U(h), 0\} \tag{3.8}$$

where U is the value of the mean wind velocity. The relative error of lidar measurements of the mean wind velocity is defined as

$$\varepsilon_{xy} = \sqrt{\langle (\hat{V}_{xy} - U)^2 \rangle}/U \tag{3.9}$$

Assuming that $\left| \hat{V}_x - U \right|, \left| \hat{V}_y \right| \ll U,$ from (3.2), (3.6), (2.18), (3.8), and (3.9) we find [10, 11, 15, 28]

$$\varepsilon_{xy} = \frac{2}{\cos\varphi UT} \sqrt{\int_0^T dt' \cos\omega_0 t' \int_0^T dt'' \cos\omega_0 t'' \int_0^\infty dz' Q_s(z') \int_0^\infty dz'' Q_s(z'') K_V(z',z'',t',t'')}$$

$$\tag{3.10}$$

where $K_V(z',z'',t',t'') = \langle V_r'(r(z',t'),t)V_r'(r(z'',t''),t)\rangle,$ $V_r' = V_r - \langle V_r \rangle,$ and $r(z',t') = z'\sin\varphi, z'\cos\varphi\cos(\omega_0 t'), z'\cos\varphi\sin(\omega_0 t')\}.$ After application of the Taylor hypothesis of frozen turbulence [40, 42], we obtain

$$\begin{aligned}
K_V(z',z'',t',t'') = {} & \cos^2\varphi[K_{xx}(p)\cos(\omega_0 t')\cos\omega_0 t'' \\
& + K_{yy}(p)\sin\omega_0 t'\sin\omega_0 t'' \\
& + K_{xy}(p)(\cos\omega_0 t'\sin\omega_0 t'' + \sin\omega_0 t''\cos\omega_0 t')] \\
& + \sin\varphi\cos\varphi[K_{xz}(p)(\cos\omega_0 t' + \cos\omega_0 t'') \\
& + K_{yz}(p)(\sin\omega_0 t' + \sin\omega_0 t'')] + \sin^2\varphi K_{zz}(p)
\end{aligned} \tag{3.11}$$

where $p = r(z',t') - r(z'',t'') + \langle V \rangle(t'-t''),$ and $K_{lk}(p) = \langle V_l'(r+p)V_k'(r)\rangle$ is the spatial correlation tensor of wind velocity fluctuations where $l, k = z, x, y.$ To simplify the analysis, we assume in the following that the turbulence anisotropy insignificantly influences the error ε_{xy} and we can use the equation [40]

$$K_{lk}(p) = K_V(p)\delta_{l-k} + \frac{1}{2}p\,\frac{dK_V(p)}{dp}\left(\delta_{l-k} - \frac{p_l p_k}{p^2}\right) \tag{3.12}$$

where $K_V(p)$ is the longitudinal correlation function of wind velocity fluctuations and $p = |p|$.

Thus, to calculate the measurement error ε_{xy}, it is necessary to know the function $K_V(p)$. The simplest model for $K_V(p)$ is the exponential one:

$$K_V(p) = \sigma_V^2 \exp(-p/L_V) \tag{3.13}$$

which will be used for calculations by (3.10) through (3.12).

In the atmospheric boundary layer, the values of U, σ_V^2, L_V, and ε depend for the most part on two dynamic parameters—the roughness of the underlying Earth's surface z_0 and the geostrophic wind velocity $G_V = |\nabla P|/(f_C \rho_a)$, where ∇P is the horizontal pressure gradient, ρ_a is the air density, and f_C is the Coriolis parameter—and one thermal parameter—the vertical turbulent flux of the heat $H_T = C_p \rho_a K_T \cdot (\gamma - \gamma_a)$, where C_p is the air heat capacity, K_T is the turbulent exchange coefficient, $\gamma = -dT_a/dz$ is the vertical gradient of mean temperature T_a, and γ_a is the adiabatic gradient [12, 43–45]. In the near-surface layer ($h < 20$m to 100m), the turbulent flux H_T is independent of height, but above the surface layer H_T varies with height due to diurnal variations of the radiation conditions of heating of the Earth's surface and air [43]. The thermal stratification in the atmospheric boundary layer is usually characterized by the Monin-Obukhov scale [45]:

$$L_{MO} = -u_*^3 T_a \rho_a C_p /(g_a \kappa H_T) \tag{3.14}$$

where $\kappa \approx 0.4$ is the von Karman constant, g_a is the free-fall acceleration, and u_* is the friction velocity. The equation for the mean velocity in the atmospheric surface layer has the form [45]

$$\frac{dU(z)}{dz} = \frac{u_* \varphi_u(\zeta)}{\kappa z} \tag{3.15}$$

where φ_u is the universal function of the dimensionless parameter $\zeta = z/L_{MO}$. For the neutral stratification ($\gamma = \gamma_a$, $H_T = 0$, $L_{MO} = \infty$), the function $\varphi_u(0) = 1$. To find $U(z)$, we use the empirical equation

$$\varphi_u(\zeta) = \begin{cases} 1 + 5\zeta, & \zeta \geq 0 \\ (1 - 15\zeta)^{1/3}, & \zeta < 0 \end{cases} \tag{3.16}$$

the results of calculation by which are in a good agreement with the experimental data [46]. Then, taking into account that $z_0 \ll L_{MO}$, from (3.15) we can derive an approximate equation for the mean wind velocity [11]

$$U(z) = \frac{u_*}{\kappa} \begin{cases} \ln\left(\dfrac{z}{z_0}\right) + 5\zeta, & \zeta \geq -\dfrac{1}{15} \\[3mm] \ln\left(\dfrac{z}{z_0}\dfrac{1}{15|\zeta|}\right) - \dfrac{1}{3} + 3[1 - (15|\zeta|)^{1/3}], & \zeta < -\dfrac{1}{15} \end{cases}$$

(3.17)

The parameters ε and σ_V^2 are calculated as [45]

$$\varepsilon = (u_*^3/\kappa z)[\varphi_u(\zeta) - \zeta]$$

(3.18)

$$\sigma_V^2 = u_*^2 \{C_v^2 \sqrt{1 - \zeta/\varphi_u(\zeta)} + (C_u^2 - C_v^2)/\sqrt{1 - \zeta/\varphi_u(\zeta)}\}$$

(3.19)

where the empirical constants C_v and C_u take the values $C_v = 1.3$–2.2 and $C_u = 2.1$–2.9 [12]. The friction velocity u_* depends on the geostrophic wind velocity, roughness parameter, Coriolis parameter, and thermal stratification. It can be calculated theoretically (see, for example, [43, 45, 47]) if we know the parameters mentioned above, or measured directly near the Earth's surface by using well-known techniques [43]. The integral correlation scale of wind speed L_V is calculated by (3.18), (3.19), and (2.71).

The models of (3.16) through (3.19) and (2.71) allow us to calculate U, σ_V^2, ε, and L_V for arbitrary heights in the atmospheric boundary layer on the assumption of a stationary and horizontally homogeneous atmosphere, if we know the parameters z_0, L_{MO}, and u_*. Theoretical models of these turbulent characteristics for the entire atmospheric boundary layer in the general case can be based on the numerical solution of the corresponding equations (see, for example, [43, 45, 48, 49]). However, in the case of neutral thermal stratification, simple empirical dependences can be used [41, 43, 47–49]:

$$\sigma_V(z) = \sigma_{VS} \exp(-C_1 z/h_B)$$

(3.20)

$$L_V(z) = L_{VS}(z)/[1 + C_2 L_{VS}(z)/h_B]$$

(3.21)

Here, σ_{VS}^2 and L_{VS} denote the variance and correlation scale of the longitudinal component of wind velocity calculated by equations for the surface layer [(3.16) through (3.19), where $\zeta = 0$], $h_B \approx \kappa u_*/|f_C|$ is the effective thickness of the atmospheric boundary layer ($h_B \sim 1$ km) [45], and C_1 and C_2 are empirical constants equal to 0.8 and 2.5, respectively, as shown by experimental data [41].

Figure 3.2 shows theoretical results for the error in mean wind velocity measurements by a cw CDL ($\lambda = 10.6$ μm) in the surface layer at a height of $h = 60$m. The calculations were performed for $t_r = 2\pi/\omega_0 = 12$s, $U = 10$ m/s, and $\sigma_V/U = 0.15$. Figure 3.2(a) illustrates the dependence of ε_{xy} on the number of rotations N for the elevation angle $\varphi = 45°$ at different values of the correlation scale $L_V = 300$m (curve 1), 200m (curve 2), and 100m (curve 3). The solid curves show the results of calculation by (3.10) [11], while the dashed curves are for the results calculated by the equation

$$\varepsilon_{xy} = (\sigma_V/U)\sqrt{L_V \omega_0/(\pi U N)}$$

(3.22)

which was obtained in [13] under the condition $N \gg L_V/(Ut_r)$. It can be seen that (3.22) yields results close to the calculations of (3.10) only at $N > 10$. Different values for the L_V scale can be interpreted as corresponding to different types of thermal stratification. The error of measurement of the mean wind velocity satisfies the condition $\varepsilon_{xy} \le 10\%$ when $N \ge 2$ at $L_V = 100$m, $N \ge 5$ at $L_V = 200$m, and $N \ge 8$ at $L_V = 300$m.

Figure 3.2(b) depicts the calculated dependence of ε_{xy} on the elevation angle φ at a fixed measurement height. Because in this case the radius of the scanning cone base a_c and the longitudinal dimension of the sensing volume Δz vary with variation of the angle φ, the results shown in this figure provide a good illustration of the influence of the spatial averaging on the error in the measurement of mean wind velocity using a coherent Doppler lidar. An increase in the angle φ leads to a decrease of the averaging volume and, consequently, to an increase in ε_{xy}. In particular, at $\varphi = 60°$, $N = 1$ (curve 1), the relative error of lidar measurements becomes greater than the intensity of turbulent fluctuations of the wind velocity $\sigma_V/U(\varepsilon_{xy} > \sigma_V/U = 0.15)$. For the limiting case $\varphi \to 90°$, the error $\varepsilon_{xy} \sim \tan\varphi$, that is, increases without limit, whereas at $\varphi \to 0°$, the volume increases and ε_{xy} decreases.

Curve 1' (dot-and-dash curve, $N = 1$) in Figure 3.2(b) corresponds to the results of calculating ε_{xy} in the case of neglected spatial averaging along the probing beam axis ($\Delta z = 0$). One can see that the longitudinal averaging makes a negligibly small contribution compared to the horizontal averaging over the circle of the scanning cone base. Curve 2 in this figure is the result of calculating ε_{xy} at $N = 10$.

To study the representativeness of lidar measurements of mean wind velocity under various atmospheric conditions, experiments were conducted near Lichtenau in the south of Germany during a spring–summer period [11, 15]. The lidar measurements were accompanied by simultaneous measurements of the wind velocity by a cup anemometer at a height of 60m. The distance between the lidar and the tower with the cup anemometer was 50m. The roughness parameter near the measurement point was $z_0 = 0.24$m. The type of thermal stratification was determined from measurements of temperature and wind velocity by cup anemometers and

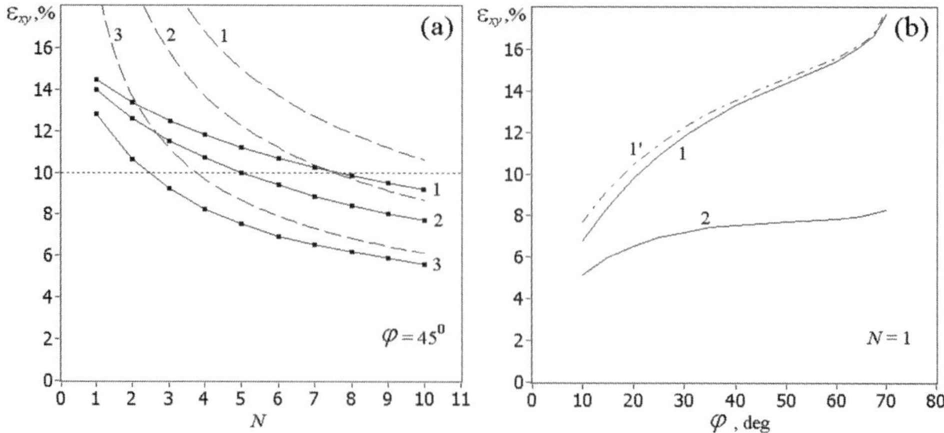

Figure 3.2 Relative error of mean wind velocity measurement by a lidar as a function of (a) the number of scans N and (b) the elevation angle φ. (© 1995 Optical Society of America. From [11].)

thermometers at five heights from 0.3m to 6m. The cup anemometer data were also used to estimate the friction velocity u_* and the intensity of turbulent fluctuations of the wind velocity σ_V/U.

The results of the wind velocity V_{xy} measurements are shown in Figure 3.3. Dots show the lidar data obtained at $N = 1$, while crosses and squares show the data of the cup anemometer averaged over periods of 2 and 10 min, respectively. The experimental data correspond to the neutral (Figures 3.3(a), 3.3(b), and 3.3(d)), stable (Figure 3.3(c)), weak unstable (Figure 3.3(e)), and unstable (Figure 3.3(f)) stratification. In all of the measurements, the elevation angle was $\varphi = 30°$, except for the data shown in Figure 3.3(b), where $\varphi = 60°$.

The data depicted in Figure 3.3 were used to calculate the dependence of the error of measurement of the mean wind velocity ε_{xy} on the number of full scans N. Figure 3.4 compares the experimental values of the error ε_{xy} in the atmospheric surface level with the theoretically predicted values. The curves are for the ε_{xy} values calculated theoretically by (3.10). The signs correspond to the error values calculated from the experimental data. One can see that the theoretical and experimental results are in good agreement. This allows us to explain the behavior of the relative error ε_{xy} as a function of the number of scans N for atmospheric conditions having different stabilities. As the stability increases ($h/L_{MO} > 0$), the turbulence intensity σ_V/U and the correlation scale L_V decrease and, consequently, the accuracy of lidar measurements of mean wind velocity increases. As can be seen from Figure 3.4, in the case of stable stratification, one scan ($N = 1$) appears to be sufficient to measure the mean wind velocity with an error $\varepsilon_{xy} < 10\%$. The measurement accuracy of the

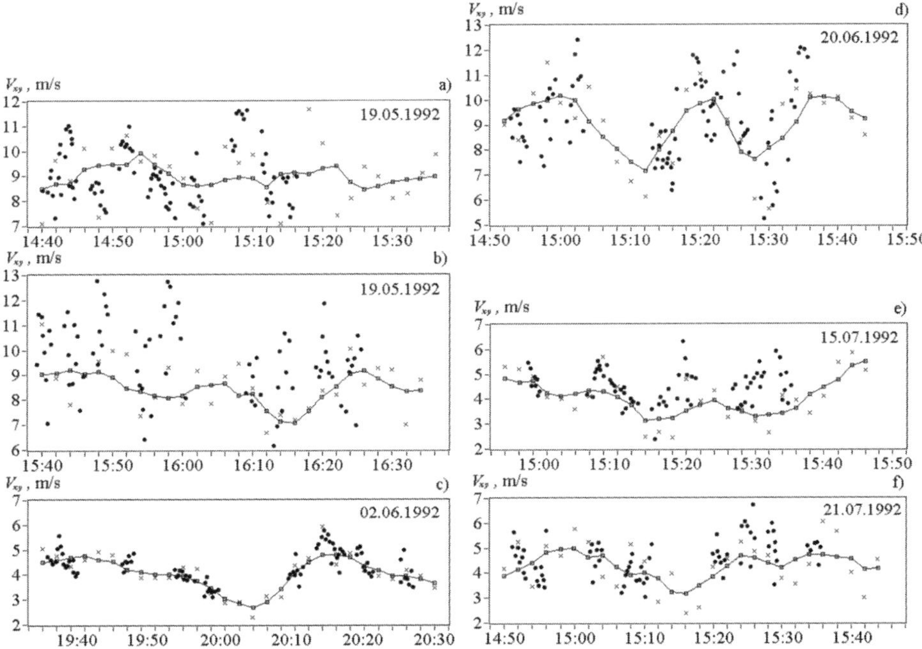

Figure 3.3 Wind velocity measured by lidar and by cup anemometer: (a) neutral stratification, $\varphi = 30°$; (b) neutral stratification, $\varphi = 60°$; (c) stable stratification, $\varphi = 30°$; (d) neutral stratification, $\varphi = 30°$; (e) weak unstable stratification, $\varphi = 30°$; (f) unstable stratification, $\varphi = 30°$. (© 1995 Optical Society of America. From [11].)

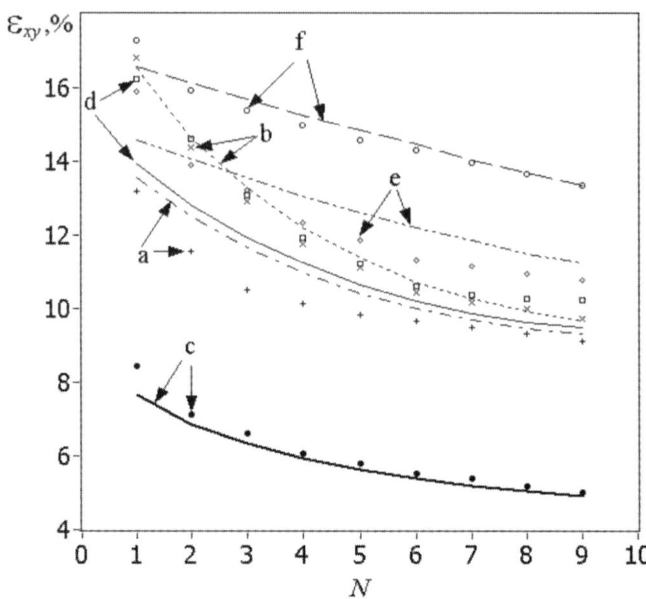

Figure 3.4 Relative error of lidar measurements ε_{xy} as a function of the number of scans N. Curves and signs marked by the letters a through f correspond to the data shown in Figures 3.3(a) through 3.3(f). (© 1995 Optical Society of America. From [11].)

mean wind velocity decreases with an increase in the parameters σ_V/U and L_V, when atmospheric instability takes place. Analysis shows [11] that for unstable stratification the duration of measurement of the mean wind velocity with an error not exceeding 10% for the lidar ($\varphi = 30°$, $h = 60$m) is a little bit shorter than that for measurement by a cup anemometer. In this case, the measurement duration should be no shorter than 10 min. It can be seen from Figure 3.4 that in the cases of neutral and weak unstable stratification five scans ($N = 5$) are already required at the elevation angle $\varphi = 30°$ to measure the mean wind velocity with a relative error of $\varepsilon_{xy} \leq 10–12\%$.

Figure 3.5 depicts the results of mean wind velocity measurements V_{xy} at $N = 1$ (dots) and $N = 10$ (dashed curve) and the relative error ε_{xy} (dots, $N = 1$). The solid curve shows the results of ε_{xy} calculations using (3.10) through (3.13), (3.20), and (3.21). Figure 3.5(b) illustrates a decrease in the measurement error for the mean wind velocity with height both theoretically and experimentally. This is explained by two reasons. On the one hand, the turbulence intensity σ_V/U decreases with an increase in h (U increases, while σ_V decreases). On the other hand, spatial averaging becomes more efficient, because the ratios $R\cos\varphi/L_V$ and $\Delta z/L_V$ increase with an increase in height (at the fixed angle φ).

The results presented above correspond to the case in which the velocity vector is estimated from the lidar data measured at the integer number N of full scans by the probing beam ($0° \leq \theta_i \leq 360°$). In practice, it is not always possible to use full scans, for example, because of buildings or trees near the lidar container. Therefore, sector scanning with the sector angle $\theta_s < 360°$ is used. In this case, the estimate of the wind velocity vector $\hat{V} = \{\hat{V}_z, \hat{V}_x, \hat{V}_y\}$ is obtained from the solution of the following system of equations [50]:

$$A \bullet \tilde{V} = B \tag{3.23}$$

where the elements of matrix $A = \{A_{kl}\}$ have the form $A_{11} = n$, $A_{12} = A_{21} = \sum_{i=1}^{n} \cos \theta_i$, $A_{22} = \sum_{i=1}^{n} \cos^2 \theta_i$, $A_{23} = A_{32} = \sum_{i=1}^{n} \sin \theta_i \cos \theta_i$, and $A_{33} = \sum_{i=1}^{n} \sin^2 \theta_i$.

In (3.23), $\tilde{V} = \{\hat{V}_z \sin \varphi, \hat{V}_x \cos \varphi, \hat{V}_y \cos \varphi\}$, $B = \left\{ \sum_{i=1}^{n} \hat{V}_{ri}, \sum_{i=1}^{n} \hat{V}_{ri} \cos \theta_i, \sum_{i=1}^{n} \hat{V}_{ri} \sin \theta_i \right\}$, and the azimuth angles θ_i take values within the scanning cone θ_s. When $\theta_s \to 0$, the determinant of the matrix $\det\{A_{kl}\} \to 0$. Consequently, at small angles θ_s the solution of the system of equations (3.23) is unstable with respect to even small fluctuations of the measured radial velocity $\hat{V}_{ri}$.

In [10], the wind velocity and direction were estimated from the raw experimental data of lidar measurements at different angles of scanning sector θ_s. Then the value of $E_{xy} = \sqrt{\langle [V_{xy}(360°, 9) - V_{xy}(\theta_s, N)]^2 \rangle}$ was calculated, where $V_{xy}(\theta_s, N)$ is the estimate of the wind velocity at angles $\theta_s = 360°$, $180°$, and $90°$ and the varying number of scans $N = 1, 2, \ldots, 9$. The relative ε_{xy} and absolute $U\varepsilon_{xy}/100\%$ errors of the estimate $V_{xy}(360°, 9)$ were, respectively, 8% and 0.7 m/s.

Figure 3.6 shows E_{xy} as a function of the scan number N at $\varphi = 30°$ and different θ_s. It can be seen that as the angle of scanning sector θ_s decreases, the error in the estimation of wind speed rapidly increases. At $\theta_s = 90°$ and $N = 1$, the value of E_{xy} becomes close to 5 m/s. An increase in the measurement time (scan number N) leads to a decrease in E_{xy}. For $\theta_s = 180°$ and $\theta_s = 90°$, the ratio $E_{xy}|_{N=1}/E_{xy}|_{N=9} \approx 3$, that is, the error, decreases proportionally to $N^{-1/2}$. Consequently, for the value of E_{xy} at $\theta_s = 90°$ to be smaller than 0.5 m/s, $N \approx 100$ scans by the probing beam (measurement time ~20 min) is required.

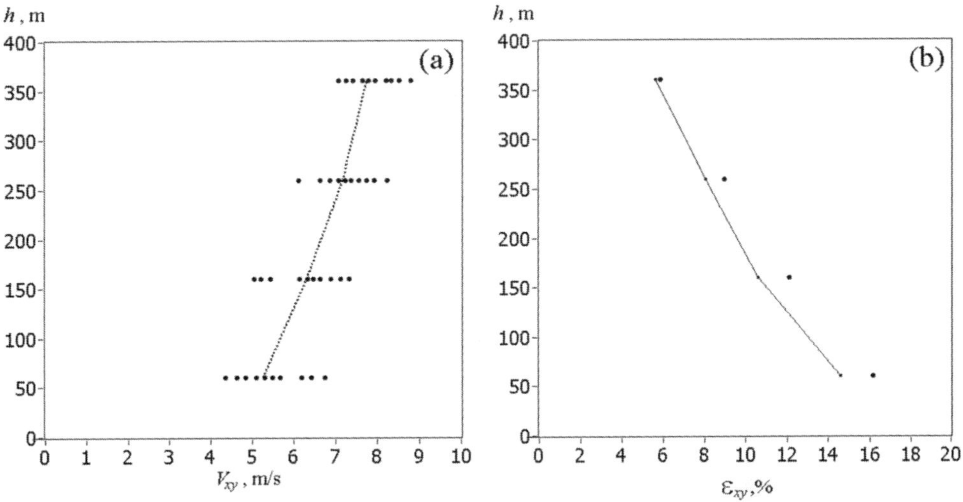

Figure 3.5   (a) Vertical profiles of the wind velocity and (b) errors in lidar measurements of mean wind velocity. (© 1995 Optical Society of America. From [11].)

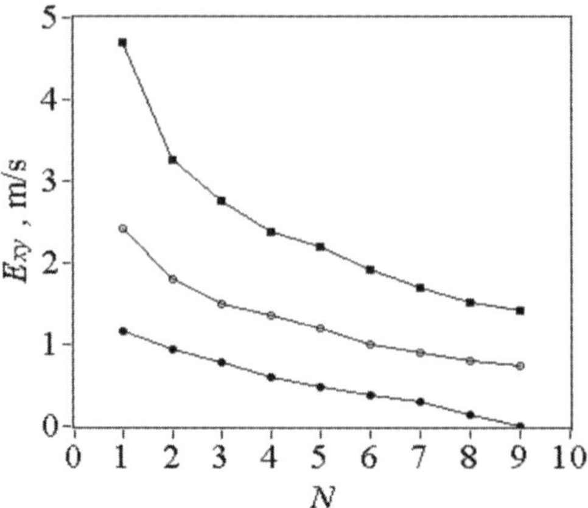

Figure 3.6 Dependence of E_{xy} on the number of scans N at $\theta_s = 360°$ (closed circles), 180° (open circles), and 90° (squares).

3.3 Methods for Estimating the Wind Velocity Vector from Pulsed CDL Data

As in the case of cw lidar, to measure wind velocity and direction using pulsed CDLs, conical scanning by the probing beam is usually used (see Figure 3.1). To obtain the vertical profile of wind velocity and direction, in the case of cw CDLs it is necessary to change the focal length of the probing beam F for each height $h = R\sin\varphi + h_L$, because the range R, according to (1.54), depends on F (h_L is the height of the lidar position). For pulsed lidar, however, there is no need to do that, because R is determined by the time interval between the launch of the probing pulse into the atmosphere (t_0) and the recording of the lidar echo signal (t_R); that is, $R = (t_R - t_0)$ $c/2$. The maximal sensing range of cw CO_2 CDLs with acceptable spatial resolution is determined by the condition $\Delta z/2 \ll R$, whereas for pulsed lidars the maximal sensing range is determined by the level of the lidar echo signal (SNR) at which wind information can be acquired from lidar data.

For many methods of estimating the radial wind velocity from a lidar echo signal, the probability density function of a single estimate $\hat{V}_{ri}$ obtained at the azimuth angle θ_i is described by the model of (2.134), which with allowance for (3.2) has the form

$$p(\hat{V}_{ri}|V) = \frac{1 - b_e}{\sqrt{2\pi}\sigma_g}\exp\left[-\frac{(\hat{V}_{ri} - S_i \cdot V)^2}{2\sigma_g^2}\right] + \frac{b_e}{B_V} \tag{3.24}$$

If individual estimates of the radial velocity are statistically independent, then the joint probability density function $p(\hat{V}_{r1}, \hat{V}_{r2}, ..., \hat{V}_{rn}|V)$ allows the representation in the form

$$p(\hat{V}_{r1}, \hat{V}_{r2}, ..., \hat{V}_{rn}|V) = \prod_{i=1}^{n} p(\hat{V}_{ri}|V) \tag{3.25}$$

The wind velocity vector $\hat{V}$ can be estimated by the maximum likelihood method [52]:

$$\max\{p(\hat{V}_{r1}, \hat{V}_{r2}, \ldots, \hat{V}_{rn} | V)\} = p(\hat{V}_{r1}, \hat{V}_{r2}, \ldots, \hat{V}_{rn} | \hat{V}) \tag{3.26}$$

Under the conditions $b_e \ll 1$ (when the number of bad estimates is negligibly small), from (3.24) and (3.25) we have the approximate formula

$$p(\hat{V}_{r1}, \hat{V}_{r2}, \ldots, \hat{V}_{rn} | V) \approx \frac{1}{\sqrt{2\pi}\sigma_g} \exp\left[-\frac{1}{2\sigma_g^2} \sum_{i=1}^{n} (\hat{V}_{ri} - S_i \cdot V)^2\right] \tag{3.27}$$

From (3.26) and (3.27) it follows that the estimate of the wind velocity vector $\hat{V}$ can be found from the solution of the linear equation set

$$\nabla \sum_{i=1}^{n} (\hat{V}_{ri} - S_i \cdot V)^2 \Big|_{V=\hat{V}} = 0 \tag{3.28}$$

where $\nabla = \{\partial/\partial V_z, \partial/\partial V_x, \partial/\partial V_y\}$. It can be shown that the solution of this linear equation set coincides with (3.3) and is the result of sine wave fitting based on the least-squares technique. In what follows, this procedure for estimating the wind velocity vector is referred to as direct sine wave fitting (DSWF). With (3.24), from (3.3) we obtain for the mean value of the estimate [18]:

$$\langle \hat{V} \rangle = (1 - b_e)V \tag{3.29}$$

It can be seen that the estimate of the velocity vector by the DSWF method at $b_e \neq 0$ and $V \neq 0$ is biased, that is, $\langle \hat{V} \rangle \neq V$. For a very low SNR, when $b_e \to 1$, the DSWF method gives the biased estimate with $\langle \hat{V}_z \rangle \to 0$, $\langle \hat{V}_x \rangle \to 0$ and $\langle \hat{V}_y \rangle \to 0$.

3.3.1 Method of Filtered Sine Wave Fitting (FSWF)

When the SNR is very low and $(1 - b_e)/(\sqrt{2\pi}\sigma_g) \ll b_e / B_V$, from (3.24) and (3.25) we obtain the approximate equation in the form

$$p(\hat{V}_{r1}, \hat{V}_{r2}, \ldots, \hat{V}_{rn} | V) \approx \left(\frac{b_e}{B_V}\right)^n + \left(\frac{b_e}{B_V}\right)^{n-1} \frac{1 - b_e}{\sqrt{2\pi}\sigma_g} \sum_{i=1}^{n} \exp\left[-\frac{(\hat{V}_{ri} - S_i \cdot V)^2}{2\sigma_g^2}\right] \tag{3.30}$$

In accordance with (3.26) and (3.30), the wind velocity vector is estimated from the maximum of the function [18]

$$Q(V) = \sum_{i=1}^{n} \exp\left[-\frac{(\hat{V}_{ri} - S_i \cdot V)^2}{2\sigma_g^2}\right] \tag{3.31}$$

that is, for $\hat{V}$ we can write

$$\max\{Q(V)\} = Q(\hat{V}) \tag{3.32}$$

To use the algorithm described by (3.31) and (3.32), it is necessary to have the information about the variance of good estimates σ_g^2. It is known that at low SNR the variance σ_g^2 nearly coincides with the squared width of the Doppler spectrum σ_S^2 [51]. Consequently, if we have *a priori* information about the width of the Doppler spectrum, then σ_g in (3.31) can be replaced with σ_S. In contrast to DSWF, the procedure for estimating the velocity vector by (3.31) and (3.32) allows us to remove contributions of poor estimates of the radial velocity $\hat{V}_{ri}$ falling beyond the interval $[|S \cdot \hat{V}| - \sigma_g, |S \cdot \hat{V}| + \sigma_g]$ from $Q(\hat{V})$ (provided that the estimate $\hat{V}$ is true); that is, the filtering of good estimates $\hat{V}_{ri}$ takes place. This approach was called filtered sine wave fitting (FSWF) in [18].

FSWF can be applied at any SNR, if we take $\sigma_g = \sigma_S$ in (3.31). At high SNR, the variance is $\langle (\hat{V}_{ri} - S_i \cdot \hat{V})^2 \rangle \ll 2\sigma_S^2$ and we can expand the exponents in (3.31) into the Taylor series, restricting our consideration to only the first two terms. Then the maximum point of the function $Q(\hat{V})$ can be determined in the form (3.3); in this case the results obtained with DSWF and FSWF coincide.

At low SNR, the number of bad estimates $\hat{V}_{ri}$ can be much larger than that of good estimates [$Q(\hat{V})$ is approximately equal to the number of good estimates, if the estimate $\hat{V}$ is true]. In such a case, the accuracy of the estimates of the wind velocity vector decreases drastically. An efficient method for decreasing the probability of obtaining bad estimates of the radial velocity $\hat{V}_r$ from lidar data is to apply an accumulation (averaging) of Doppler spectra [see (2.42) and Figure 2.15 at SNR = 0.01 and 0.1]. It turns out that an accumulation of Doppler spectra measured at different azimuth scan angles can be used to increase the accuracy of estimation of the wind velocity vector $\hat{V}$ [16–21].

3.3.2 Maximum of the Function of Accumulated Spectra (MFAS) Method

According to (2.42), the Doppler spectrum $\hat{S}_i(V_k)$ [with allowance for the Doppler relationship $V = (\lambda/2)(f - f_I)$], measured at the azimuth angle θ_i can be written in the form

$$\hat{S}_i(V_k) = \frac{1}{L}\sum_{l=1}^{L}\left[\frac{1}{M}\left|\sum_{m=0}^{M-1} Z(mT_s, l + (i-1)L)\exp\left(-2\pi j \frac{mk}{M}\right)\right|^2\right] \tag{3.33}$$

where $V_k = (k - M/2)\Delta V$, $k = 0, 1, ..., M - 1$, $\Delta V = (\lambda/2)\Delta f$ is the velocity resolution, $\Delta f = 1/(MT_s)$ is the frequency resolution, and $i = 1, 2, ..., n$. The radial velocities V_k fall within the range $[-B_V/2, B_V/2]$, where $B_V = (\lambda/2)B_F$ and $B_F = 1/T_s$ is the spectral bandwidth. The spectrum $\hat{S}_i(V_k)$ is a function of a discrete argument. We replace V_k with $S_i \cdot V$ (components of the vector V are continuous values) and use the linear interpolation [18]

$$\hat{S}_i^L(V_r) = \hat{S}_i(k'\Delta V)(1 - V_r/\Delta V + k') + \hat{S}_i((k' + 1)\Delta V)(V_r/\Delta V - k') \quad (3.34)$$

where $V_r = S_i \cdot V$, $k' = [V_r/\Delta_V]$ and $[...]$ is an integer part. Then we can introduce the function of accumulated spectra in the form [18]

$$F_a(V) = \frac{1}{n}\sum_{i-1}^{n}\hat{S}_i^L(S_i \cdot V) \quad (3.35)$$

and the estimate of the wind velocity vector $\hat{V}$ can be obtained at the point of maximum of this function:

$$\max\{F_a(V)\} = F_a(\hat{V}) \quad (3.36)$$

In [18], this approach was called the method of estimation of the wind velocity vector from the maximum of the function of accumulated spectra (MFAS).

Thus, the simple algorithm of (3.34) through (3.36) allows us to determine the wind velocity vector without obtaining individual estimates of the radial wind velocity $\hat{V}_{ri}$.

3.3.3   Wind Vector Maximum Likelihood (WVML) Method

Direct estimation of the wind velocity vector from measured Doppler spectra is also possible with the use of the maximum likelihood method (WVML) [18].

Assuming statistically independent estimates of the Doppler spectrum $\hat{S}_{ik} = \hat{S}_i(V_k)$ at different i and k, the log-likelihood function has the form

$$\Phi(V) = \ln\left[\prod_{i=1}^{n}\prod_{k=0}^{M-1}p(\hat{S}_{ik}|V)\right] \quad (3.37)$$

where, according to (2.50), the probability density function $p(\hat{S}_{ik}|V)$ is described by the equation

$$p(\hat{S}_{ik}|V) = (L^L/\Gamma(L))(\hat{S}_{ik}/\hat{S}_{ik})^{L-1}\exp(-L\hat{S}_{ik}/S_{ik})/S_{ik} \quad (3.38)$$

For the normalized Doppler spectrum $S_{ik} = S_{ik}(V)$, we use the model analogous to (2.53), that is,

$$S_{ik}(V) = \alpha_1\exp\left[\frac{(V_k - S_i \cdot V)^2}{2\sigma_S^2}\right] + 1 \quad (3.39)$$

where $\alpha_1 = \dfrac{SNR \cdot B_V}{\sqrt{2\pi}\sigma_S}$. As a result, from (3.37) through (3.39), we have [18]

$$\Phi(V) = nM\ \ln[L^L/\Gamma(L)] + \sum_{i=1}^{n}\sum_{k=0}^{M-1}[(L-1)\ln(\hat{S}_{ik}) - L\ln(S_{ik}(V)) - L\hat{S}_{ik}/S_{ik}(V)]$$

$$(3.40)$$

The estimate of the wind velocity vector $\hat{V}$ is determined at the maximum point of this function:

$$\max\{\Phi(V)\} = \Phi(\hat{V}) \tag{3.41}$$

If SNR and σ_S are unknown, then, in contrast to the function of accumulated spectra $F_a(V)$, $\Phi(V)$ is a function of five variables: V_z, V_x, V_y, σ_S, and SNR. In many situations taking place in the atmosphere, we can set $V_z = 0$. In the case of weak turbulence, the width of the Doppler spectrum σ_S is determined mostly by the duration of the probing pulse and the width of the time window, which are both known. The information about the vertical profile of the SNR can be obtained from lidar measurements at a fixed position of the probing beam (without scanning) and with the use of accumulation of Doppler spectra sufficient for obtaining a stable estimate of $\hat{\text{SNR}}$. Thus, the problem reduces to the determination of horizontal components of wind velocity V_x and V_y from the measured function $\Phi(V)$.

3.3.4 Cramer-Rao Lower Bound

The accuracies of the wind velocity estimates by the different methods vary. It is well known [52] that the smallest possible error of estimation of the wind velocity is the Cramer-Rao lower bound $\sigma_{CR,WV}$, which in the case of an unbiased estimate ($\langle \hat{V}_x \rangle = U$, $\langle \hat{V}_y \rangle = 0$) is defined as

$$\sigma_{CR,WV}^2 = \left\{ \left\langle \left[\frac{\partial \Phi(V_x)}{\partial V_x} \right]^2 \right\rangle \right\}^{-1} \tag{3.42}$$

where $\Phi(V_x)$ is described by (3.40) at $V = \{V_x, 0\}$ [52]. Substituting (3.40) into (3.42) and averaging under the condition $n \gg 1$, for the Cramer-Rao lower bound of the wind velocity estimate, we have [18]

$$\sigma_{CR,WV}^2 = \frac{2}{n\cos^2\varphi}\sigma_{CR,WV}^2 \tag{3.43}$$

where

$$\sigma_{CR,RV}^2 = \frac{\sigma_S^2 \Delta V}{L}\left\{ \int_{-B_V/2}^{B_V/2} dV\, \frac{(V/\sigma_S)^2}{[1+(\alpha_1\exp[-V^2/(2\sigma_S^2)])^{-1}]^2} \right\}^{-1} \tag{3.44}$$

is the Cramer-Rao lower bound (CRLB) for the estimate of radial wind velocity [53].

3.3.5 Analysis of the Accuracy of Wind Velocity Vector Estimation Techniques Based on Numerical Simulations

To compare the accuracy of the wind velocity vector estimates by the four methods discussed in the preceding section (DSWF, FSWF, MFAS, and WVML), we use the numerical simulation data from scanning pulsed CDL operations. The algorithm for simulation of lidar echo signals is described in Section 2.6. The simulation was performed for a lidar with $\lambda = 2$ μm at the fixed angle $\varphi = 60°$ and one full scan ($0 \leq \theta_i \leq 360°$). The estimates of Doppler spectra $\hat{S}_{ik}$ were obtained from simulated echo signals $Z(mT_s, i)$ at one laser shot, that is, in (3.33) $L = 1$. It was assumed that for one full scan the lidar emits 100 pulses into the atmosphere ($n = 100$), the bandwidth is 50 MHz ($B_V = M\Delta V = 50$ m/s), $M = 64$, the velocity resolution is $\Delta V \approx 0.78$ m/s, and the Doppler spectrum width is $\sigma_S = 1$ m/s. In the simulation, the wind velocity vector $\mathbf{V}$ was considered to be constant (that is, it was assumed that the wind is homogeneous and its turbulent fluctuations can be neglected) and $V_z = 0$, $V_x = U = 20$ m/s, and $V_y = 0$. Only two components of the wind velocity vector were estimated from the simulated data: $\hat{V}_x$ and $\hat{V}_y$. This means that $\mathbf{V}$ and $\mathbf{S}_i$ are two-dimensional vectors: $\mathbf{V} = \{V_x, V_y\}$ and $\mathbf{S}_i = \{\cos\varphi\cos\theta_i, \cos\varphi\sin\theta_i\}$. The functions $Q(V_x, V_y)$, $F_a(V_x, V_y)$, and $\Phi(V_x, V_y)$ were calculated at the nodes of the two-dimensional grid $\{\Delta V'k, \Delta V'l\}$ with the cell size $\Delta V' < \Delta V$ and -50 m/s $\leq V_{x,y} \leq 50$ m/s.

In the case of DSWF and FSWF, the estimates $\hat{V}_{ri}$ can be obtained by various techniques. For example, two estimators can be used for $\hat{V}_{ri}$: the periodogram maximum estimator (PM) [53] and Levin maximum likelihood estimator (ML) [see (2.51) and (2.52)] [66]. We introduce the designations PM DSWF and ML DSWF meaning the direct sine wave fitting method is used with the $\hat{V}_{ri}$ PM and ML estimators, respectively. Analogous designations can also be introduced for the case of filtered sine wave fitting: PM FSWF [$Q(V_x, V_y) \equiv Q_1(V_x, V_y)$] and ML FSWF [$Q(V_x, V_y) \equiv Q_2(V_x, V_y)$]. We assume that, when using the ML estimator and WVML, there is *a priori* information about SNR and the width of the Doppler spectrum σ_S, and then the problem reduces to obtaining the estimates $\hat{V}_{ri}$ and $\hat{\mathbf{V}}$, respectively.

Figure 3.7(a) shows the example for the function $Q_1(V_x, V_y)$ at SNR $= 0$ dB. The maximum point for $(\hat{V}_x, \hat{V}_y)$ coincides with the point of intersection of the horizontal and vertical lines in the figure. The estimates of radial velocities $\hat{V}_{ri}$ and the result of sine wave fitting $\tilde{V}_r = \mathbf{S} \cdot \hat{\mathbf{V}}$ are shown in Figure 3.7(b). Similar results are also obtained for the functions $Q_2(V_x, V_y)$, $F_a(V_x, V_y)$, and $\Phi(V_x, V_y)$, if SNR $= 0$ dB or higher. In this case all three considered techniques provide nearly identical accuracies for wind vector estimation.

Figure 3.8(a) shows the examples for the functions $Q_1(V_x, V_y)$ (PM FSWF), $Q_2(V_x, V_y)$ (ML FSWF), $F_a(V_x, V_y)$, and $\Phi(V_x, V_y)$ obtained from the same realization of the simulated spectra $\hat{S}_{ik}$ at SNR $= -20$ dB. In this example, $V_x = 20$ m/s and $V_y = 0$ m/s. It can be seen that in the case of PM FSWF and ML FSWF the estimates of the velocity vector $\{V_x, V_y\}$ are obtained with a large error. At the same time, the MFAS and WVML techniques give highly accurate results. It is shown below that at SNR $= -20$ dB the probability of estimation of the wind velocity with an error of less than 2 m/s is 15% in the case of MFAS and 25% in the case of WVML. Figure 3.8(b) depicts the radial velocities $\hat{V}_{ri}$, $\tilde{V}_r$, and V_{rt}, where $V_{rt} = \mathbf{S} \cdot \mathbf{V}$ is the true radial wind velocity, as functions of the azimuth angle θ. Individual estimates $\hat{V}_{ri}$ have a nearly uniform distribution in the range from -25 to 25 m/s due to the very low

SNR. Nevertheless, the MFAS and WVML techniques (in which noise fluctuations are partially suppressed due to averaging of spectral estimates $\hat{S}_{ik}$) give accurate estimates of the wind velocity vector. The probability of obtaining accurate estimates by the MFAS and WVML techniques in this case is much higher than in the case of using the filtered sine wave fitting methods (PM FSWF and ML FSWF) [18].

The bias $B_U = \langle \hat{U} \rangle - U$ and error $E_U = [\langle (\hat{U} - U)^2 \rangle]^{1/2}$ of wind velocity estimates were calculated with the use of numerical simulation data at various SNRs [18]. For the calculation of each of these characteristics, 2,000 independent realizations of estimates $\hat{U} = [\hat{V}_x^2 + \hat{V}_y^2]^{1/2}$ were used. Figure 3.9 shows the results of calculation of E_U (solid curves) and $|B_U|$ (dashed curves) as functions of SNR. The bias of the velocity estimate is shown only for the PM DSWF and ML DSWF cases. It can be seen that, when these techniques are used at SNR < -5 dB, the bias is the major contributor to the error, since $V_x = U = 20$ m/s and $V_y = 0$ m/s. As SNR decreases, however, the number of bad estimates $\hat{V}_{ri}$ increases, and $\langle \hat{V}_x \rangle \to 0$, $\langle \hat{V}_y \rangle \to 0$ at $b \to 1$. For PM FSWF and ML FSWF at SNR ≥ -14 dB and MFAS and WVML at SNR ≥ -16 dB, the bias of the wind velocity estimate B_U is negligibly small. Compared to the direct sine wave fitting method, the filtered sine wave fitting and the methods of estimation from the maximum of the function of accumulated spectra $F_a(V_x, V_y)$ and log-likelihood $\Phi(V_x, V_y)$ allow the wind velocity vector to be estimated at far lower SNR values. As expected, the WVML method provides the best estimation accuracy.

In Figure 3.9, the bold curve is the Cramer-Rao lower bound calculated by (3.43) and (3.44). One can see that only the use of ML FSWF and WVML allows us to achieve the CRLB, respectively, at SNR ≥ -9 dB and SNR ≥ -14 dB.

Let the wind vector estimate $\hat{V} = \{\hat{V}_x, \hat{V}_y\}$ be efficient, if the absolute random errors $\hat{E}_x = |\hat{V}_x - V_x|$ and $\hat{E}_y = |\hat{V}_y - V_y|$ are less than 2 m/s. That is, if $V_x = U = 20$ m/s and $V_y = 0$ m/s, the relative error $\hat{E}_x / V_x$ does not exceed 10% and the error of estimation of the wind direction is not more than 7°. The precision of ±2 m/s satisfies the requirements

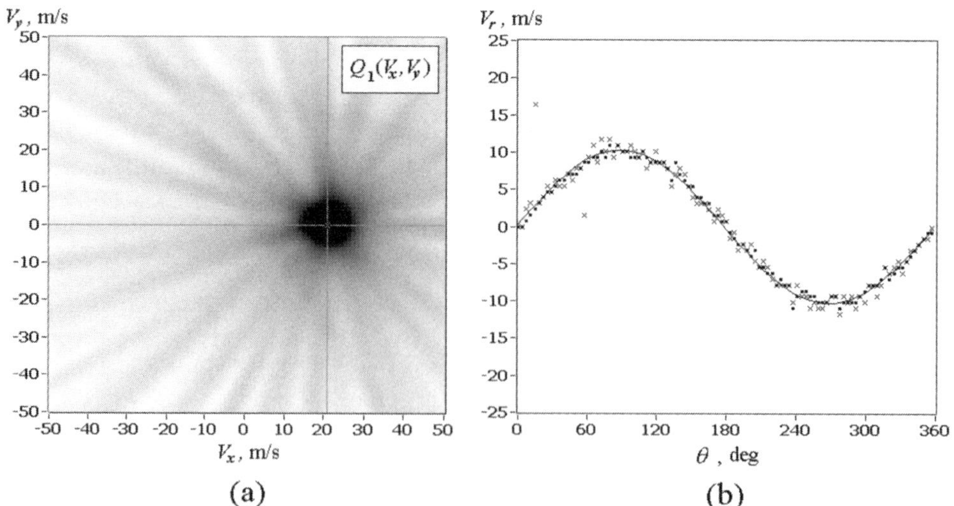

Figure 3.7 Examples of (a) the function $Q_1(V_x, V_y)$ and (b) the radial wind velocity as a function of the azimuth angle at SNR $= 0$ dB. Crosses: estimates by the PM; dots: estimates by the ML; solid curve: $\tilde{V}_r = S \cdot \hat{V}$, where $\hat{V}$ is the maximum point in part (a). (© 2003 American Meteorological Society. From [18].)

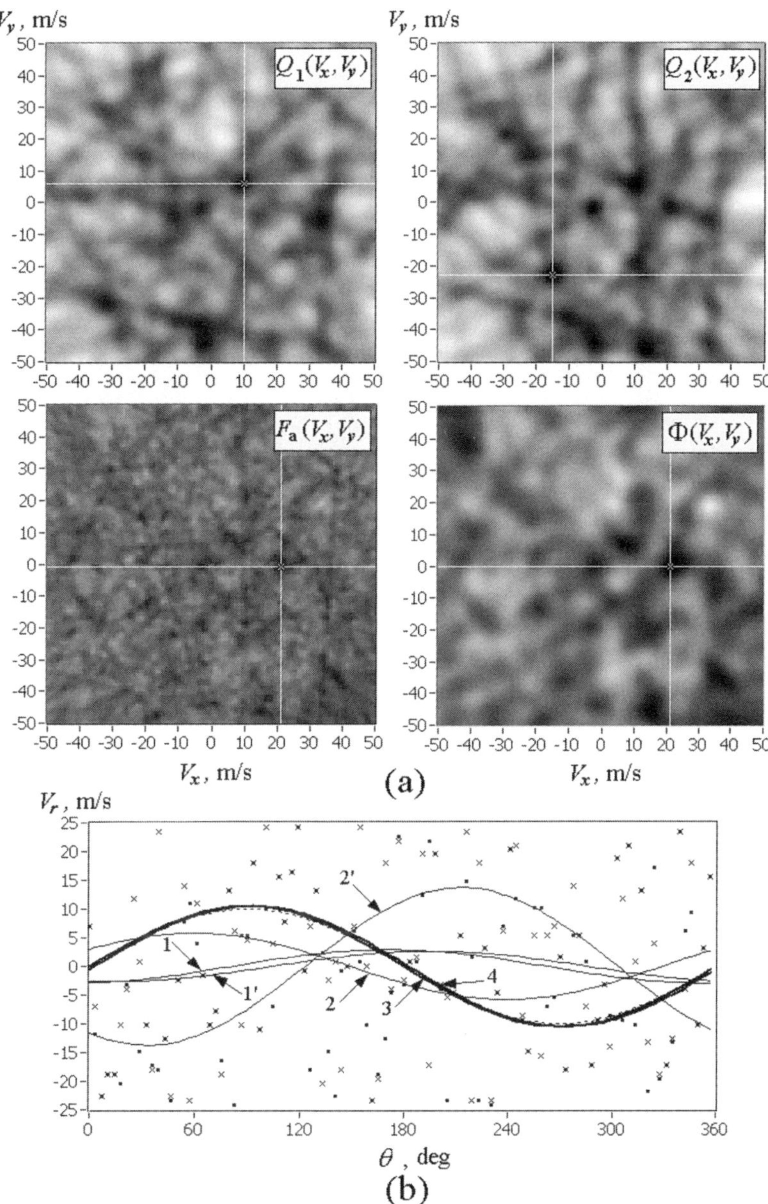

Figure 3.8   Example of the functions $Q_1(V_x, V_y)$ (for PM FSWF), $Q_2(V_x, V_y)$ (for ML FSWF), $F_a(V_x, V_y)$, and $\Phi(V_x, V_y)$ obtained from the same realization of simulated spectra at (a) SNR = −20 dB and (b) the radial wind velocity. Crosses: $\hat{V}_r$ estimated by the PM; dots: $\hat{V}_r$ estimated by the ML technique; solid curves: $\tilde{V}_r$ calculated with the use of estimates $\hat{V}$ obtained by the PM DSWF (1), ML DSWF (1′), PM FSWF (2), ML FSWF (2′), MFAS (3), and WVML (4) techniques; dashed curve: V_{rt}. (© 2003 American Meteorological Society. From [18].)

of many practical tasks. Figure 3.10 shows the probability of efficient estimate P_{ae} as a function of SNR. The wind velocity vector was estimated from the data of numerical simulation by these four techniques: PM FSWF, ML FSWF, MFAS, and WVML. It can be seen that an efficient estimate with a probability no lower than 50% is possible in the case of PM FSWF at SNR $\geq$ −16 dB and in the case of WVML at SNR $\geq$ −19 dB.

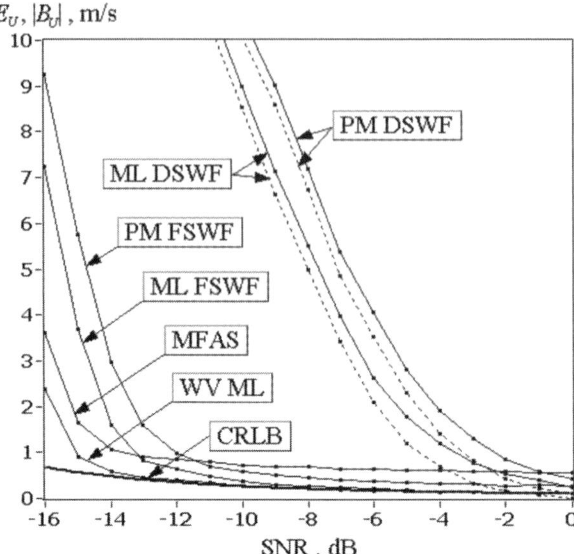

Figure 3.9 Error E_U (solid curve) and absolute value of the bias $|B_U|$ (dashed curves) of the lidar estimate of wind velocity as functions of SNR. Bold curve shows the Cramer-Rao lower bound. (© 2003 American Meteorological Society. From [18].)

The accuracy of the techniques for estimating the wind velocity vector from the data of pulsed CDLs at $L = 1$ has been analyzed above. It has been shown that at low SNR the MFAS method provides better accuracy than does the FSWF method, and WVML is more accurate than MFAS. However, keep in mind that WVML assumes the presence of *a priori* information about SNR. If this information is absent, then the practical implementation of this method is significantly complicated and becomes computationally expensive. In addition, the advance in accuracy above FSWF and MFAS is not so significant as in the case of known SNR. The accumulation of spectra at $L \gg 1$ also decreases the difference in accuracy between FSWF, MFAS, and WVML (at SNR $\ll 1$) compared to $L = 1$. The numerical analysis shows that the number L should be chosen so that an increase in the width of the Doppler spectrum due to the scanning by the probing beam does not lead to a more than 10% decrease in the maximum of the spectral peak.

The results of numerical simulation of the operation of a scanning pulsed CDL at $\lambda = 2~\mu$m, $\sigma_P = 120$ ns, and $T_s = 20$ ns with the use of the algorithms described by (2.130) and (2.131) with neglected wind turbulence ($\sigma_V = 0$) are presented in [27, 39]. The simulation was performed in accordance with field experiments [27, 39], whose results are discussed in Section 3.4. In the calculations, one full scan ($0° \leq \theta_i \leq 360°$) with 12,000 laser shots was considered. The Doppler spectra were calculated from the array of simulated signals $Z(mT_s, l)$ by (3.33), where $L = 50$ and $M = 10$. To pass on from the velocity resolution $\Delta V = (\lambda/2)/(MT_s) = 5$ m/s to the velocity bin $\delta V = (\lambda/2)/(M'T_s) = 0.1$ m/s, in the exponent in (3.33) we should replace $M = 10$ with $M' = 500$. In this case, $V_k = (k - M'/2)\delta V$ and $k = 0, 1, 2, ..., M' - 1$. The radial velocity was estimated from the position of the spectral peak (the PM estimator was used); that is, $\hat{V}_{ri} = (k_{max} - M'/2)\delta V$. To obtain estimates of the wind velocity vectors $\hat{V}_x$ and $\hat{V}_y$, the DSWF and FSWF methods were used (the parameter σ_g was considered equal to 2 m/s).

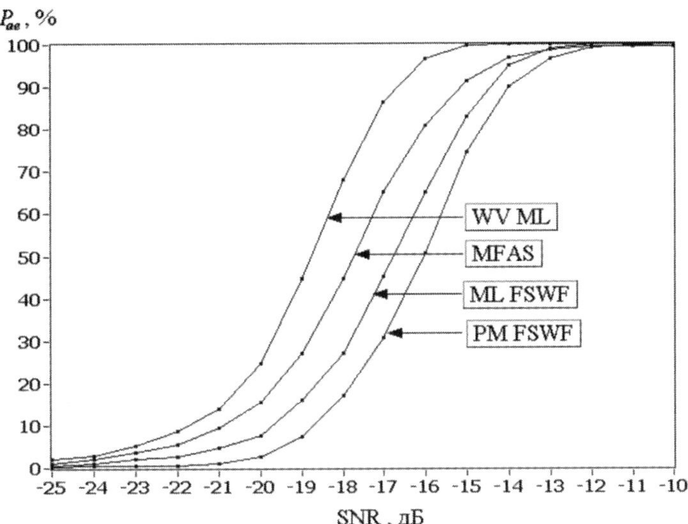

Figure 3.10 Probability of efficient estimate of the wind velocity vector from scanning CDL data and the use of different estimation techniques. (© 2003 American Meteorological Society. From [18].)

The simulation was conducted at different SNRs. The error of estimation of the wind velocity $E_U = \sqrt{\langle(\hat{U} - U_t)^2\rangle}$ and the error of estimation of the wind direction $E_\theta = \sqrt{\langle(\hat{\theta}_V - \theta_{Vt})^2\rangle}$, where $\hat{U} = |\hat{V}_x + j\hat{V}_y|$ and $\hat{\theta}_V = \arg[\hat{V}_x + j\hat{V}_y]$ are lidar estimates of the wind velocity and wind direction, respectively, while U_t and θ_{Vt} are true values of the wind velocity and direction, were calculated with the use of 2,000 independent realizations of the estimates $\hat{V}_x$ and $\hat{V}_y$ for each SNR value.

Figure 3.11(a) shows calculated E_U and Figure 3.11(b) shows calculated E_θ as functions of SNR at an elevation angle of $\varphi = 15°$. One can see that for the conditions $E_U < 1$ m/s and $E_\theta < 5°$ to be true, the SNR should be no lower than −20 dB (or 0.01). Numerical experiments have shown that the variation of angle φ from 0° to 30° results in only a slight difference from the results shown in Figure 3.11.

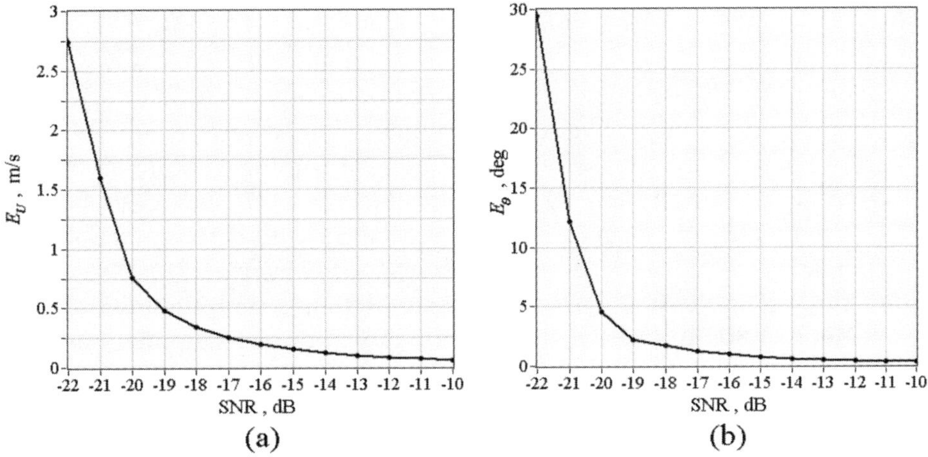

Figure 3.11 Errors of lidar estimate of (a) wind velocity and (b) wind direction obtained with the use of FSWF at $L = 50$ as functions of the SNR.

It follows from the data shown in Figure 3.9 (PM FSWF) and Figure 3.11 that the 120-fold increase of the number of laser shots and the use of spectral accumulation with $L = 50$ allows us to obtain lidar estimates of wind velocity with an error smaller than 1 m/s at far lower (by an order of magnitude) SNR values than in the case of 100 shots and $L = 1$.

3.4   Experimental Testing of the FSWF and MFAS Techniques

To retrieve vertical profiles of wind velocity $U(h)$ and direction $\theta_V(h)$, we have used raw data from an experiment conducted in September 2003 in southeast Colorado (USA) within the framework of the Lamar Low-Level Jet Project [54, 55]. The main goal of this experiment was to study the nocturnal atmospheric boundary layer at the stable thermal stratification, when a jet flow is formed at relatively low levels. The experiment involved four sonic anemometers installed on a 120m meteorological tower (at heights of 54m, 67m, 85m, and 116m) and a 2-μm pulsed coherent Doppler lidar [56]. The distance between the tower and the lidar container was 167m.

The lidar measurements were conducted with conical scanning by the probing beam at different elevation angles φ. Estimates of the Doppler spectra $\hat{S}_i(V_k)$ were obtained by using (3.33) and 50 pulses for accumulation, that is, $L = 50$. These spectra were used to find, by (2.47) and (2.48), signal-to-noise ratio $\hat{\mathrm{SNR}}(\theta_i, R_l)$ and radial wind velocity $\hat{V}_r(\theta_i, R_l)$ as functions of the azimuth angle θ_i and the distance from the lidar to the center of the sensing volume $R_l = R_0 + l\Delta R$, where $R_0 = 190$m, $\Delta R = 30$m, and $l = 1, 2, 3, \ldots$. The resolution in the azimuth angle was $\Delta\theta = 1.5°$. Consequently, for one turn ($0° \leq \theta_i \leq 360°$) made for 1 min, $n = 240$ velocity estimates for every R_l were obtained. The data on $\hat{\mathrm{SNR}}(R_l, \theta_i)$ were used to calculate the vertical profiles of $\mathrm{SNR}(h_l) = n^{-1}\sum_{i=1}^{n}\hat{\mathrm{SNR}}(R_l, \theta_i)$, where $h_l = h_0 + R_l \sin\varphi$.

The components of the wind velocity vector $\hat{V}_x$ and $\hat{V}_y$ were determined from estimates of the radial velocity $\hat{V}_r(R_l, \theta_i)$ for heights h_l by two techniques: DSWF and FSWF.

During the experiment on September 15, 2003, there was strong wind with a velocity within the jet flow that sometimes exceeded 20 m/s. The wind turbulence varied from moderate to strong. That is why the parameter σ_g was taken equal to 2 m/s in the filtered sine wave fitting procedure used in (3.31).

Figure 3.12 exemplifies vertical profiles of wind velocity and direction retrieved from the data of lidar measurements at different angles φ. It can be seen that at $\varphi = 30°$ up to a height of ~ 800m, where SNR ≈ -10dB, the DSWF and FSWF methods give nearly identical results. With an increase in the height h above 800m, SNR decreases, and the increase in the number of bad estimates of $\hat{V}_{ri}$ causes an increase of the bias in the wind velocity estimate obtained with the use of DSWF. Up to a height of 1,800m, where SNR ≈ -20 dB, FSWF, according to the results of numerical simulations, should give a practically unbiased estimate of the wind velocity vector despite the increase in the estimate variance.

Measurements at angles $\varphi = 7°$ and $\varphi = 2°$ (Figures 3.12(b) and 3.12(c)) have allowed the comparison of lidar data with the results of simultaneous measurements by four acoustic anemometers. At $\varphi = 7°$ (see Figure 3.12(b)) and heights h where

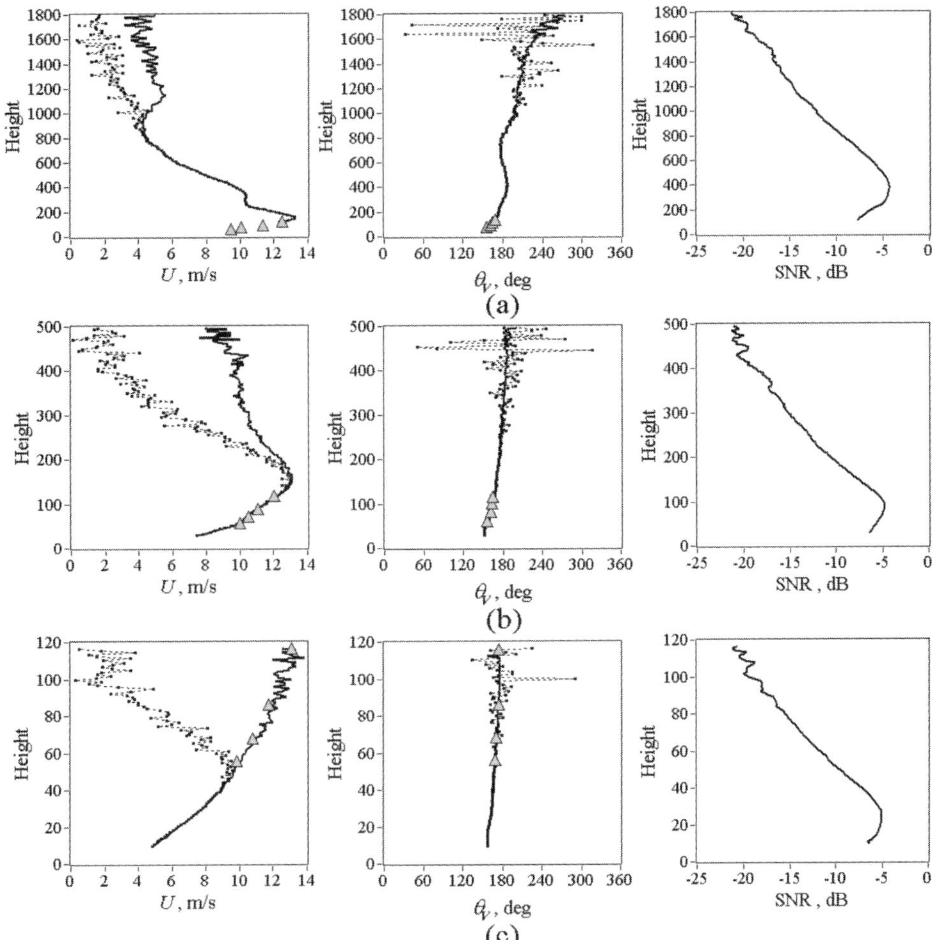

Figure 3.12 Results of retrieval of vertical profiles of wind velocity U, wind direction θ_V, and SNR from the data of a conically scanning 2-μm pulsed CDL, measured on September 15, 2003, (a) at 01:58 at $\varphi = 30°$, (b) 02:02 at $\varphi = 7°$, and (c) 02:32 at $\varphi = 2°$. The wind profiles are obtained from lidar data with the use of DSWF (dashed curves) and FSWF (solid curves). The triangles are the results of simultaneous measurements by sonic anemometers with the use of 5-min averaging.

SNR $\geq$ −10 dB, the results of DSWF and FSWF application also nearly coincide, and the lidar measurement data are in satisfactory agreement with the results of measurements by sonic anemometers. In the case of $\varphi = 2°$ (see Figure 3.12(c)) only the results of FSWF application agree satisfactorily with the sonic anemometer data, since already at those heights the sonic anemometers were installed at SNR < −10 dB and the DSWF estimates of the wind velocity appear to be significantly biased. At the same time, the FSWF technique even at SNR = −21 dB gives estimates of the wind velocity components that are in a good agreement with the acoustic anemometer data.

Figure 3.13 shows the temporal profile of wind velocity measured by the sonic anemometers at different heights (curves) along with the results of lidar measurements of the wind velocity at $\varphi = 2°$ (symbols). As for the data shown in Figure 3.12(c), at the heights of installation of the sonic anemometers the SNR values ranged within −10 dB $\leq$ SNR $\leq$ −20 dB, where DSWF gives a biased estimate of the wind velocity. Therefore, the lidar measurement data were processed using the FSWF method. One can see the satisfactory agreement between the results of wind velocity measurements by the sonic anemometers and CDL. The largest deviation does not exceed 0.5 m/s, despite the nonstationarity and inhomogeneity of the wind flow. It is interesting to note that at height $h = 116$m, where SNR ≈ -20 dB, the results of measurements by the sonic anemometer and the lidar nearly coincide. Thus, we can conclude that at $L = 50$, as in the numerical experiment (see Figure 3.11), the FSWF method allows us to estimate the wind velocity and wind direction with a good accuracy from lidar measurements under conditions of weak echo signal up to SNR $= -20$ dB. According to the results of the numerical simulation shown in Figure 3.11, at SNR $= -20$ dB the errors of estimation of the wind velocity and wind direction are, respectively, 0.7 m/s and 4°. The more accurate measurements of the wind velocity in the field experiment considered here are explained by the fact that, in contrast to the numerical experiment, the field experiment included the additional averaging of the velocity measured by the lidar over five neighboring vertical levels [27].

Figure 3.14 shows the results of retrieval of vertical profiles of the wind velocity and wind direction from the lidar data with the use of the FSWF techniques [27]. The figure illustrates well the evolution of the jet flow, which could be observed with the lidar in the atmospheric boundary layer starting from about 20:00 LT. The use of FSWF allows one to double the maximal height of the lidar measurements of wind velocity and wind direction when compared to DSWF and to observe the variation of the velocity and direction of the wind flow in lower layers of the free atmosphere up to a height of ~2 km.

In [57, 58], to retrieve vertical wind profiles from data measured by airborne 2-μm pulsed CDLs, the MFAS technique was used to process the raw lidar data. Figure 3.15 exemplifies the vertical profile of the wind velocity retrieved from the lidar data (dots). The curve shows the vertical profile of the wind velocity measured by a wind sensor installed on a dropsonde. The altitude of the airplane with the

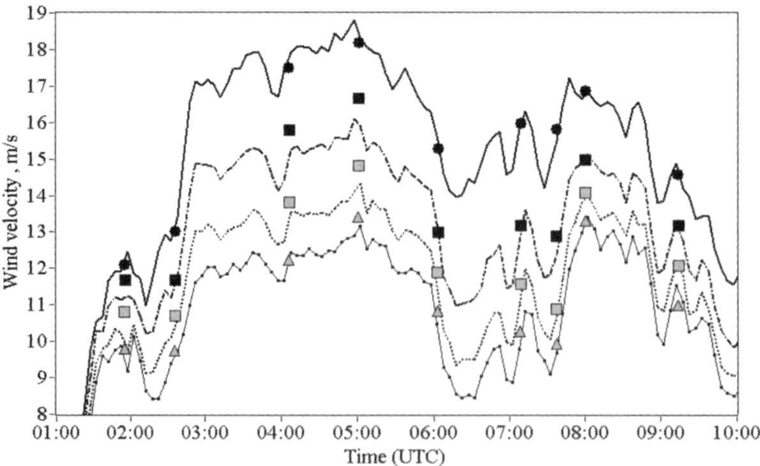

Figure 3.13 Time profiles of wind speed measured by sonic anemometers at a height of 54m (dots connected by lines), 67m (dashed curve), 85m (dot-and-dash curve), and 116m (solid curve) and the corresponding results of lidar measurements at $\varphi = 2°$ at a height of 54m (triangles), 67m (gray squares), 85m (black squares), and 116m (circles).

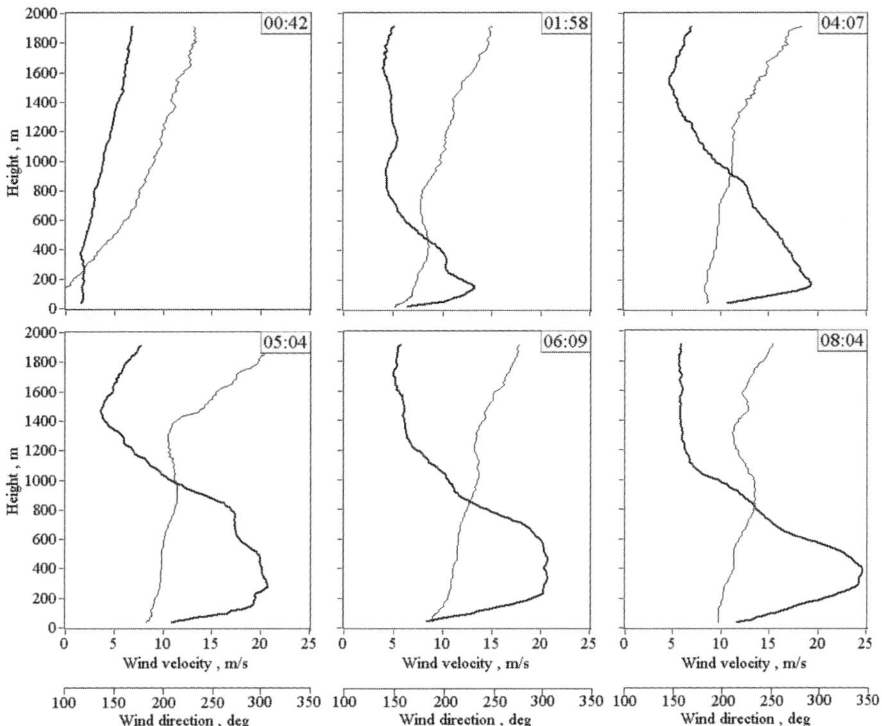

Figure 3.14 Vertical profiles of wind velocity (bold curves) and wind direction (thin curves) retrieved from the lidar data with FSWF. The measurement time (UTC) is shown in every panel.

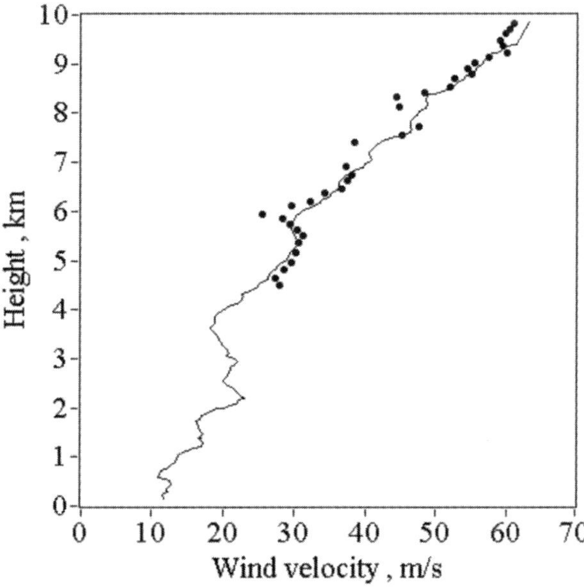

Figure 3.15 Vertical profiles of the wind speed retrieved from the data of simultaneous measurements by 2-μm CDL from aboard an aircraft (dots) and wind sensor installed at a dropsonde (curve). The lidar profile of wind was obtained with the MFAS method.

lidar was 10.3 km. The maximal range of measurement of lidar echo signals was 6 km. The measurements were conducted at low SNR up to SNR ~ −23 dB. The lidar wind profile shown in the figure was obtained for one full scan by the probing beam at 24,000 laser shots (measurement time of 48s). The use of MFAS allowed the retrieval of the vertical profile of wind velocity in the layer from 4.5 to 10 km, which is in good agreement with measurements by the dropsonde wind sensor.

3.5 Simulation of Retrieval of Vertical Wind Profiles from Measurements by Spaceborne CDLs

An increase in the reliability of weather forecasts depends to a large extent on our ability to obtain information about wind on a global scale. An efficient instrument for the solution of this problem is a spaceborne pulsed CDL [59–64]. In this regard, it becomes necessary to study methodically the problems of retrieval of vertical profiles of wind velocity and wind direction from lidar data. One of the possible approaches here is the numerical simulation of the operation of spaceborne lidar imitating the processes of propagation of the probing radiation in the atmosphere and sampling of the echo signal with the following extraction of the information about wind from simulated data [60, 65].

To study the possibility of retrieving the wind velocity field from measurements by a spaceborne Doppler lidar, the Lidar Group of the German Aerospace Center (DLR) developed the ALIENS software [65], which imitates wind measurements by spaceborne CDLs. This software uses data on the global distribution of atmospheric parameters for the January 19–30, 1998, period furnished by the German Weather Service (DWD). The Laboratory of Wave Propagation of the Institute of Atmospheric Optics SB RAS (IAO) developed the software unit, which simulates turbulent fluctuations of wind velocity in accordance with the DWD data. The cooperation between DLR, DWD, and IAO has allowed for the development of a virtual Doppler lidar instrument to study the possibilities of wind sensing from space under realistic conditions [25].

The ALIENS software consists of four main units. The first unit sets the lidar parameters (probing pulse wavelength, energy, and duration, pulse repetition frequency, diameter of transceiver telescope, parameters of analog-to-digital converter, receiver bandwidth, and so on), geometrical parameters of sensing and satellite position (nadir, angular rate of scanning, satellite altitude), and atmospheric parameters (vertical profiles of backscatter and extinction coefficients, DWD data on the global distribution of vertical profiles of the meridional and zonal wind velocities, temperature, cloudiness, and turbulent exchange coefficients).

The second unit successively simulates the process: backscattering of the probing pulse by aerosol particles—optical shift of the backscattered and reference waves—detection—and data digitizing. This unit outputs temporal profiles of the photocurrent arising in the circuit of the lidar receiving system. The numerical simulation is based on the algorithm described in Section 2.6.

In the third unit, the simulated data are processed to obtain vertical profiles of the radial component of wind velocity with the preset height resolution. Then, in the case of conical scanning by the probing beam, the array of radial velocities can be used to retrieve vertical profiles of wind velocity and direction in the considered sensing region (for example, with a height resolution of 1 km and 100×100 km horizontal resolution) along the satellite trajectory.

The obtained vertical profiles of wind velocity and direction are transmitted into the output (fourth) unit, where they are compared with the initial DWD wind data. From the results of this comparison, one can judge the efficiency of lidar measurements of wind fields.

Figure 3.16 shows the geometry of sensing by the scanning Doppler lidar. The arrow shows the direction of motion of the satellite at which the lidar is installed. Dots show the distribution of single probing pulses in the horizontal plane. The size of grid cells corresponds to the resolution of DWD data: $1.125°$ in longitude and $1.121°$ in latitude; that is, about 100×100 km near the equator. The relationship between radial velocities, propagation directions (nadir, azimuth scan angle), and the wind velocity vector (two horizontal components, the vertical one is assumed to be zero) is used on the assumption that the wind remains unchanged inside every cell at the chosen altitude.

In the numerical simulation, the following parameters were set: satellite altitude of 400 km; probing pulse repetition frequency of 10 s^{-1}; pulse duration of 1 μs; pulse energy of 0.5J; wavelength of 2.02184 μm; telescope diameter of 70 cm; sampling frequency of 200 MHz; bandwidth of 50 MHz; vertical resolution of 1,258m; horizontal resolution of $1.125°$ in longitude and $1.121°$ in latitude.

Figure 3.17 shows examples of the vertical profiles of wind velocity U and wind direction θ_V retrieved from the simulated lidar data with noise taken into account and neglected. The values of wind velocity and direction were calculated from the estimates of the radial wind velocity $\hat{V}_{ri}$ for probing pulses falling within the selected cell of DWD data. The radial velocities were estimated from the power

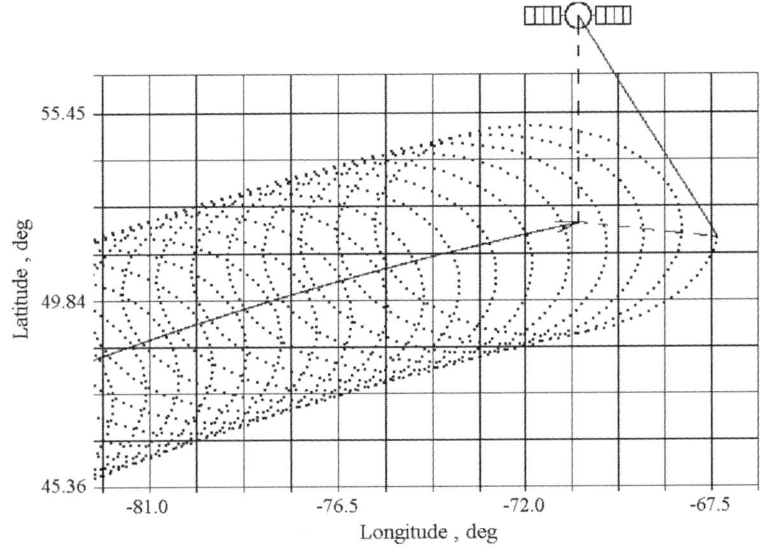

Figure 3.16 Geometry of measurements by satellite scanning CDLs. (© 2005 Society of Photo-Optical Instrument Engineering. From [67].)

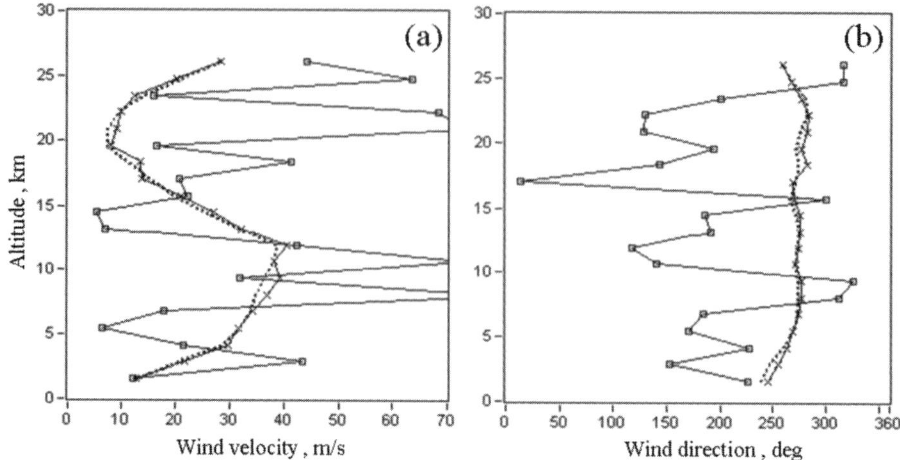

Figure 3.17 Examples of retrieval of the vertical profiles of (a) wind velocity and (b) wind direction from simulated lidar data with (squares) and without (crosses) noise; dashed curves are for the DWD data. (© 2005 Society of Photo-Optical Instrument Engineering. From [67].)

spectra of simulated signals from the position of a peak in the spectral distribution. The dashed curves show the DWD data. One can see that with no noise the wind is reconstructed with good accuracy. However, if noise is present, due to the low SNR, the useful component of the spectrum is hidden among noise peaks and the error of wind retrieval for the above examples is very large (in this example, even at a height of ~3 km).

The most reliable way to estimate correctly the radial wind velocity under conditions of low SNR values is through the accumulation of spectra measured from single probing pulses, as a result of which noise peaks are averaged and the signal peak carrying the information about the radial wind velocity is separated. Spectra can be accumulated with the use of only the data measured successively (from pulse to pulse) within the considered 100×100 km cell. As a result, N accumulated Doppler spectra (at different azimuth angles, average for scan sectors) are obtained for every cell, where N is determined by how many times the probing beam crosses the cell (see Figure 3.16). Then, after the estimation of radial velocities from the accumulated spectra, the wind velocity and direction are calculated.

According to the data shown in Figure 3.16, a maximum of 53 pulses (on average ~26) can fall within one cell. It follows from the results of numerical simulation that this number of pulses is sufficient for the retrieval of wind profiles having a height of up to ~7 to 8 km. The analysis of DWD data for wind at high altitudes has shown that to increase the number of probing pulses used for the wind profile retrieval, we can increase the number of cells from one to nine without fear of excessive averaging of wind over a large area (~300×300 km). However, when the probing beam is scanning within this area, the azimuth scan angle changes significantly. That is why the MFAS method (see Section 3.3) was used to obtain the information about the wind velocity and direction, because in this method all measured Doppler spectra are accumulated and the estimates of the radial wind velocity are not required.

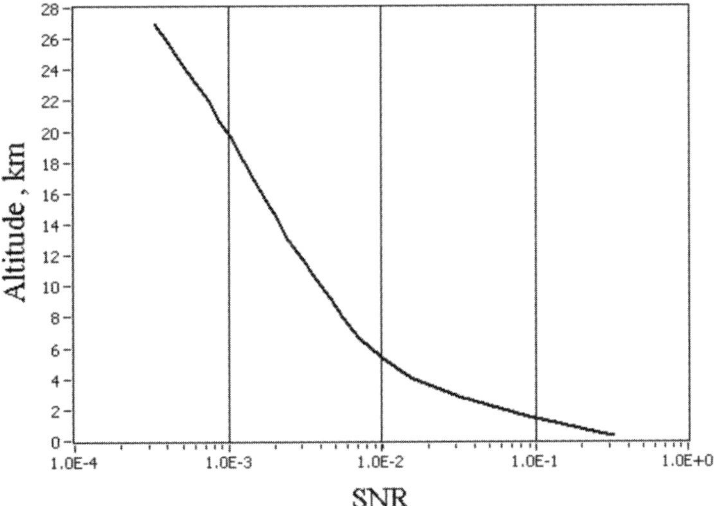

Figure 3.18 Vertical profile of the SNR.

Figure 3.18 shows the vertical profile of the SNR for the above lidar parameters in the case of a cloudless atmosphere [16]. At altitude $h > 20$ km, SNR becomes smaller than 10^{-3}. The results presented below suggest that the retrieval of the wind field with the required accuracy at this SNR becomes impossible.

The MFAS method was used in numerous numerical experiments on lidar retrieval of the wind field in different regions of the globe. Figure 3.19 depicts some examples of reconstruction of the wind velocity and direction in the cloudless atmosphere and in the presence of clouds along with the DWD data [16, 19]. One can see that without clouds the wind profiles are reconstructed with an acceptable accuracy up to heights of ~18 to 20 km. In the presence of clouds, a large error appears in the layer below ~8 km due to the strong extinction of the probing and backscattered radiation in clouds.

The vertical wind profiles retrieved by the MFAS method were used to calculate the absolute errors for the wind velocity $\varepsilon_U = [\langle(U_L - U_W)^2\rangle]^{1/2}$ and direction $\varepsilon_\theta = [\langle(\theta_L - \theta_W)^2\rangle]^{1/2}$, where U_L and θ_L are the lidar estimates, and U_W and θ_W are the DWD data. Figure 3.20 shows the region of the globe along the satellite track for which the DWD data were used in the numerical simulation. The errors ε_U and ε_θ were calculated with the use of 746 retrieved vertical profiles of wind velocity and direction falling within separate 1.125° (longitude) × 1.121° (latitude) cells of the computational grid. The wind profiles were retrieved on the assumption of no clouds in the whole sensing area and with regard for the DWD data on the cloud distribution in the atmosphere for January 20, 1998. First, the errors ε_U and ε_θ were calculated for individual wind profiles, and then they were averaged over the sensing area demonstrated in Figure 3.20. In the cross section of the satellite track, the size of the sensing area was ~600 km.

Figure 3.21 depicts the altitude dependence of the errors ε_U and ε_θ calculated on the assumption of no clouds along the satellite track and with regard for the cloud

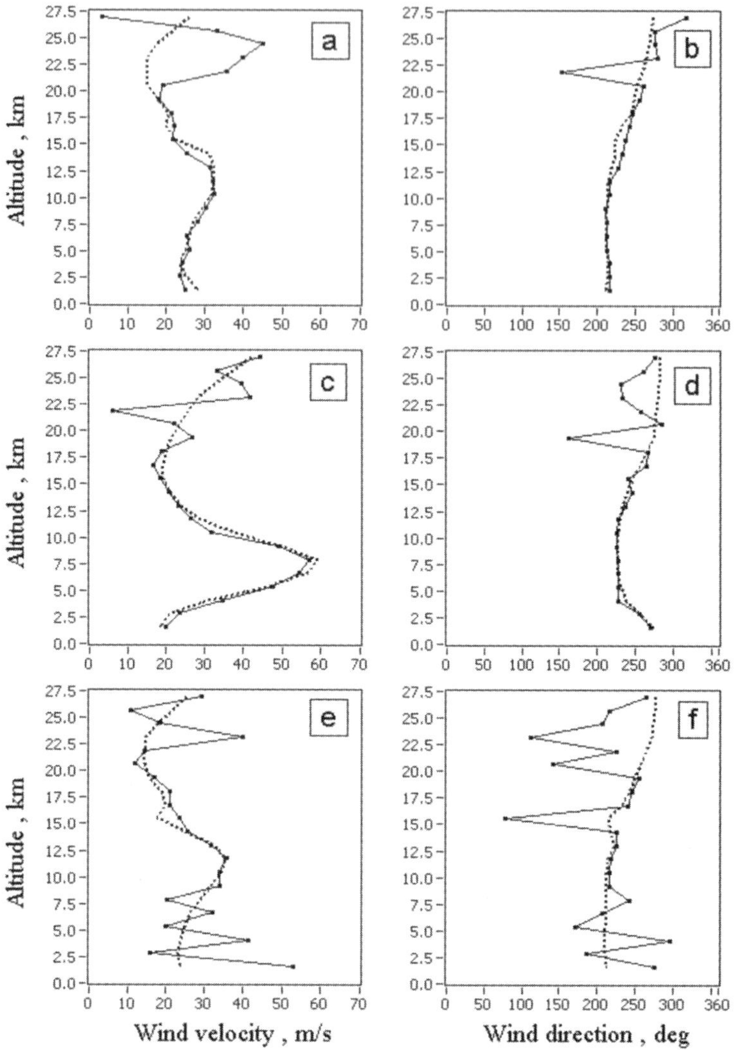

Figure 3.19 Examples of retrieval of (a, c, e) wind velocity and (b, d, f) wind direction in the (a–d) cloudless atmosphere and (e, f) in the presence of clouds. Solid curves: lidar data; dashed curves: DWD data. (© 2005 Society of Photo-Optical Instrument Engineering. From [67].)

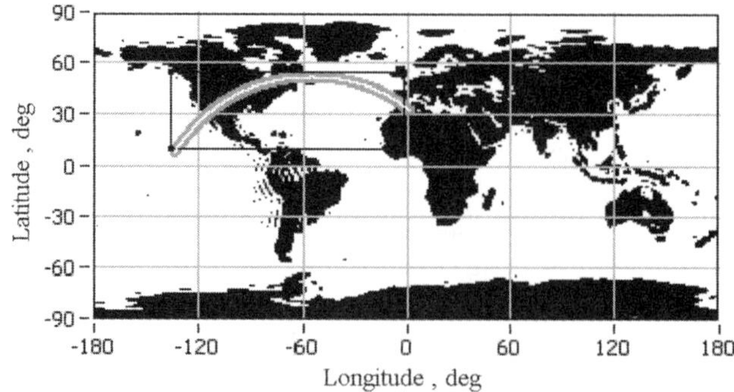

Figure 3.20 Region of the globe along the satellite track for which the DWD data were used in the analysis of wind retrieval errors. (© 2005 Society of Photo-Optical Instrument Engineering. From [67].)

distribution (according to the DWD data) in the sensing area [16]. It can be seen that the use of the MFAS method allows for the retrieval of wind velocity with an error $\leq$ 2 m/s and wind direction with an error $\leq 20°$ in the cloudless atmosphere up to altitudes of ~17 km. If clouds are present in the sensing area, then, as follows from Figures 3.21(c) and 3.21(d), the lidar retrieval of wind is possible only in the layer from about 12 to 15 km. The accuracy of $\pm$2 m/s and $\pm20°$ quite satisfies the requirements specified by meteorological services on wind sensing data for their use in weather forecast models.

The results presented indicate that CDLs can be used principally for the retrieval of wind velocity fields on a global scale if the lidar has the parameters mentioned above and a specialized procedure (for example, MFAS) is used for the processing of raw measurement data.

3.6   Conclusions

1. Wind turbulence is the main source of error in measurements of the mean wind velocity and direction by cw CO_2 CDLs conducting conical scanning by the probing beam in the atmospheric boundary layer. Because lidar measurements imply the spatial averaging of turbulent fluctuations of wind velocity along the probing beam axis and over the base of the scanning cone, a shorter averaging time is needed for obtaining estimates of the wind velocity and direction than in measurements by a point sensor. The larger the ratio of the radius of the scanning cone base to the outer scale of turbulence, the shorter the lidar measurement time needed to achieve the necessary accuracy.

2. In the case of stable thermal stratification, measurements of the mean velocity in the atmospheric surface and boundary layers by a cw CO_2 lidar at an elevation angle of 30° are representative (relative error no higher than 10%) even after just one complete scan made for 12s. At the neutral stratification and an elevation angle of 30°, one scan is also sufficient if the measurement height is no lower than 150m. The duration of lidar measurements for obtaining a representative estimate of the mean wind velocity ($\varphi = 30°$) in the atmospheric surface layer at an unstable stratification is only a little shorter than the duration of measurements by a cup anemometer, which is 10 min.

3. An increase in the elevation angle φ and a decrease in the scan sector angle θ_s lead to an increase in the measurement error of the mean wind velocity by a cw CO_2 lidar. Already at the angle $\theta_s = 90°$, the number of scans by the probing beam necessary for the representative measurements of the mean velocity is at least 10-fold larger than in the case of $\theta_s = 360°$.

4. The FSWF, MFAS, and WV ML techniques for the estimation of wind velocity and wind direction from the data measured by pulsed CDLs using conical scanning by the probing beam significantly extend the capabilities of lidar wind sensing. Estimations of the mean wind velocity vector by these techniques remain unbiased at SNRs an order of magnitude smaller than the limiting SNR, at which an unbiased estimate by the DSWF method is still possible. Theoretically, the WVML method allows for a more accurate estimate. However, if there is no *a priori* information about the SNR, then

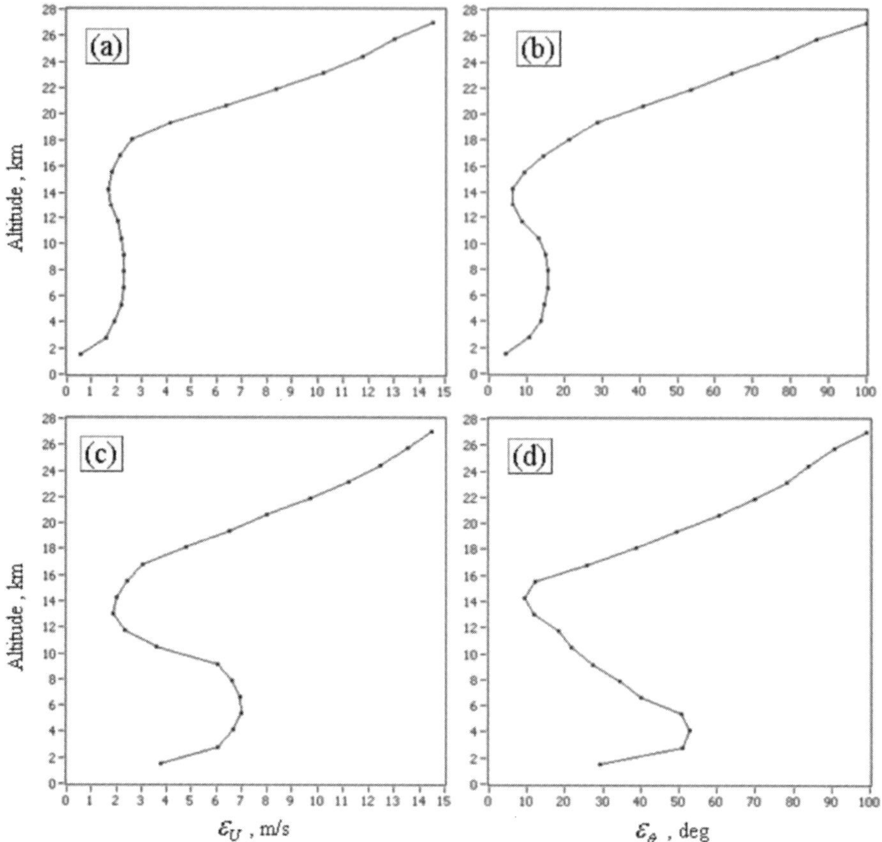

Figure 3.21 Errors of estimation of the wind velocity ε_U and wind direction ε_θ as functions of altitude, calculated on the assumption of (a, b) no clouds and (c, d) with regard for the cloud distribution according to the DWD data for January 20, 1998. (© 2005 Society of Photo-Optical Instrument Engineering. From [67].)

the application of WVML becomes less efficient compared to the FSWF and MFAS methods. At low SNR, MFAS allows for a more accurate estimate of the mean velocity vector than FSWF does. At SNR > −11 dB, the FSWF method is more accurate.

5. A comparative analysis of the results of simultaneous measurements of vertical profiles of the wind velocity by 2-μm pulsed CDL and by direct sensors (sonic anemometers at a meteorological tower and a dropsonde sensor) has demonstrated the efficiency of using the FSWF and MFAS methods in the case of very low SNR (in the experiments, SNR took values from −10 to −23 dB).

6. The computer simulation shows that the use of the MFAS method allows for the retrieval of vertical profiles of the wind velocity and direction from satellite CDL measurements under conditions of a cloudless atmosphere with a height resolution of 1 km and errors of ±2 m/s for the wind velocity and ±20° for the wind direction in the lower atmospheric layer, which is ~17 km thick. The simulation was carried out for a 2-μm CDL with a pulse energy of 0.5J, pulse duration of 1 μs, pulse repetition frequency of 10 Hz, and telescope diameter of 70 cm.

References

[1]   Lhermitte, R.M., and Atlas, D., "Precipitation motion by pulse Doppler," *Proc. 9th Weather Radar Conference*, Kansas City, MO, USA, 1961, pp. 218–223.

[2]   Huffaker, R.M., "Laser Doppler detection systems for gas velocity measurement," *Applied Optics*, Vol. 9, No. 5, 1970, pp. 1026–1039.

[3]   Lawrence, T.R., et al., "A laser velocimeter for remote wind sensing," The Review of Scientific Instruments, Vol. 43, No. 3, 1972, pp. 512–518.

[4]   Bilbro, J.M., "Atmospheric laser Doppler velocimetry: An overview," *Optical Engineering*, Vol. 20, No. 12, 1980, pp. 2048–2054.

[5]   Köpp, F., Schwiesow, R.L., and Werner, Ch., "Remote measurements of boundary layer wind profiles using a cw Doppler lidar," *Journal of Climate Applied Meteorology*, Vol. 23, No. 1, 1984, pp. 148–158.

[6]   Köpp, F., Bachstein, F., and Werner, Ch., "On-line data system for a cw laser Doppler anemometer," *Applied Optics*, Vol. 23, 1984, pp. 2488–2491.

[7]   Werner, Ch., "Fast sector scan and pattern recognition for a cw laser Doppler anemometer," *Applied Optics*, Vol. 24, No. 21, 1985, pp. 3557–3564.

[8]   Schwiesow, R.L., Köpp, F., and Werner, Ch., „Comparison of cw-lidar-measured wind values by full conical scan, conical sector scan and two-point techniques," *Journal of Atmospheric and Oceanic Technology*, Vol. 2, No. 1, 1985, pp. 3–14.

[9]   Doviak, R.J., and Zrnic, D.S., *Doppler Radar and Weather Observations*, Academic Press, San Diego, 1984, p. 458.

[10] Köpp, F., et al., "Laser Doppler wind measurements in the planetary boundary layer," *Contributions to Atmospheric Physics*, Vol. 67, No. 4, 1994, pp. 269–286.

[11] Banakh, V.A., et al., "Representativeness of wind measurements with a cw Doppler lidar in the atmospheric boundary layer," *Applied Optics*, Vol. 34, No. 12, 1995, pp. 2055–2067.

[12] Lumley, J.L., and Panofsky, H.A., *The Structure of Atmospheric Turbulence*, Interscience Publishers, New York, 1964.

[13] Werner, Ch., et al., *WIND: An Advanced Wind Infrared Doppler Lidar for Mesoscale Meteorological Studies*, Phase O/A-Study, 1989, p. 211.

[14] Werner, Ch., Köpp, F., and Schwiesow, R.L., "Influence of clouds and fog on LDA wind measurements," *Applied Optics*, Vol. 23, 1984, pp. 2482–2487.

[15] Banakh, V.A., et al., "Effect of dynamic turbulence of the atmospheric boundary layer on the accuracy of Doppler lidar wind-velocity measurements," *Atmos. Oceanic Opt.*, Vol. 6, No. 11, 1993, pp. 786–793.

[16] Banakh V.A., et al., "Modeling of wind reconstruction from measurements with a spaceborne coherent Doppler lidar," *Atmos. Oceanic Opt.*, Vol. 14. No. 10, 2001, pp. 848–855.

[17] Banakh, V.A., et al., "Accuracy of estimates by the method of variational spectrum accumulation of the wind velocity in turbulent atmosphere from lidar data," *Atmos. Oceanic Opt.*, Vol. 16, No. 8, 2003, pp. 658–662.

[18] Smalikho, I.N., "Techniques of wind vector estimation from data measured with a scanning coherent Doppler lidar," *Journal of Atmospheric and Oceanic Technology*, Vol. 20, No. 2, 2003, pp. 276–291.

[19] Banakh, V.A., et al., "Simulation of retrieval of wind height profile from data of scanning spaceborne coherent Doppler lidar," *Proc. 11th Coherent Laser Radar Conference*, Malvern, Worcestershire, UK, July 2001, pp. 216–219.

[20] Banakh, V.A., et al., "Method for reconstruction of the height wind profiles from the data of space Doppler lidar," *Digest of the VIII Joint International Symposium on Atmospheric and Ocean Optics, Atmospheric Physics*, Irkutsk, Russia, 25–29 June 2001, pp. 168.

[21]   Banakh, V.A., et al., "Dependence of the accuracy of Doppler lidar spectra accumulation method on wind turbulence," *SPIE Proc. 9th International Symposium on Remote Sensing of Clouds and the Atmosphere VII*, Crete, Greece, 24–27 September 2002, Vol. 4882, pp. 211–217.

[22]   Banakh, V.A., et al., "Effect of turbulence on accuracy of wind retrieval from Doppler lidar data by spectra accumulation method," *SPIE Proc. of IX Joint International Symposium on Atmospheric and Ocean Optics. Atmospheric Physics*, Tomsk, Russia, 2–5 July 2002, Vol. 5027, pp. 98–105.

[23]   Banakh, V.A., et al., "Computer simulation of cw Doppler wind lidar operation in the turbulent atmosphere," *Atmos. Oceanic Opt.*, Vol. 12, No. 10, 1999, pp. 905–911.

[24]   Werner, Ch., et al., "Update on the Results of DLR/DWD/IAO Impact Study," presented at NOAA Working Group Meeting on Space-Based Lidar Winds, Key West, FL, USA, 19–22 January1999.

[25]   Leike, I., et al., "Virtual Doppler lidar instrument," *Journal of Atmospheric and Oceanic Technolog*, Vol. 18, No. 9, 2001, pp. 1447–1456.

[26]   Smalikho, I.N., et al., "Laser remote sensing of the mean wind," *Atmos. Oceanic Opt.*, Vol. 15, No. 8, 2002, pp. 607–614.

[27]   Banakh, V.A., et al., "Wind velocity and direction measurement with a coherent Doppler lidar under conditions of weak echo signal," *Atmos. Oceanic Opt.*, Vol. 23, No. 5, 2010, pp. 333–340.

[28]   Banakh, V.A., et al., "Representativity of the wind measurements by a laser Doppler anemometer (LDA) in the boundary layer of the atmosphere," *SPIE Proc. Atmospheric Propagation and Remote Sensing II*, Orlando, FL, USA, 14–15 April 1993, Vol. 1968. pp. 483–493.

[29]   Banakh, V.A., and Smalikho, I.N., "Laser Doppler wind measurements in the planetary boundary layer," in *Doppler Lidar*, DLR Institute of Optoelectronics, Oberpfaffenhofen. 1994, pp. 63–75.

[30]   Banakh, V.A., et al., "Laser Doppler wind sensors in atmospheric boundary layer," *SPIE Proc. Lidar and Atmospheric Sensing*, Munich, Germany, 19 June 1995, Vol. 2505, pp. 142–151.

[31]   Banakh, V.A., et al., "Measurements of wind velocity by cw coherent lidar using sector conical scanning," *SPIE Proc. Lidar Techniques for Remote Sensing II*, Paris, France, 25–26 September 1995, Vol. 2581, pp. 234–242.

[32]   Werner, Ch., et al., "Wind profiler for the atmospheric boundary layer," *Proc. 8th Coherent Laser Radar Conference*, Keystone, CO, USA, July 1995, pp. 5–8.

[33]   Banakh, V.A., et al., "Conical sector scan in wind velocity measurements by cw Doppler lidar," *Proc. 8th Coherent Laser Radar Conference*, Keystone, CO, USA, July 1995, pp. 120–123.

[34]   Banakh, V.A., I et al., *Laser Remote Sensing of the Mean Wind in the Atmospheric Boundary Layer*, Deutsche Forschungsanstalt fur Luft and Raumfart e.v. Forschungsbericht 95-13, Köln, 1995, p. 80.

[35]   Werner, Ch., et al., "Spaceborne Doppler lidar perspectives," *SPIE Proc. Fifth International Symposium on Atmospheric and Ocean Optics*, Tomsk, Russia, 15–18 June 1998, Vol. 3583, pp. 350–359.

[36]   Werner, Ch., et al., "ALADIN Impact study," *SPIE Proc. Laser Radars Techniques (Ranging and Atmospheric Lidar) II*, Barcelona, Spain, 21–24 September 1998, Vol. 3494, pp. 259–269.

[37]   Werner, Ch., et al., "ALADIN impact study," *Proc. 10th Coherent Laser Radar Conference*, Mount Hood, OR, USA, 28 June–2 July 1999, pp. 294–297.

[38]   Werner, Ch., et al., "Virtual instrument for wind and atmospheric turbulence," *SPIE Proc. Atmospheric Propagation, Adaptive Systems, and Laser Radar Technology*, Barcelona, Spain, 25–28 September 2000, Vol. 4167, pp. 259–269.

[39] Banakh, V.A., Smalikho, I.N., and Werner, Ch., "Aerosol particle microstructure dependence of accuracy of cw Doppler estimate of wind velocity," *SPIE Proc. Atmospheric Propagation, Adaptive Systems, and Laser Radar Technology for Remote Sensing*, Barcelona, Spain, 25–28 September 2000, Vol. 4167, pp. 270–280.

[40] Monin A.S., and Yaglom, A.M., *Statistical Fluid Mechanics, Volume II: Mechanics of Turbulence*, MIT Press, Cambridge, MA, 1971.

[41] Byzova, N.L., Ivanov, V.N., and Garger, E.K., *Turbulence in Atmospheric Boundary Layer*, Gidrometeoizdat, Leningrad, 1989, p. 263.

[42] Tatarskii, V.I., *Wave Propagation in a Turbulent Medium*, McGraw-Hill, New York, 1961.

[43] Zilitinkevich, R.R., *Dynamics of Atmospheric Boundary Layer*, Gidrometeoizdat, Leningrad, 1970, p. 292.

[44] Laikhtman, D.L., *Physics of Atmospheric Boundary Layer*, Gidrometeoizdat, Leningrad, 1970, p. 342.

[45] Monin, A.S., and Yaglom, A.M., *Statistical Fluid Mechanics*, Vol. 2, MIT Press, Cambridge, MA, 1975.

[46] Hogstrom, U., "Non-dimensional wind and temperature profiles in the atmospheric surface layer—A re-evaluation," *Boundary Layer Meteorology*, Vol. 42, 1988, pp. 55–78.

[47] Nieuwstadt, F.T.M., and Van Dop, H., *Atmospheric Turbulence and Air Pollution*, course held at the Hague, The Netherlands, 21–25 September, 1981, p. 351.

[48] Vager, B.G., and Nadezhina, E.D., *Atmospheric Boundary Layer under Conditions of Horizontal Inhomogeneity*, Gidrometeoizdat, Leningrad, 1979, p. 136.

[49] Blackadar, A.K., "The vertical distribution of wind and turbulent exchange in a neutral atmosphere," *Journal of Geophysical Research*, Vol. 67, No. 8, 1962, pp. 3095–3102.

[50] Werner, Ch., et al., "Intercomparison of laser Doppler wind measurements with other methods and forecast model," *Journal of Optics A: Pure and Applied Optics*, Vol. 7, No. 12, 1998, pp. 1473–1487.

[51] Frehlich, R.G., and Yadlowsky, M.J., "Performance of mean-frequency estimators for Doppler radar and lidar," *Journal of Atmospheric and Oceanic Technology*, Vol. 11, No. 5, 1994, pp. 1217–1230.

[52] Van Trees, H.L., *Detection, Estimation, and Modulation Theory*, Wiley, New York, 1968.

[53] Ray, B.J., and Hardesty, R.M., "Discrete spectral peak estimation in incoherent backscatter heterodyne lidar. I: Spectral accumulation and Cramer-Rao lower bound," *IEEE Trans. on Geoscience and Remote Sensing*, Vol. 31, No. 1, 1993, pp. 16–27.

[54] Kelley, N., et al., *Lamar Low-Level Jet Program—Interim report*, NREL Report TP-500-34593, National Renewable Energy Laboratory, Golden, CO, 2004, p. 216.

[55] Pichugina, Y.L., et al., "Nocturnal boundary layer height estimate from Doppler lidar measurements," *Proc. 18th Symposium on Boundary Layer and Turbulence*, Stockholm, Sweden, June 2008, 7B.6.

[56] Grund, C.J., et al., "High-resolution Doppler lidar for boundary layer and cloud research," *Journal of Atmospheric and Oceanic Technology*, Vol. 18, No. 3, 2001, pp. 376–393.

[57] Rahm, S., Simmet, R., and Wirth, M., "Airborne two micron coherent lidar wind profiles," *Proc. 12th Coherent Laser Radar Conference*, Bar Harbor, ME, USA, 15–20 June 2003, pp. 94–97.

[58] Weissmann, M., et al., "Targeted observations with an airborne wind lidar," *Journal of Atmospheric and Oceanic Technology*, Vol. 22, No. 10, 2005, pp. 1706–1719.

[59] Huffaker, M.R., et al., "Feasibility studies for global wind measuring satellite system (Windsat): Analysis of simulated performance," *Applied Optics*, Vol. 23, 1984, pp. 2523–2536.

[60] Baker, W.F., et al., "Lidar-measured wind from space: A key component for weather and climate prediction," *Bull. Amer. Meteorol. Soc.*, Vol. 76, No. 6, 1995. pp. 869–888.

[61] Hardesty, R.M., et al., "A potential NPOESS winds mission," *Proc. 13th Coherent Laser Radar Conference*, 2005, pp. 45–48.

[62] Menzies, R.T., "Doppler lidar atmospheric wind sensors: A comparative performance evaluation for global measurement applications from earth orbit," *Applied Optics*, Vol. 25, 1986, pp. 2546–2553.

[63] Petheram, J.C., Frohbeiter, G., and Rosenberg, A., "Carbon dioxide Doppler lidar wind sensor on a space station polar platform," *Applied Optics*, Vol. 28, 1989, pp. 834–839.

[64] Endemann, M., and Ingmann, P., "The European Spaceborne Doppler Wind Lidar ADM-Aeolus," *Proc. 13th Coherent Laser Radar Conference*, 2005, pp. 49–52.

[65] Leike, I., Streicher, J., and Werner, Ch., "ALIENS: Atmospheric lidar end-to-end simulator," *SPIE Proc. Atmospheric and Ocean Optics*, Vol. 3583, 1998, pp. 380–386.

[66] Levin, J.M., "Power spectrum parameter estimation," *IEEE Trans. on Information and Theory*, Vol. IT-11, 1965, pp. 100–107.

[67] Banakh, V.A., and Werner, Ch., "Computer Simulation of Coherent Doppler Lidar Measurement of Wind Velocity and Retrieval of Turbulent Wind Statistics," *Optical Engineering*, Vol. 44, No. 7, 2005, pp. 1–19.

CHAPTER 4

Estimation of Atmospheric Turbulence Parameters from Wind Measurements with Coherent Doppler Lidars

4.1 Introduction

The analytical equations for determining the statistical characteristics of lidar estimates of the radial velocity and the Doppler spectrum width presented in Chapter 2, as well as the results of investigation of their features depending on atmospheric factors, form the theoretical basis for methods of estimations of the turbulence parameters in the atmosphere from wind measurements by both cw and pulsed coherent Doppler lidars.

The vertical profiles of the variance of wind velocity and the momentum flux retrieved from wind data measured by pulsed CO_2 CDLs are reported in [1, 2]. In [3], the raw data measurements from 2-μm pulsed CDLs were used to obtain the vertical profiles for the variance of wind velocity. In these works the variance of wind velocity was evaluated from lidar estimates of the radial velocity without regard for the spatial averaging over the sensing volume. As is shown in Chapter 2, neglecting to average the turbulent fluctuations of velocities of aerosol particles in the sensing volume underestimates the result, especially when the longitudinal dimension of the sensing volume exceeds the outer scale of turbulence. To correctly determine the variance of wind velocity, in addition to the variance of lidar estimates of the radial velocity, it is necessary to calculate the mathematical expectation of the squared width of the Doppler spectrum estimated from the same raw lidar data (see Sections 2.3 and 2.6).

The dissipation rate of the kinetic energy of turbulence ε is one of the key parameters of the turbulent wind fields in the atmosphere. It characterizes the intensity of turbulent processes and determines the rapidity of atmospheric exchange phenomena, diffusion of atmospheric pollutants, aircraft wake vortex evolution, and so on. A large part of the investigations into the possibility of measuring wind turbulence with coherent Doppler lidars is devoted just to the problem of measurement of the dissipation rate.

References [4] and [5] proposed that the method for estimation of the dissipation rate ε be based on determination of the width of a cw CDL's Doppler spectrum. The applicability of this method is restricted by the condition of small longitudinal sizes of the lidar sensing volume, which must be much smaller than the outer scale of turbulence. With an increase in the measurement range, the sensing volume formed by a cw CDL increases rapidly (see Section 1.3). Therefore, as shown by theoretical calculations and field experiments [6, 7], the applicability of the method of estimation of the dissipation rate ε from the Doppler spectrum width for ground-based (cw CO_2) lidars is limited to the heights of the atmospheric surface layer (less than 100m). To determine ε at higher altitudes, in addition to knowing the width of the Doppler spectrum we must also know the outer scale of turbulence L_V. The method for estimating ε taking L_V into account is considered in [8]. This method requires the estimation be made using the raw lidar data of not only the Doppler spectrum width, but also the wind velocity variance (see Section 4.2).

Reference [9] proposed to acquire information about dissipation rate ε from the temporal structure function or the temporal spectrum of fluctuations of the radial velocity measured by a cw CDL, calculated for time intervals (frequencies) corresponding to the inertial subrange of the turbulence spectrum. In contrast to a point sensor (for example, a cup or sonic anemometer), here it is necessary to take into account the averaging of the velocity measured by the lidar over the sensing volume (see Section 2.3). It follows from theoretical analysis [9] that this approach allows one to estimate ε for arbitrary dimensions of the sensing volume Δz, in particular, at $\Delta_z > L_V$. However, as follows from field experiments [10–12], this is not always the case. As mentioned earlier in Section 2.3, possible reasons for the discrepancy between the theoretical predictions and the experimental data may be the inapplicability of the Taylor hypothesis of frozen turbulence for calculation of the temporal structure function and temporal spectrum of radial velocity measured by lidar at large sensing volumes or the influence of refractive turbulence on the temporal variations of the lidar estimates of radial velocity (see Section 2.5).

To avoid this drawback in the method of estimating ε from the temporal statistics of radial velocity, researchers proposed [13] to estimate the dissipation rate based on the transverse spatial structure function of radial velocity measured by a cw CDL using conical scanning by the probing beam. This approach is considered in Section 4.3. Later, the idea of ε determination from data from a conically scanning lidar was applied by R. Frehlich to pulsed CDLs [14]. The issue of the accuracy of this method was touched on in [15, 16] (Section 4.7).

The problem of estimation of the dissipation rate ε from the longitudinal spatial structure function of the radial wind velocity measured by a pulsed CDL was considered in [17, 18]. In contrast to earlier attempts to estimate the dissipation rate (see, for example, [2, 19]) through the fitting of the "two-thirds dependence" of the structure function on the distance between observation points (Kolmogorov law) to the lidar-measured structure function, in [17, 18] the dissipation rate is estimated from the structure function calculated with regard to the spatial averaging of the radial velocity, measured by lidar, over the sensing volume, which causes a significant deviation in the structure function from the "two-thirds" Kolmogorov law.

For the measurement of aircraft wake vortex parameters by a coherent Doppler lidar, scanning by the probing beam in the vertical plane across the wake is used [20–26]. The pulsed CDL data obtained in this case carry the information about wake vortices and about wind turbulence in the vicinity of wake vortices. This allows us to study the effect of atmospheric turbulence on wake vortex evolution, using only one lidar [25]. The possibility of estimating wind turbulence parameters from data measured by pulsed CDLs conducting vertical scanning were studied for the first time in [8, 27, 28]. The results obtained are reported in Sections 4.5 and 4.6.

In this chapter, we consider various methods for estimating wind turbulence parameters (variance of wind velocity, outer scale of turbulence, and rate of dissipation of turbulent energy) from data measured by cw and pulsed CDLs. The conditions in which these methods are applicable are also determined. The results of retrieval of vertical profiles of turbulence parameters from lidar data are presented. The data from simultaneous measurements of the turbulence energy dissipation rate by lidars and sonic anemometers are comparatively analyzed. Detecting clear air turbulence zones using an airborne coherent Doppler lidar is discussed.

The chapter is based on the results reported in [6–11, 13, 15–17, 27–46].

4.2   Estimation of Wind Turbulence Parameters from Doppler Spectrum Width and Temporal Statistics of Radial Velocity Measured with Continuous-Wave CDLs

Assume that the measurements by cw CDLs (without scanning by the probing beam) for the time T yield an array of Doppler spectra $\hat{S}(f_k;t_i)$, where $t_i = t_0 + i\Delta t$, $i = 0, 1, 2, ..., M - 1$, Δt is the time for measurement of one spectrum, and $M\Delta t = T$. Using (2.47) through (2.49), we obtain estimates of the radial velocity $\hat{V}_r(t_i)$ and the Doppler spectrum width $\hat{\sigma}_S(t_i)$. Then we calculate the following characteristics:

$$\langle \hat{V}_r \rangle_E = M^{-1} \sum_{i=0}^{M-1} \hat{V}_r(t_i),\ \langle \hat{V}_r^2 \rangle_E = M^{-1} \sum_{i=0}^{M-1} \hat{V}_r^2(t_i),\ \langle \hat{\sigma}_S^2 \rangle_E = M^{-1} \sum_{i=0}^{M-1} \hat{\sigma}_S^2(t_i)$$

and

$$\hat{\sigma}_e^2 = 0.5(M - 1)^{-1} \sum_{i=0}^{M-2} [\hat{V}_r(t_{i+1}) - \hat{V}_r(t_i)]^2$$

In the case of an unbiased estimate of the radial velocity and under stationary atmospheric conditions at $M \to \infty$, we have $\langle \hat{V}_r \rangle_E \to \langle V_r \rangle$, $(\langle \hat{V}_r^2 \rangle_E - \langle \hat{V}_r \rangle_E^2) \to \sigma_{\hat{V}}^2$, $\langle \hat{\sigma}_S^2 \rangle_E \to \langle \hat{\sigma}_S^2 \rangle$, and $\hat{\sigma}_e^2 \to \sigma_e^2 + D_{\bar{V}}(\Delta t)/2$ (see Section 2.3). At $\Delta t = 50$ ms, the condition $\sigma_e^2 \gg D_{\bar{V}}(\Delta t)/2$ is usually true. Therefore, $\hat{\sigma}_e^2$ is practically an unbiased estimate of the variance σ_e^2. The estimate $\hat{\sigma}_e^2$ can also be obtained from measurements of the spectrum of wind velocity (see discussion later in this section).

According to (2.79), even at the statistical homogeneity of the wind flow (when $\hat{\sigma}_{\langle V \rangle}^2 = 0$) it is quite problematic to retrieve the information about wind turbulence from $\langle \hat{\sigma}_S^2 \rangle_E$, because $\langle \hat{\sigma}_S^2 \rangle_E$ includes also a contribution from the error $\langle E_\sigma \rangle$, the method of whose determination from experimental data is unknown. The random error E_σ is mostly caused by noise of the lidar signal (even at high SNR). With an increase in the order of accumulation of Doppler spectra L, the variance of fluctuations of the noise component of the Doppler spectrum estimate decreases, and as a result the error $\langle E_\sigma \rangle$ decreases [8]. To decrease $\langle E_\sigma \rangle$ (to reduce it to nearly zero), the following procedure was proposed in [7].

First, the averaged normalized spectrum (having the dimension of the probability density of velocity) centered about the first spectral moment of velocity is calculated:

$$S_D(V_k') = \frac{1}{\Delta V} \cdot \frac{\tilde{S}(V_k')}{\displaystyle\sum_{k=k_1}^{k_2-1} \tilde{S}(V_k')} \tag{4.1}$$

where

$$\tilde{S}(V'_k) = \frac{1}{M} \sum_{i=1}^{M} \hat{S}(f_{ki}, t_i) - 1 \tag{4.2}$$

$$f_{ki} = (2/\lambda)\Delta V \left[\frac{V'_k + \hat{V}_r(t_i)}{\Delta V} \right] \tag{4.3}$$

Note that the operator [...] denotes the rounding to the nearest integer, $V'_k = \Delta V k$, and $k = 0, \pm1, \pm2, \dots$. Then, $\langle \hat{\sigma}_S^2 \rangle_E$ is calculated according to (2.49) and (4.1) as

$$\langle \hat{\sigma}_S^2 \rangle_E = \Delta V \sum_{k=k_1}^{k_2-1} (V'_k)^2 S_D(V'_k) \tag{4.4}$$

One can show [47] that at a sufficiently large number M (when $\langle \hat{\sigma}_S^2 \rangle_E \approx \langle \hat{\sigma}_S^2 \rangle$) and for statistical homogeneity of the wind flow ($\sigma_{\langle V \rangle}^2 = 0$), the use of the procedure in (4.1) through (4.4) does not yield (2.79) for $\langle \hat{\sigma}_S^2 \rangle_E$, but

$$\langle \hat{\sigma}_S^2 \rangle_E = \sigma_t^2 + \sigma_{VI}^2 + \sigma_e^2 + (\Delta V)^2/12 \tag{4.5}$$

Figure 4.1 shows the data for a field experiment [7] for the averaged normalized Doppler spectra $S_D(V')$ obtained with the use of (4.1) through (4.3) for the case where $\Delta z \ll L_V$ (curves 1 and 2) and $\Delta z \sim L_V$ (curve 3). The dashed curves demonstrate the results of calculation of the Gaussian distributions $(\sqrt{2\pi}\sigma_s)^{-1} \exp[-V'^2/(2\hat{\sigma}_S^2)]$, where $\sigma_S^2 \equiv \langle \hat{\sigma}_S^2 \rangle_E$ were calculated from the spectra $S_D(V')$ by (4.5). One can see a significant deviation of $S_D(V')$ from the Gaussian distribution at $\Delta z \ll L_V$, which is in agreement with the theoretical results obtained in [7].

Under the condition $\Delta z \ll L_V$, the turbulent broadening of the Doppler spectrum σ_t^2 is described by (2.69). Then, with regard to (4.5), the estimate of the turbulent energy dissipation rate can be obtained as

$$\hat{\varepsilon} = \frac{(\hat{\sigma}_t^2)^{3/2}}{C_K^{3/2}(2/\pi)\Delta z} \tag{4.6}$$

where

$$\hat{\sigma}_t^2 = \langle \hat{\sigma}_S^2 \rangle_E - [\sigma_{VI}^2 + \hat{\sigma}_e^2 + (\Delta V)^2/12] \tag{4.7}$$

is the estimate of turbulent broadening of the Doppler spectrum.

The turbulent energy dissipation rates estimated from the raw data of simultaneous measurements by the sonic anemometer and cw CO_2 CDLs are compared in [10]. The dissipation rate ε was estimated from the temporal structure function of

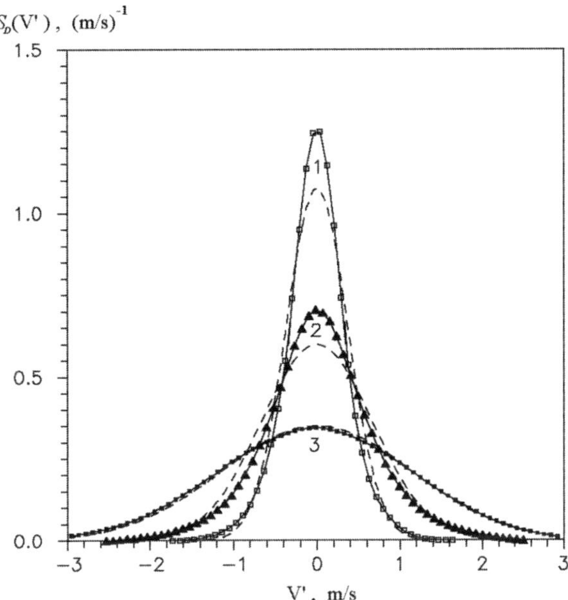

Figure 4.1   Normalized power spectra of lidar echo signal (Doppler spectra) measured at weak turbulence with $\Delta z = 2.3$m (curve 1) and strong turbulence with $\Delta z = 4.5$m $\ll L_V$ (curve 2) and $\Delta z = 100$m $\sim L_V$ (curve 3). Dashed curves are the calculated corresponding Gauss distributions. (© 1999 American Meteorological Society. Used with permission. From [7].)

wind velocity measured by the sonic anemometer and from the Doppler spectrum width. In the estimation, the terms in the square brackets in (4.7) were set equal to zero. The analysis shows that for this experiment the terms σ_e^2 and $(\Delta V)^2/12$ (velocity resolution $\Delta V = 0.0848$ m/s) actually can be neglected. However, consideration of the instrumental broadening $\sigma_{VI}^2 = 0.0015$ (m/s)^2 has allowed some correction of the lidar data.

Table 4.1 summarizes the results of estimation of the turbulent energy dissipation rate from the data of simultaneous measurements by cw CDL at $\Delta z = 2.3$ m (ε was estimated by (4.6)) and sonic anemometer installed at a height of 7 m [10]. One can see a good agreement between ε values estimated from the sonic anemometer and lidar data.

If the condition $\Delta z \ll L_V$ is not true, then the value $\hat{\varepsilon}$ obtained with (4.6) is understated. The outer scale of turbulence can be taken into account through the use of some model spectrum of turbulent fluctuations of the wind velocity, for example, the von Karman model [see (2.69) through (2.71)]. In this case, in addition to σ_t^2, lidar measurements of the variance of wind velocity σ_V^2 are necessary. With (2.56),

Table 4.1   Estimation of the Turbulent Energy Dissipation Rate

	Local time (hour:min)		
Estimate of (m^2s^{-3})	*14:51*	*15:25*	*15:29*
From lidar data	$5 \cdot 10^{-3}$	$5.5 \cdot 10^{-3}$	$7 \cdot 10^{-3}$
From sonic anemometer data	$4.7 \cdot 10^{-3}$	$5.8 \cdot 10^{-3}$	$7.4 \cdot 10^{-3}$

(2.61), (2.63), (2.70), (2.71), and (2.78), the dissipation rate can be estimated as follows [8].

With (4.7), we determine the turbulent broadening of the Doppler spectrum $\hat{\sigma}_t^2$. The estimate of the wind velocity $\hat{\sigma}_V^2$ is obtained as

$$\hat{\sigma}_V^2 = \langle \hat{V}_r^2 \rangle_E - \langle \hat{V}_r \rangle_E^2 - \sigma_e^2 + \hat{\sigma}_t^2 \tag{4.8}$$

The outer scale of turbulence is determined from the solution of the following equation:

$$\hat{\sigma}_t^2 / \hat{\sigma}_V^2 = F_K(\hat{L}_V / \Delta z) \tag{4.9}$$

where $F_K(x)$ is a universal function, which in the case of the von Karman model has the form

$$F_K(x) = x \int_0^\infty d\xi \frac{1 - \exp(-\xi)}{[1 + (2.105 x \xi)^2]^{5/6}} \tag{4.10}$$

Finally, the turbulent energy dissipation rate is calculated as

$$\hat{\varepsilon} = (1.972 / C_K^{3/2}) \hat{\sigma}_V^3 / \hat{L}_V \tag{4.11}$$

The algorithm described by (4.7) through (4.11) allows, in principle, the estimation of the turbulent energy dissipation rate at high altitudes in the atmospheric boundary layer from measurements of the radial wind velocity by ground-based cw CDLs, when the longitudinal dimension of the sensing volume Δz becomes larger than the outer scale of turbulence L_V. However, if within the sensing volume the mean radial velocity varies along the optical axis, then the inhomogeneity of the mean wind [according to (2.79), $\sigma_{\langle V \rangle}^2 \neq 0$] contributes additionally to the broadening of the Doppler spectrum. In the case of strong wind shear (or jet flow), this contribution can dominate. Now it seems impossible to take this factor into account in the processing of cw CDL data.

An alternative to the approach described above is to estimate the dissipation rate ε from the temporal spectrum (or structure function) of wind velocity. In [29], the following algorithm was proposed for the estimation of ε.

The one-sided spectral density function is calculated from a series of values of the wind velocity radial component $\hat{V}_r(t_i)$ measured for time $T = \Delta t M$ with the use of the discrete (fast) Fourier transform [48, 49]:

$$\hat{S}_V(f_k) = \frac{2\Delta t}{M} \left| \sum_{i=0}^{M-1} \hat{V}_r(t_i) \exp\left(-2\pi j \frac{ik}{M}\right) \right|^2 \tag{4.12}$$

where $f_k = k/T$, $k = 0, 1, \ldots, M/2$. At sufficiently large T, when $T \gg \tau_V$ (τ_V is the wind velocity correlation time), we take the frequency range $[f_{k1}, f_{k2}]$, within which the estimate $\hat{S}_V(f_k)$ is unbiased [that is, $\langle\hat{S}_V(f_k)\rangle = S_V(f_k)$], and its mean value is determined by (2.94) in the case of point measurements (for example, by sonic anemometer) or by (2.93) in the case of measurement by a Doppler lidar. We introduce the designation: $S_V(f_k) = \varepsilon^{2/3}Q(f_k)$. The frequency $f_{k1} = k_1/T$ is the lower bound of the inertial interval, whereas $f_{k2} = k_2/T$ corresponds to the highest frequency, at which the noise contribution to the measured spectrum can be neglected.

The estimate of the dissipation rate $\hat{\varepsilon}$ can be obtained only within the considered frequency range $[f_{k1}, f_{k2}]$ including $n = (f_{k0} - f_{k2})T$ (where $f_{k0} = f_{k1} - 1/T$) spectral channels. The complex random variable under the modulus sign in (4.12), due to the condition $T \gg \tau_V$, obeys the normal distribution law of probability densities with zero mean and equal variances for the real and imaginary parts proportional to $S_V(f_k)$. Consequently, the probability density of the spectrum estimate $\hat{S}_{k'} = \hat{S}_V(f_{k0} + k'/T)$ has the exponential distribution. In this case, the estimates $\hat{S}_{k'}$ and $\hat{S}_l$ (at $k' \neq l$) are statistically independent. Consequently, the probability density of the vector $\hat{S} = \{\hat{S}_1, \hat{S}_2, ..., \hat{S}_n\}$ has the form

$$p(\hat{S}) = \left(\prod_{k'=1}^{n}\frac{1}{S_{k'}}\right)\exp\left[-\sum_{k'=1}^{n}\frac{\hat{S}_{k'}}{S_{k'}}\right] \tag{4.13}$$

where $S_{k'} = \varepsilon^{2/3}Q_{k'}$ and $Q_{k'} = Q(f_{k0} + k'/T)$. Assuming that the values of $Q_{k'}$ are known, we can find the estimate of the dissipation rate $\hat{\varepsilon}$ from the measured vector $\hat{S}$ by the maximum likelihood [50]. Toward this end, it is necessary to differentiate the log-likelihood function

$$\Phi(\varepsilon) = \ln p(\hat{S}) = -\sum_{k'=1}^{n}\left[\ln(\varepsilon^{2/3}Q_{k'}) + \frac{\hat{S}_{k'}}{\varepsilon^{2/3}Q_{k'}}\right] \tag{4.14}$$

with respect to ε and to solve the equation

$$\left.\frac{d\Phi(\varepsilon)}{d\varepsilon}\right|_{\varepsilon=\hat{\varepsilon}} = 0 \tag{4.15}$$

As a result, we have

$$\hat{\varepsilon} = \left(\frac{i}{n}\sum_{k'=1}^{n}\frac{\hat{S}_{k'}}{Q_{k'}}\right)^{3/2} \tag{4.16}$$

In [29], it is shown that if the condition $n \gg 1$ is true, the estimate of the dissipation rate by (4.16) is unbiased ($\langle\hat{\varepsilon}\rangle = \varepsilon$), and its error can be calculated by the equation

$$E_\varepsilon = \left[\frac{9}{4n} + \alpha^2 2\frac{\sigma_V^2}{U^2}\frac{\tau_V}{T}\right]^{1/2} \times 100\% \tag{4.17}$$

where the second term in square brackets is connected with the accuracy of measurement of the mean wind velocity U. The parameter $\alpha = 1$ in the case of a point sensor and $\alpha = 5/2$ for lidar measurements at large sensing volume and strong lateral wind, when (2.99) is valid. Thus, if the contribution from the measurement error of the mean wind velocity to the estimation error of the dissipation rate is significant, then the value of ε calculated from sonic anemometer data is more accurate than that calculated from lidar data.

It is obvious that the number n (and, consequently, the error E_ε) depends on the level of the noise component of the measured spectrum S_e. The lower the S_e value, the higher the frequency $f_{k2} = k_2/T$ and the larger the number of spectral channels n that can be used for estimation of dissipation rate ε. Taking into account that S_e is white noise, we can obtain its estimate $\hat{S}_e$ from the measured spectrum $\hat{S}_V(f_k)$ by the equation

$$\hat{S}_e = \frac{1}{M/2 + 1 - k_3} \sum_{k=k_3}^{M/2} \hat{S}_V(f_k) \tag{4.18}$$

where the condition $S_{\bar{V}}(f_{k_3}) \ll S_e$ (realizable at $k_3 \gg k_2$) should be true at the frequency f_{k_3}. Then, according to (2.39), we can calculate the variance of error of radial wind velocity estimates as

$$\hat{\sigma}_e^2 = \hat{S}_e/(2\Delta t) \tag{4.19}$$

Figure 4.2 shows as dots the temporal spectrum (Figure 4.2(a)) obtained with the averaging over 24 degrees of freedom and the structure function (Figure 4.2(b)) of the radial wind velocity measured by the Doppler lidar at a height of 10m, focal length of 50m ($\Delta z = 2.3$m), and weak wind [7]. The data of the unsmoothed spectrum [see (4.12)] were used to estimate ε and S_e by (4.16) and (4.18). The obtained estimates were, respectively, $4.2 \cdot 10^{-3}$ m^2/s^3 and $1.6 \cdot 10^{-4}$ m^2/s. According to (4.19), where $\Delta t = 50$ ms, the variance of the random error of measurement of the radial velocity averaged over the sensing volume is $\sigma_e^2 = 1.6 \cdot 10^{-3}$ (m/s)^2. The obtained values of ε, S_e, and σ_e^2, as well as the information about the mean wind velocity and direction, allow us to calculate the structure function $D_{\hat{V}}(\tau)$ and the spectrum $S_{\hat{V}}(f)$ by (2.83), (2.88), (2.100), and (2.93) through (2.95). The calculated results are shown in Figure 4.2 as solid curves. One can see a good agreement between theory and experiment in this case of a small sensing volume ($\Delta z = 2.3$m). The estimate of the variance of error σ_e^2 obtained from the temporal spectrum with the use of (4.19) nearly coincides with $D_{\hat{V}}(\Delta t)/2$. This means that $D_{\hat{V}}(\Delta t) \gg D_{\bar{V}}(\Delta t)$ and, with regard to (2.83), the variance of error σ_e^2 can also be determined from the structure function of the radial wind velocity measured by lidar at a minimal nonzero temporal shift $\tau = \Delta t$ (in our case $\Delta t = 50$ ms).

Field experiments [6, 7, 10, 11] show that the results of estimation of the dissipation rate ε from the temporal spectrum of the wind velocity measured by lidar and from the Doppler spectrum width at a small sensing volume ($\Delta z \sim 2$–10 m) are rather

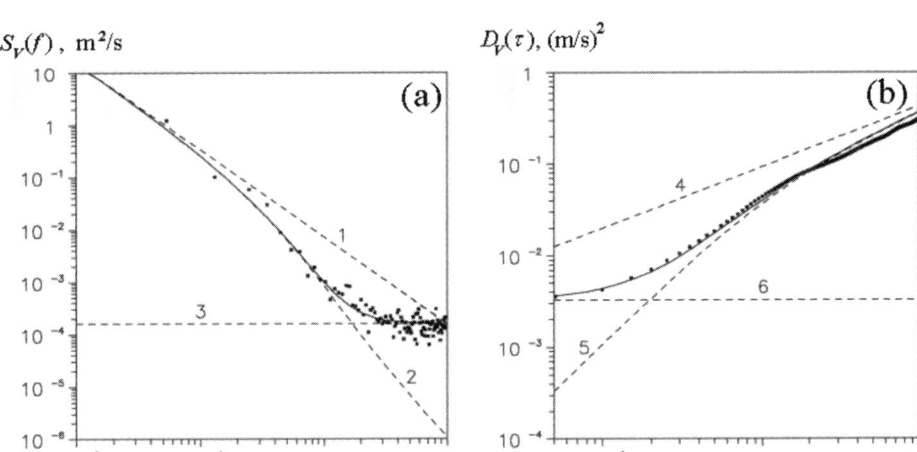

Figure 4.2   (a) Temporal spectrum and (b) structure function of wind velocity measured by lidar. Dots: experiment; solid curves: theoretical calculations of $S_{\hat{V}}(f)$ and $D_{\hat{V}}(\tau)$; dashed curves: (1) S_V $(f) \sim f^{-5/3}$, (2) $S_{\bar{V}}(f) = S_{\hat{V}}(f) - S_e$, (3) S_e, (4) $D_V(\tau) \sim \tau^{2/3}$, (5) $D_{\bar{V}}(\tau) = D_{\hat{V}}(\tau) - 2\sigma_e^2$, and (6) $2\sigma_e^2$. (© 1999 American Meteorological Society. Used with permission. From [7].)

close. Thus, the value of the dissipation rate estimated from the lidar spectrum of wind velocity ($4.2 \cdot 10^{-3}$ m^2/s^3) differs from that in Table 4.1 by no more than 15%.

The dissipation rate ε can be estimated with a required accuracy from the temporal spectrum (structure function) of the wind velocity at large longitudinal dimensions of the sensing volume Δz (in particular, at $\Delta z > L_V$) as well, but, as shown in Sections 2.3 and 2.5, only at the rather strong lateral wind, which is not always observed in practice. However, the conditions corresponding to the strong lateral wind can be established artificially by means of fast scanning being done by the probing beam.

4.3   Determination of the Turbulent Energy Dissipation Rate from the Transverse Spatial Structure Function of Radial Velocity Measured by Conically Scanning Continuous-Wave CDLs

In [13], a method was proposed for the estimation of the dissipation rate ε from data measured by a cw Doppler lidar with conical scanning being done by its probing beam. This approach is free of the requirement that the Taylor hypothesis of frozen turbulence (transfer of turbulent vortices by the mean wind with the vortex shape unchanged) be applicable, since even at a very weak lateral wind the scanning done by the probing beam at a sufficiently high rate provides the conditions of frozen turbulence.

Section 2.5 showed that turbulent fluctuations of the air refractive index can affect significantly the temporal spectrum (and, consequently, the structure function) of fluctuations of the wind velocity measured by cw CDLs at frequencies corresponding to the inertial interval of the turbulence spectrum. The effect is stronger for a weak wind. The smaller the lateral wind velocity, the larger the difference between the measured spectrum and the calculations of (2.93) through (2.95). Numerical experiments show that if the correlation time of random displacements of the sensing volume along the beam axis τ_z equal to the ratio of the initial radius of the probing beam a_0 to the lateral wind velocity $U = |V_\perp|$ (see Section 2.5) is much shorter than the time for measurement of the Doppler spectrum Δt, then the lidar spectrum of wind velocity is close in shape to the total spectrum, being a sum of the spectrum of velocity averaged over the sensing volume $S_{\bar{V}}(f)$ and the noise spectrum S_e. Figure 2.13 illustrates this (see, for example, curves 3 and 6 in Figure 2.13(a)).

According to (2.35), (2.107), and (2.108), we present the unbiased lidar estimate of the radial wind velocity $\hat{V}_r(t_i)(t_i = t_0 + i\Delta t, i = 0,1,2,...)$ in the form

$$\hat{V}_r(t_i) = \langle V_r \rangle + \int_0^\infty dz \langle Q_s(z) \rangle V_r'(z,t_i) + \int_0^\infty dz Q_s'(z,t_i) V_r'(z,t_i) + V_e(t_i) \qquad (4.20)$$

where $V_r' = V_r - \langle V_r \rangle$, $\langle Q_s(z) \rangle \equiv \langle Q_s(z,t_i) \rangle$, and $Q_s'(z,t_i) = Q_s(z,t_i) - \langle Q_s(z) \rangle$. The random variables V_r', Q_s', and V_e are statistically independent with zero means. With regard to this, from (4.20) we have for the structure function $D_{\hat{V}}(\tau) = \langle [\hat{V}_r'(t_i + \tau) - \hat{V}_r'(t_i)]^2 \rangle$ $(\tau = l\Delta t, l = 1, 2, 3, ...)$:

$$D_{\hat{V}}(\tau) = D_{\bar{V}}(\tau) + 2[K_n(0) - K_n(\tau)] + 2\sigma_e^2 \qquad (4.21)$$

where $D_{\bar{V}}(\tau)$ is the structure function of the radial velocity averaged over the sensing volume, and

$$K_n(\tau) = \int_0^\infty dz_1 \int_0^\infty dz_2 \langle Q_s'(z_1,t_i + \tau)Q_s'(z_2,t_i) \rangle \langle V_r'(z_1,t_i + \tau)V_r'(z_2,t_i) \rangle \qquad (4.22)$$

is the correlation function of fluctuations of the lidar estimate of radial velocity caused by refractive turbulence.

The scanning being done by the probing beam allows us to fulfill the condition $\tau_z \ll \Delta t$. Thus, we can take $K_n(\tau) = 0$ in (4.21), and as a result

$$D_{\hat{V}}(\tau) = D_{\bar{V}}(\tau) + 2\sigma_n^2 \qquad (4.23)$$

where the following designation is introduced:

$$\sigma_n^2 = K_n(0) + \sigma_e^2 \qquad (4.24)$$

From the condition $\tau_z \ll \Delta t$ it also follows that during the time of measurement of one Doppler spectrum Δt the sensing volume can increase its "effective" longitudinal size due to random displacements of the sensing volume along the optical axis caused by refractive turbulence. At the same time, the results shown in Figures 2.11(a) through 2.11(c) demonstrate that for typical atmospheric refractive turbulence strength this increase in the longitudinal size of the sensing volume is insignificant. Therefore, we believe that under the condition $\tau_z \ll \Delta t$ the function $D_{\tilde{V}}(\tau)$ is also determined by (2.84), (2.85), (2.62), (2.55), and (1.57) where Δz is used as the longitudinal size of the sensing volume.

Figure 3.1 shows schematically the geometry of measurements by a ground-based cw CDL at the conical scanning location. The probing beam focused at a distance F from the lidar is tilted at an angle φ to the horizontal plane and rotates with angular speed ω_0 around vertical axis z. During the scanning, Doppler spectra are measured for equal time intervals $\Delta t = \Delta\theta/\omega_0$, where $\Delta\theta$ is the azimuth angle resolution. With regard to (3.2), (2.18), and (4.20), we can represent the estimate of the radial velocity in the form

$$\hat{V}_r(\theta) = \int_0^\infty dz\, Q_s(z) S(\theta) \cdot V(zS(\theta)) + V_n(\theta) \tag{4.25}$$

where $\theta = \theta_0 + i\Delta\theta$, $S(\theta) = \{\sin\varphi,\, \cos\varphi\cos\theta,\, \cos\varphi\sin\theta\}$, $V = \{V_z, V_x, V_y\}$ is the vector of instantaneous wind velocity at point $r = zS(\theta)$, and

$$V_n(\theta) = \int_0^\infty dz\, Q_s'(z,\theta/\omega_0) V_s'(z,\theta/\omega_0) + V_r(\theta/\omega_0) \tag{4.26}$$

is the noise component of the estimate with $\langle V_n \rangle = 0$ and $\langle V_n(\theta + i'\Delta\theta)V_n(\theta)\rangle = \sigma_n^2 \delta_{i'}$.

According to (3.5) through (3.7), the estimate of the mean wind velocity $\hat{V}$ can be written in the form

$$\hat{V} = \frac{1}{2\pi} \int_0^{2\pi} d\theta\, \hat{V}_r(\theta) A(\theta) \tag{4.27}$$

where $A(\theta) = \{1/\sin\varphi,\, 2\cos\theta/\cos\varphi,\, 2\sin\theta/\cos\varphi\}$. The estimate of the mean radial wind velocity can be represented as

$$\langle \hat{V}_r(\theta)\rangle = S(\theta) \cdot \hat{V} \tag{4.28}$$

We assume that the scanning rate is high enough to neglect variations of the wind velocity during the period of azimuth scanning between the angles θ_1 and θ_2. Then, for the analysis of the spatial structure of fluctuations of the radial wind velocity, we can use the transverse structure function of the velocity measured by lidar

$$D_V(\theta_1,\theta_2) = \langle [\hat{V}_r'(\theta_1) - \hat{V}_r'(\theta_2)]^2 \rangle \tag{4.29}$$

where $\hat{V}_r'(\theta) = \hat{V}_r(\theta) - \langle \hat{V}_r(\theta)\rangle$.

Substituting (4.25), (4.27), and (4.28) into (4.29) and averaging over the ensemble of random realizations, for the conditions $L_V \ll R$ and $|\theta_1 - \theta_2| \ll \pi/2$, when the spatial structure of velocity difference is determined by the two-thirds Kolmogorov law, we obtain [13]

$$D_V(\theta_1 - \theta_2) = C_K \varepsilon^{2/3} \int_0^\infty dz_1 Q_s(z_1) \int_0^\infty dz_2 Q_s(z_2)$$

$$\times \left\{ \left(1 + \frac{1}{3}\frac{B_2^2}{B_1^2 + B_2^2}\right)(B_1^2 + B_2^2)^{1/3} - B_1^{2/3} \right\} + 2\sigma_n^2 \qquad (4.30)$$

where $Q_s(z)$ is given by (2.55), $B_1 = |z_1 - z_2|$, $B_2^2 = (\theta_1 - \theta_2)^2 z_1 z_2 \cos^2\varphi$, and the angles $\theta_1 \neq \theta_2$ are specified in radians. In the particular case of small dimensions of the sensing volume, when $\Delta z \to 0$, (4.30) reduces to the well-known equation for the transverse structure function of the wind velocity in the inertial interval of turbulent inhomogeneities of the wind [51, 52]:

$$D_V(\theta_1 - \theta_2) = \frac{4}{3} C_K \varepsilon^{2/3} \left(|\theta_1 - \theta_2| F \cos\varphi\right)^{2/3} + 2\sigma_n^2 \qquad (4.31)$$

Equation (4.31) is applicable under the condition that the sector arc length in the base of the scanning cone $|\theta_1 - \theta_2| F \cos\varphi$ does not exceed the low-frequency bound of the inertial interval:

$$|\theta_1 - \theta_2| F \cos\varphi \ll L_V \qquad (4.32)$$

Thus, the structure function $D_V(\theta_1 - \theta_2)$ calculated from lidar data (measured at the conical scanning) can be used to estimate the turbulent energy dissipation rate ε with the use of (4.30). In this case, the only restriction is condition (4.32) without a limitation on the longitudinal size of sensing volume Δz, which can be arbitrary, in particular, $\Delta z \gg L_V$.

Figure 4.3 exemplifies the estimates $\hat{V}_r(\theta)$, $\langle \hat{V}_r(\theta) \rangle = S(\theta) \cdot \hat{V}$, and $\hat{V}_r'(\theta) = \hat{V}_r(\theta) - \langle \hat{V}_r(\theta) \rangle$ derived from the field experiment data [13] obtained at one full scan. The estimates are shown by dots, a dashed curve, and two solid curves in the regions of positive and negative values of $\hat{V}_r(\theta)$. In the both regions, approximately 60 points have nearly equal angular distances between the points $\Delta\theta = 2.7°$.

From the data obtained at N full scans, in angular ranges $[0°, 180°]$ and $[180°, 360°]$, we have $2N$ series of $\hat{V}_r'(i\Delta\theta)$. Using all points of each series, we can obtain $2N$ estimates of the structure function at $|\theta_1 - \theta_2| \ll 180°$ and then average them at the corresponding angle increments $|\theta_1 - \theta_2|$. The values of ε and σ_n^2 can be determined with the use of these equations [13]:

$$\varepsilon = \left[\frac{1}{n-1} \sum_{k=1}^{n-1} \frac{1}{n-k} \sum_{i=1}^{n-k} \frac{D_V((i+k)\Delta\theta) - D_V(i\Delta\theta)}{G_V((i+k)\Delta\theta) - G_V(i\Delta\theta)} \right]^{3/2} \qquad (4.33)$$

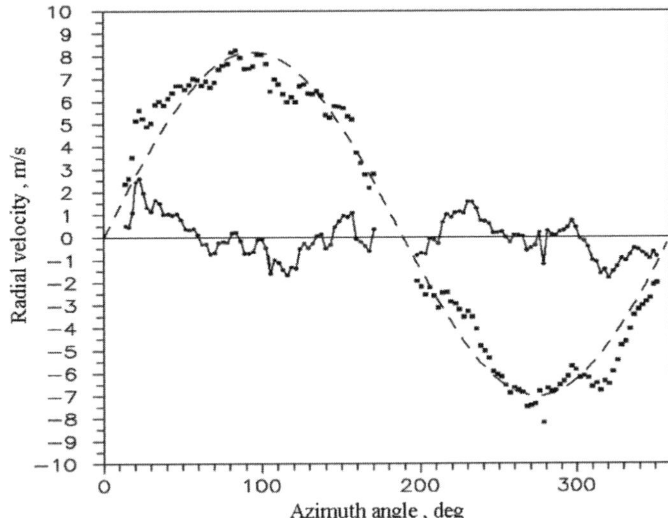

Figure 4.3   Estimates $\hat{V}_r$ (squares), $\langle\hat{V}_r\rangle = S \cdot \hat{V}$ (dashed curve), and $\hat{V}_r' = \hat{V}_r - \langle\hat{V}_r\rangle$ (solid curves) as functions of the azimuth scanning angle θ. (© 1999 American Meteorological Society. Used with permission. From [7].)

$$\sigma_n^2 = \frac{1}{2n}\sum_{i=1}^{n}[D_V(i\Delta\theta) - \varepsilon^{2/3}G_V(i\Delta\theta)] \tag{4.34}$$

where the $D_V(i\Delta\theta)$ are experimental values of the structure function and $G_V(i\Delta\theta)$ is the factor of $\varepsilon^{2/3}$ in the right-hand side of (4.30). According to the requirement of applicability of the theory $n\Delta\theta \ll \pi/2$ and $n\Delta\theta F\cos\varphi \ll L_V$, the number n was equal to 10 in the experiments, whose results are shown below.

Two examples of experimental values of structure functions $D_V(i\Delta\theta)$ obtained from lidar measurement data at heights $h = 50$m ($\Delta z = 9.2$m, closed squares) and $h = 550$m ($\Delta z = 557$ m, open squares) [13] are shown in Figure 4.4. With the use of (4.33), the following estimates of the dissipation rate were obtained for these data: $\varepsilon = 0.039$ m^2/s^3 for $h = 50$m and $\varepsilon = 0.012$ m^2/s^3 for $h = 550$m. The results of calculations using (4.30) are shown as solid curves in Figure 4.4. One can see good agreement between the theory and experiment at both small ($\Delta z = 9.2$m) and large ($\Delta z = 557$m) sizes for the sensing volume. The results of calculation of the transverse structure functions of wind velocity by (4.31) at $\sigma_n = 0$ with the use of the obtained ε values are shown as dashed curves in this figure.

4.4   Retrieval of Vertical Profiles of the Turbulent Energy Dissipation Rate from Continuous-Wave CDL Data

Measurements were conducted with a cw CO$_2$ coherent Doppler lidar of the German Aerospace Center (DLR, Oberpfaffenhofen) in southern Germany at locations with different terrain (Earth's surface roughness parameters) under different thermodynamic conditions in the atmospheric boundary layer. Experimental series were

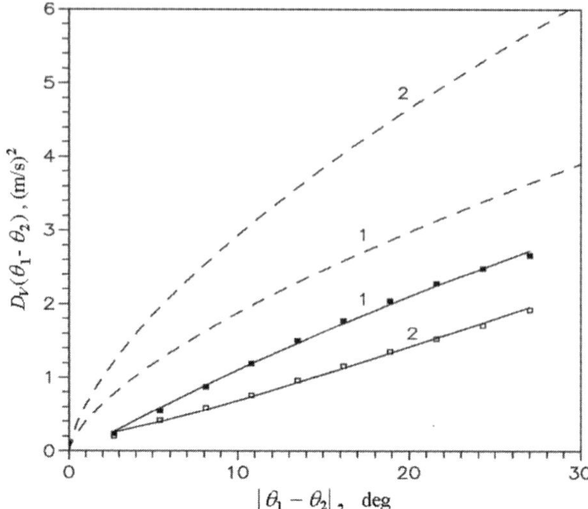

Figure 4.4 Examples of structure functions of radial velocity fluctuations drawn from the data of a scanning lidar on radial velocity fluctuations (closed and open squares, respectively) at heights $h = 50m$ (1) and 550m (2). (© 1999 American Meteorological Society. Used with permission. From [7].)

carried out in the summer of 1992 near Lichtenau, in fall 1993 in a narrow alpine valley near Garmisch Partenkirchen, and at the airfield of the DLR airdrome (Oberpfaffenhofen) in fall 1994 and 1995 and summer 1996. The lidar measurements of 1992 were accompanied by simultaneous measurements of wind and temperature by cup anemometers and thermometers at different heights in the atmospheric boundary layer, which provided the information about temperature stratification of the atmosphere. In 1994, the lidar measurements were conducted simultaneously with measurements by a sonic anemometer [53]. Nearly all lidar measurements were characterized by a very high SNR.

To retrieve the vertical profile of the turbulent energy dissipation rate from the cw CDL data, it is necessary to conduct measurements successively at fixed heights h_l ($l = 1, 2, 3, ...$), repeating them many times to obtain stable estimates of $\varepsilon(h_i)$. If ε is estimated from the Doppler spectrum width $\sigma_S = \sqrt{\langle \sigma_S^2 \rangle}$, it is possible to use raw data measured at both the fixed position of the probing beam and at the scanning location. If ε is determined from the temporal spectrum of wind velocity $S_{\hat{V}}(f)$, the probing beam should remain fixed throughout the measurement at a chosen height h_i.

Figure 4.5 shows the retrieved vertical profiles of the dissipation rate ε as symbols connected by solid lines. The depicted results correspond to the calculations from lidar data at different heights: σ_S^2 (1, 2), $S_V(f)$ (3), and $D_V(\theta_1 - \theta_2)$ (4–6). Profiles 1 and 5 were obtained during a weak wind ($U < 3$ m/s), whereas profiles 2, 3, 4, and 6 correspond to a strong wind ($U > 10$ m/s). Numerical simulation revealed that the relative errors of the ε estimates are 15% to 20% [30].

The weather conditions during the measurement of profiles 2 through 4 were characterized by fitful wind and strong turbulence of airflow (measurements were conducted prior to rain), which can explain the relatively high ε values observed in these experiments. Profile 5, in contrast, was measured during a very weak wind not exceeding 1.5 m/s and weak turbulence. As a rough approximation, one can take that ε is proportional to the mean wind velocity to the third power [51, 54–56].

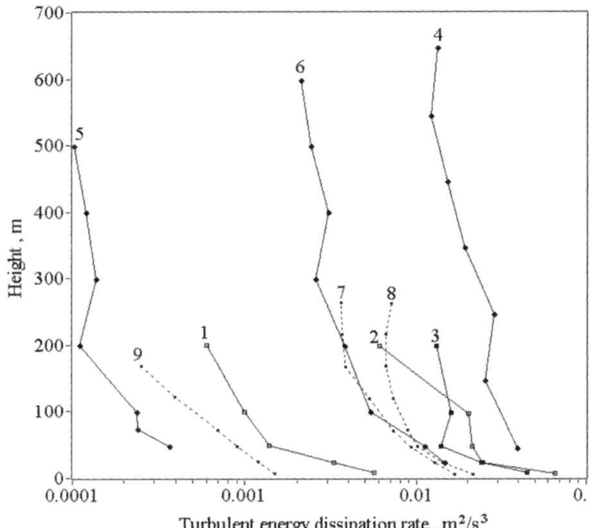

Figure 4.5 Vertical profiles of the turbulent energy dissipation rate (TEDR) retrieved from CDL measurements (1–6) and measurements at a meteorological mast (7–9): results of ε retrieval (curves 1 and 2) from the Doppler spectrum width σ_S, (curve 3) from the temporal velocity spectrum $S_{\hat{v}}(f)$, and (curves 4–6) from the transverse structure function of velocity $D_V(\theta_1 - \theta_2)$.

Correspondingly, at the weak wind, small values of the turbulent energy dissipation rate should be observed.

It is interesting to compare results obtained by different methods. Profiles 2 and 3 are retrieved from the same raw data. It can be seen that the methods provide close results except for the highest level. The comparison of the profiles retrieved from analogous data at the weak wind (curve 1) has shown that up to a height $h = 50$m (the size Δz is small) the estimates of ε from and $S_{\hat{v}}(f)$ differ within the random error. Above this level, the estimates of ε from σ_S $S_{\hat{v}}(f)$ become overstated, which is connected with significant deviations of the spectrum measured at high altitudes from the spectrum calculated by (2.93) through (2.95) for two possible reasons: (1) inapplicability of the Taylor hypothesis of frozen turbulence or (2) the influence of refractive turbulence (see Figure 2.13). If the dissipation rate ε is estimated from the measurements of the Doppler spectrum width σ_S or measurements (at the rather fast conical scanning speeds of the probing beam) of the transverse structure function of wind velocity $D_V(\theta_1 - \theta_2)$, these two factors are not significant.

Figure 4.6 compares the $\varepsilon(h)$ profiles retrieved from σ_S (curve 1) and from $D_V(\theta_1 - \theta_2)$ (curve 2, which corresponds to curve 5 in Figure 4.5) calculated from the same scanning lidar data. The dissipation rate estimates from σ_S were obtained by using (4.6), ignoring the outer scale of turbulence L_V. It can be seen that at lower heights the methods give close results. However, at high altitudes, the values of ε obtained from σ_S are clearly understated, because Δz exceeds the maximal size of turbulent wind inhomogeneities in the inertial interval. In particular, for $h = 500$m, the estimate of ε from σ_S is understated by a factor of 5 compared to the value obtained from $D_V(\theta_1 - \theta_2)$. The same difference is also observed when comparing the $\varepsilon(h)$ profiles determined from σ_S by (4.6) with those obtained from the structure function $D_V(\theta_1 - \theta_2)$ and shown as curves 4 and 6 in Figure 4.5.

Since the turbulence in the atmospheric boundary layer is anisotropic, that is, L_V depends on the probing beam direction with respect to the mean wind direction, the measurement of the integral scale of turbulence L_V from data of a conically scanning lidar causes certain difficulties. Therefore, the method of ε determination from $D_V(\theta_1 - \theta_2)$, where the information about L_V is not required, is more efficient (accurate) compared to the estimation of the dissipation rate from the Doppler spectrum width. What is more, in contrast to the method of ε estimation from $D_V(\theta_1 - \theta_2)$, the information about turbulence can be retrieved from measurements of the Doppler spectrum width only when the regular inhomogeneity of the wind can be neglected; that is, when $\sigma^2_{\langle V \rangle} \ll \sigma^2_t$ in (2.79).

A comparison of the results of simultaneous measurements of ε by cw CDLs and sonic anemometers has demonstrated the satisfactory agreement of the obtained data (see, in particular, Table 4.1 and Figure 2.2). Figure 4.5 shows the results of measurements at a meteorological mast published in [57] as dashed curves 7–9, which correspond to measurements at neutral atmospheric stratification (curve 7), stable stratification (curve 8), and unstable stratification (curve 9). The Doppler lidar data shown in Figure 4.5 were obtained from fall daytime measurements at the neutral or close to neutral stratification. The conditions of the measurements whose findings are shown by curve 6 maximally correspond to neutral stratification. It can be seen from the figure that these data are in good agreement with the results of direct measurements (curve 7).

Thus, in this section we have considered three methods of estimation of the turbulent energy dissipation rate from wind measurements by cw CDLs: (1) from the width of the Doppler spectrum σ_S, (2) from the temporal spectrum $S_{\hat{V}}(f)$ [or structure function $D_{\hat{V}}(\tau)$] of wind velocity measured by lidar at the fixed position of the probing beam, and (3) from the transverse structure function $D_V(\theta_1 - \theta_2)$ calculated

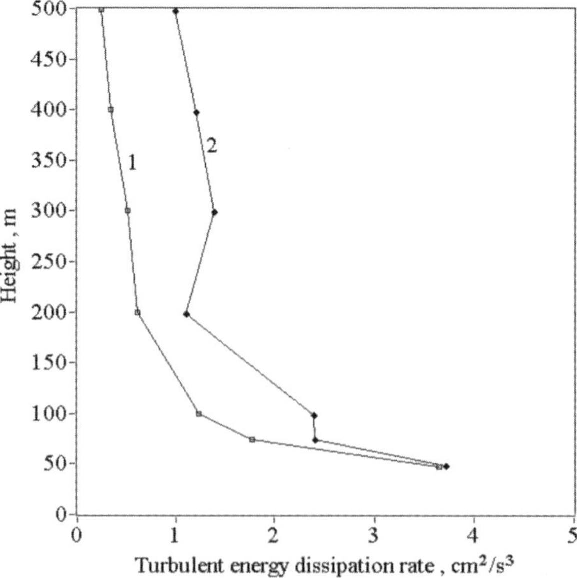

Figure 4.6 Vertical profiles of the turbulent energy dissipation rate retrieved from scanning lidar data at ε estimation from σ_S (1) and from $D_V(\theta_1 - \theta_2)$ (2). (© 1999 American Meteorological Society. Used with permission. From [7].)

from conical scanning lidar data. The results of the field experiments demonstrate the applicability of these methods for reconstruction of the vertical profile of the dissipation rate $\varepsilon(h)$ in the layer up to a height of ~200m (at least, the obtained results do not contradict the available experimental data on the vertical profile and absolute values of ε). If lidar measurements are conducted above this layer, the first and the second methods in many cases give results with high errors for the reasons discussed above.

The third method, estimation of the dissipation rate ε from $D_V(\theta_1 - \theta_2)$, does not require application of the Taylor hypothesis of frozen turbulence, and practically does not depend on the refractive turbulence because the last influences only the variance σ_n^2 of the noise component in (4.30) for $D_V(\theta_1 - \theta_2)$. Therefore, this method, in contrast to the first two methods, allows for lidar measurements of turbulence to be taken at heights much higher than 200m. In this case, the height restrictions are determined only by the spatial (vertical) resolution of the cw CDL measurements. For example, the maximal height of lidar data shown in Figure 4.5 is about 650m. An increase in the elevation angle φ allows one, in principle, to carry out measurements at heights up to h ~ 800m with the same spatial resolution.

In [30], numerical simulation was used to calculate the relative error of ε estimation from the transverse structure function $D_V(\theta_1 - \theta_2)$ for different numbers N of full scans ($0° \leq \theta \leq 360°$) by the probing beam during the measurement at a fixed height h_l. It has been shown that at $N = 10$ the error is 18%. Consequently, if one scan takes 12s, then to obtain the vertical profile of the dissipation rate $\varepsilon(h_l)$ at seven height levels ($l = 1, 2, ..., 7$) with this accuracy, the duration of lidar measurements (with regard to the time needed to change the focal length of the probing beam) should be about 15 min.

4.5   Methods for Estimating Wind Turbulence Parameters from Pulsed CDL Scanning Data in the Vertical Plane

The geometry of lidar measurements of aircraft wake vortex parameters supposes the scanning by the probing beam in the vertical plane across the aircraft wake [20–26]. Figure 4.7 shows the geometry of measurements by ground-based pulsed CDLs using vertical scanning. During the measurement, the probing beam moves alternatively up-down, that is, the elevation angle is a function of time $\varphi_{l'} \equiv \varphi(t_{l'})$, where l' is the shot number. In the experiments, whose results are presented below, the maximal elevation angle $\varphi_{\max}$ did not exceed 30° ($0° \leq \varphi_{l'} \leq 30°$) with the angular scanning rate $d\varphi/dt$ of 2°/s and 1°/s. A similar measurement geometry was used in [58] for lidar studies of the nocturnal atmospheric boundary layer.

An array of estimates of the Doppler spectra $\hat{S}_V(V_k; R_l, \varphi_i, n)$ for ranges $R_l = R_0 + l\Delta R$ ($l = 0, 1, 2, ..., L' - 1$) and elevation angles $\varphi_i = i\Delta\varphi$ ($i = 0, 1, 2, ..., I$; $\varphi_{\max} = I\Delta\varphi$) is obtained from raw data (measured array of lidar signal samples) with the use of (2.42) through (2.46). Here, according to the Doppler equation, (1.17), the frequency f_k is replaced with the velocity $V_k = (k - M'/2)\delta V$, $n = 1, 2, ..., N$ is the number of full scans ($0° \leq \varphi_i \leq \varphi_{\max}$), ΔR is a preset step along the optical axis, and $\Delta\varphi = LT_P d\varphi/dt$ is the elevation angle resolution. Then, the spectral moments are calculated by (2.47) through (2.49); that is, the normalized echo signal power

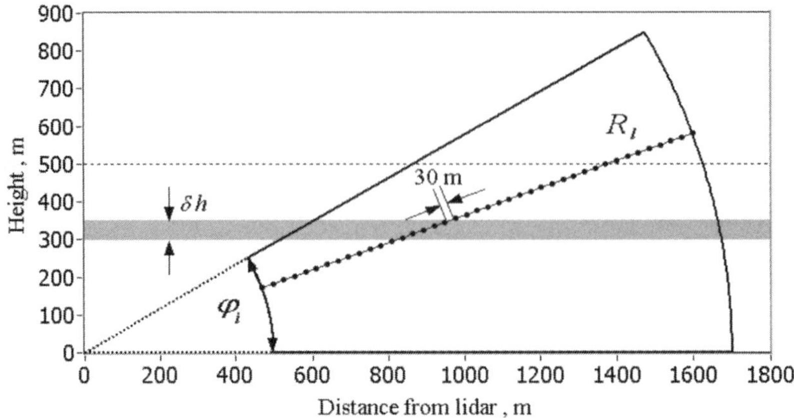

Figure 4.7 Geometry of measurement by pulsed CDL with probing beam scanning in the vertical plane. (© 2005 American Meteorological Society. Used with permission. From [8].)

(signal-to-noise ratio) $\hat{P}_S(R_l,\varphi_i,n) \equiv \text{S}\hat{\text{N}}\text{R}(R_l,\varphi_i,n)$, radial velocity $\hat{V}_r(R_l,\varphi_i,n)$, and the squared width of the Doppler spectrum $\hat{\sigma}_S^2(R_l,\varphi_i,n)$ are obtained.

The results of investigation of statistical characteristics of the estimates $\hat{V}_r$ and $\hat{\sigma}_S$ discussed in Section 2.6 form the basis for development of methods for determination of wind turbulence parameters from the mathematical expectation of the squared Doppler spectrum width $\langle\hat{\sigma}_S^2\rangle$ and the longitudinal structure function of the radial wind velocity $D_{\hat{V}}(r) = \langle[\hat{V}_r'(R + r,\varphi_i) - \hat{V}_r'(R,\varphi_i)]^2\rangle$. Because the data obtained with the probing beam scanning are used, it may be possible to apply one of the approaches discussed in previous section. This approach is based on estimation of the turbulent energy dissipation rate from the spatial structure function of the radial wind velocity measured at different elevation angles $D_{\hat{V}}(R,\varphi_{i+i'},\varphi_i) = \langle[\hat{V}_r'(R,\varphi_{i+i'}) - \hat{V}_r'(R,\varphi_i)]^2\rangle$. Taking into account the statistical independence of the errors $V_e(R, \varphi_{i+i'})$ and $V_e(R, \varphi_i)$ at $i' \neq 0$, we can write the equation for the structure function $D_{\hat{V}}(R,\varphi_{i+i'},\varphi_i)$ in the form

$$D_{\hat{V}}(R,\varphi_{i+i'},\varphi_i) = D_{\bar{V}}(R,\varphi_{i+i'},\varphi_i) + 2\sigma_e^2$$

(4.35)

where $D_{\bar{V}}(R,\varphi_{i+i'},\varphi_i) = \langle[\bar{V}_r'(R,\varphi_{i+i'}) - \bar{V}_r'(R,\varphi_i)]^2\rangle$.

The turbulence is assumed to be homogeneous and isotropic. Then, at small angular separations $|\varphi_{i+i'} - \varphi_i| \ll \pi/2$ and the condition $L_V \ll R$, for the function $D_{\bar{V}}(R,\varphi_{i+i'},\varphi_i)$ we can say that the variation of the scanning angle from φ_i to $\varphi_{i+i'}$ is equivalent to the displacement of the probing beam in the direction transverse to the beam optical axis by a distance $y = R \mid \varphi_{i+i'} - \varphi_i \mid$. Thus, $D_{\bar{V}}$ can be considered as a transverse structure function $D_{\bar{V}}(y)$. With regard to the above assumptions, we obtain from (2.29) that [28]

$$D_{\bar{V}}(y) = 2 \int\limits_{-\infty}^{+\infty} \int\limits_{+\infty}^{+\infty} d\kappa_z\, d\kappa_y S_V^{(2)}(\kappa_z,\kappa_y)H_p(\kappa_z)[1 - \cos(2\pi\, d\kappa_z y)]$$

(4.36)

where the von Karman model of (2.86) and (2.71) is used for the two-dimensional spectrum of the wind velocity $S_V^{(2)}(\kappa_z,\kappa_y)$. The function $H_p(\kappa_z)$ entering into (4.36) is described by (2.118) when a rectangular temporal window is used or by (2.119) if a Gaussian temporal window is used.

The dissipation rate ε can be determined from lidar measurements of the longitudinal structure function $D_{\hat{V}}(r)$ within the inertial interval of turbulence (with neglected outer scale of turbulence), when $r \ll L_V$. In this case, for ε to be estimated with an acceptable accuracy, the value of Δz should not exceed L_V. For a lidar with $\Delta z \sim 90\text{m}$, this condition is not always true. If the information about the dissipation rate is retrieved from the transverse structure function $D_{\hat{V}}(y)$, then ε can be estimated with the outer scale of turbulence neglected at $y \ll L_V$ even if $L_V \ll \Delta z$. The condition $y \ll L_V \ll \Delta z$ allows us to obtain a simple asymptotic equation

$$D_{\hat{V}}(y) = 4.2C_K \frac{\varepsilon^{2/3}y^{5/3}}{\Delta z} + 2\sigma_e^2 \tag{4.37}$$

where $y \neq 0$ and Δz is described by (2.26), from (4.35), (4.36), (2.71), and (2.86) if in (2.86) we use $L_V \to \infty$ and $\kappa_z = 0$ and use in (4.36) the Gaussian temporal window [(2.119) for $H_p(\kappa_z)$].

Assume that the spectra $\hat{S}(V_k; R_l, \varphi_i, n)$ were obtained with the use of the Gaussian temporal window. The mathematical expectation of the estimate of the squared Doppler spectrum width $\langle \hat{\sigma}_S^2(R,\varphi) \rangle$ can be represented, by analogy with (2.79), in the form

$$\langle \hat{\sigma}_S^2(R,\varphi) \rangle = \sigma_t^2 + \sigma_{\langle V \rangle}^2 + \sigma_{V1}^2 + \langle E_\sigma \rangle \tag{4.38}$$

where the instrumental broadening of the Doppler spectrum $\sigma_{VI}^2 = (\lambda/2)^2\sigma_{fI}^2$ is given by (2.28) and the regular spectral broadening associated with the inhomogeneity of the mean wind velocity is described by the equation

$$\sigma_{\langle V \rangle}^2 = \frac{1}{\Delta z}\int\limits_{-\infty}^{+\infty} dz' \exp\left(-\pi\frac{z'^2}{\Delta z^2}\right)\langle V_r(R + z')\rangle^2$$

$$-\left[\frac{1}{\Delta z}\int\limits_{-\infty}^{+\infty} dz' \exp\left(-\pi\frac{z'^2}{\Delta z^2}\right)\langle V_r(R + z')\rangle\right]^2 \tag{4.39}$$

If the mean radial velocity within the sensing volume ($|z'| \leq \Delta z$) can be approximated by the expansion

$$\langle V_r(R + z')\rangle \approx \langle V_r(R)\rangle + \frac{d\langle V_r(R)\rangle}{dR}z' + \frac{1}{2}\frac{d^2\langle V_r(R)\rangle}{dR^2}z'^2 \tag{4.40}$$

then [8]

$$\sigma^2_{\langle V \rangle} \approx \mu_V = \frac{1}{2\pi}\left(\frac{d\langle \bar{V}_r \rangle}{dR}\Delta z\right)^2 + \frac{1}{8\pi^2}\left(\frac{d^2\langle \bar{V}_r \rangle}{dR^2}\Delta z^2\right)^2 \tag{4.41}$$

With regard to (4.38) and (4.41), we can write $\langle \hat{\sigma}^2_S(R,\varphi) \rangle$ in the form

$$\langle \hat{\sigma}^2_S(R,\varphi) \rangle = \sigma^2_t + \mu_V + \gamma_V + \sigma^2_{VI} + \langle E_\sigma \rangle \tag{4.42}$$

where $\gamma_V = \sigma^2_{\langle V \rangle} - \mu_V$. In (4.42), the instrumental broadening σ^2_{VI} is known. The value of μ_V can be calculated by (4.41) with the use of the measured lidar profile of the mean radial velocity $\langle \bar{V}_r(R) \rangle$. In addition to the sought-after characteristic σ^2_t, the parameters γ_V and $\langle E_\sigma \rangle$ remain unknown. If approximation (4.40) is actually valid, then we can use $\gamma_V = 0$ in (4.42).

For cw CDLs, the estimate of the squared Doppler spectrum width $\hat{\sigma}^2_S$ from a single measurement of the spectrum includes an error E_σ that does not vanish upon averaging. This can lead to significant overestimation of the dissipation rate ε in the case of weak turbulence. Therefore, to minimize $\langle E_\sigma \rangle$, one should use additional accumulation of Doppler spectra similar to what is done in (4.1) through (4.3). Another approach is also possible. If the estimate $\hat{\sigma}^2_S$ is obtained with (2.49), it is necessary first to replace $\tilde{S}(f_k)/\bar{S}_N - 1$ with $\tilde{S}(f_k)/\bar{S}_N - n_{th}$, where n_{th} is a dimensionless parameter exceeding unity. The value of n_{th} was determined in [8] by numerical simulation for the conditions of the field experiment. It was found that n_{th} varies from $n_{th} = 1.12$ at SNR = 10 to $n_{th} = 1.14$ at SNR = 1.

We believe that the wind flow is statistically homogeneous in the horizontal direction, that is, the mean wind velocity and the turbulence parameters σ^2_V, L_V, and ε depend only on the height h, and the turbulence anisotropy can be neglected in the processing of lidar data. In [8, 28], the following algorithm has been proposed for obtaining the estimates of the turbulent broadening of the Doppler spectrum $\hat{\sigma}^2_t(h)$, variance of random velocity measurement error $\hat{\sigma}^2_e(h)$, variance of wind velocity $\hat{\sigma}^2_V(h)$, and the longitudinal $\hat{D}_{\bar{V}}(r_k;h)$ and transverse $\hat{D}_{\bar{V}}(y_{m'};h)$ structure functions of wind velocity.

The vertical profile of the horizontal component of the mean wind velocity vector (projection onto the scanning plane) $V_T = \langle V_r \rangle/\cos\varphi = \langle \hat{V}_r \rangle/\cos\varphi$ is determined as

$$V_T(h) = (N \cdot N_h)^{-1}\sum_{n=1}^{N}\sum_{i,l}^{N_h}\hat{V}_r(R_l,\varphi_i,n)/\cos\varphi_i \tag{4.43}$$

where summation is performed over subscripts i and l satisfying the condition $h - \delta h/2 < R_l\sin\varphi_i < h + \delta h/2$, and N_h is the number of points in the averaging layer with a thickness δh for one scan. As an example, the averaging layer with a thickness $\delta h = 50$m at $h = 325$m is shown as a gray bar in Figure 4.7. Then, we obtain arrays for the mean value

$$\langle \hat{V}_r(R_l,\varphi_i) \rangle_E = V_T(R_l\sin\varphi_i)\cos\varphi_i \tag{4.44}$$

and fluctuations

$$\hat{V}_r'(R_l,\varphi_i,n) = \hat{V}_r(R_l,\varphi_i,n) - \langle \hat{V}_r(R_l,\varphi_i)\rangle_E \tag{4.45}$$

of estimates of the radial wind velocity.

According to (4.41) and (4.42) at $\gamma_V = 0$ and $\langle E_\sigma\rangle = 0$, as well as (2.113) through (2.115), (2.127), (2.128), and (4.35), the estimates of the regular broadening of the Doppler spectrum $\hat{\mu}_V(R_l,\varphi_i)$ and statistical characteristics $\hat{\sigma}_t^2(h)$, $\hat{\sigma}_e^2(h)$, $\hat{\sigma}_V^2(h)$, $\hat{D}_{\bar{V}}(r_{k'};h)$, and $\hat{D}_{\bar{V}}(y_{m'};h)$ are calculated by the following equations:

$$\hat{\mu}_V(R_l,\varphi_i) = \frac{1}{2\pi}\left[(\langle \hat{V}_r(R_{l+1},\varphi_i)\rangle_E - \langle \hat{V}_r(R_{l-1},\varphi_i)\rangle_E)\frac{\Delta z}{2\Delta R}\right]^2$$
$$+\frac{1}{8\pi^2}\left[(\langle \hat{V}_r(R_{l-1},\varphi_i)\rangle_E - 2\langle \hat{V}_r(R_l,\varphi_i)\rangle_E + \langle \hat{V}_r(R_{l+1},\varphi_i)\rangle_E)\left(\frac{\Delta z}{\Delta R}\right)^2\right]^2 \tag{4.46}$$

$$\hat{\sigma}_t^2(h) = (N \cdot N_h)^{-1}\sum_{n=1}^{N}\sum_{i,l}^{N_h}[\hat{\sigma}_S^2(R_l,\varphi_i,n) - \hat{\mu}_V(R_l,\varphi_i) - \sigma_{VI}^2] \tag{4.47}$$

$$\hat{\sigma}_e^2(h) = (N \cdot N_h)^{-1}\sum_{n=1}^{N}\sum_{i,l}^{N_h}[\Delta V_e(R_l,\varphi_i,n)]^2/2 \tag{4.48}$$

$$\hat{\sigma}_V^2(h) = (N \cdot N_h)^{-1}\sum_{n=1}^{N}\sum_{i,l}^{N_h}[\hat{V}_r'(R_l,\varphi_i,n)]^2 - \hat{\sigma}_e^2(h) + \hat{\sigma}_t^2(h) \tag{4.49}$$

$$\hat{D}_{\bar{V}}(r_{k'};h) = (N \cdot N_{hk'})^{-1}\sum_{n=1}^{N}\sum_{i,l}^{N_{hk'}}\{[\hat{V}_r'(R_l + r_{k'},\varphi_i,n) - \hat{V}_r'(R_l,\varphi_i,n)]^2$$
$$-[\Delta V_e(R_l + r_{k'},\varphi_i,n) - \Delta V_e(R_l,\varphi_i,n)]^2/2\} \tag{4.50}$$

and

$$\hat{D}_{\bar{V}}(y_{m'};h) = (N \cdot N_{hm'})^{-1}\sum_{n=1}^{N_s}\sum_{i,l}^{N_{hm'}}[\hat{V}_r'(R_l,\varphi_{i+i'},n) - \hat{V}_r'(R_l,\varphi_i,n)]^2 - 2\hat{\sigma}_e^2(h) \tag{4.51}$$

where $\Delta V_e(R_l,\varphi_i,n) = \hat{V}_r'(R_l,\varphi_{i+i'},n) - \hat{V}_r'(R_l,\varphi_i,n)$, $r_{k'} = k'\Delta R$, $k' = 1, 2, 3, \ldots$, $y_{m'} = m'\Delta y$, $m' = 1, 2, 3, \ldots$, and $i' = 1, 2, 3, \ldots$. In (4.50) and (4.51), the summation is performed over the subscripts i and l satisfying the condition $h - \delta h/2 < (R_l + r_{k'}/2)\sin\varphi_i < h + \delta h/2$ and simultaneously two conditions: $h - \delta h/2 < R_l(\sin\varphi_{i+i'} + \sin\varphi_i)/2 < h + \delta h/2$ and $y_{m'} - \Delta y/2 < R_l|\varphi_{i+i'} - \varphi_i| < y_{m'} + \Delta y/2$.

Thus, the results of the primary processing of raw lidar data are estimates of the variance of wind velocity $\hat{\sigma}_V^2(h)$, the turbulent broadening of the Doppler spectrum $\hat{\sigma}_t^2(h)$, and the longitudinal $\hat{D}_{\bar{V}}(r_{k'};h)$ and transverse $\hat{D}_{\bar{V}}(y_{m'};h)$ structure functions of wind velocity. From the estimates of these characteristics, we can determine the turbulent energy dissipation rate ε and the outer scale L_V in three ways: (1) from the Doppler spectrum width (DSW method), where $\hat{\sigma}_t^2(h)$ and $\hat{\sigma}_V^2(h)$ are used; (2) from the longitudinal structure function of wind velocity (LSF method), when $\hat{D}_{\bar{V}}(r_{k'};h)$ is used to obtain estimates of turbulence parameters; and (3) from the transverse structure function of wind velocity (TSF method), where the turbulence parameters are estimated from $\hat{D}_{\bar{V}}(y_{m'};h)$.

For the DSW method, the estimates $\hat{L}_V$ and $\hat{\varepsilon}$ are obtained through the solution of (4.9) and by calculation using (4.11), where $F_K(\hat{L}_V/\Delta z) = (1.972)^{2/3} C_K^{-1} L_V^{-2/3} G_W(\Delta z, L_V)$ and the function $G_W(\Delta z, L_V)$ is given by (2.125) [8].

With the use of the LSF method, the estimates $\hat{L}_V$ and $\hat{\varepsilon}$ are obtained as a result of the minimization of the functional [59]

$$\rho(\varepsilon, L_V) = \sum_{k'=1}^{K} \left[\frac{\hat{D}_{\bar{V}}(r_{k'}) - \varepsilon^{2/3} G_s(r_{k'}, L_V)}{k' G_s(r_{k'}, L_V)} \right]^2 \tag{4.52}$$

where the function $G_s(r, L_V)$ is calculated by (2.121) and, depending on the chosen temporal window, by (2.118) or (2.119). The values of ΔR and K are chosen so that the minimal separation of observation points $r_1 = \Delta R$ lies within the inertial interval of turbulence, and the number K exceeds the ratio $L_V/\Delta r$ [8]. The results presented below were obtained at $\Delta R = 30\text{m}$ and $K = 16$.

The estimates of turbulent parameters by the TSF method can be obtained through the fitting (for example, by the least-squares method) of the function $D_{\bar{V}}(y_{m'})$ calculated by (4.36) to the measured transverse structure function $\hat{D}_{\bar{V}}(y_{m'})$ [28].

In Figure 2.14(a), one can see that as the SNR decreases, the error of lidar estimation of the radial wind velocity increases, and, correspondingly, the error of estimation of wind turbulence parameters should increase. The numerical simulation of raw lidar data and application of the above procedures have shown [8] that for the conditions of the field experiment (see Section 4.6) the dissipation rate ε can be estimated with the use of the DSW method at SNR > 1 and for the LST (or TSF) method at SNR > 0.3 (−5 dB). Therefore, when estimating ε based on (4.47) and (4.48) through (4.51), experimental data with, respectively, $\hat{\text{SNR}}(R_l, \varphi_i, n) \leq 1$ and $\hat{\text{SNR}}(R_l, \varphi_i, n) \leq 0.3$ were rejected. To avoid the influence of aircraft wake vortices when using (4.43) through (4.51), experimental data in the vicinity of wake vortices were rejected as well.

In [8], to calculate errors of estimation of the dissipation rate ε by the DSW and LSF methods, the data for numerical simulation of the operation of 2-μm pulsed CDLs were used in accordance with the algorithm described in Section 2.6. In the simulation and processing of the simulated raw data of the lidar signal, the following parameters were specified: $\sigma_P = \sigma_W = 0.2488\ \mu\text{m}$, $B_F = 50$ MHz, $M = 2{,}048$, pulse repetition frequency $f_P = 500$ Hz, number of probing pulses used for accumulation of spectra $L = 25$, and measurement duration = 5 min. It was assumed that SNR decreases linearly along the probing beam from 3 to 1, and that turbulent wind inhomogeneities are transferred by the lateral wind through the beam with a speed of 3 m/s. Random realizations of the two-dimensional field of the radial velocity $V_r(z, y)$ were simulated at $L_V = 150$m and different ε.

From the model lidar data, 40 estimates for each of the parameters $\hat{V}_r(R_l,t_i)$ and $\hat{\sigma}_S^2(R_l,t_i)$ with a step of 30m along the probing beam at every fixed moment $t_i = i\Delta t$ were obtained, where $\Delta t = 0.5$s and $i = 1, 2, ..., 600$. Thus, to obtain one estimate of the dissipation rate $\hat{\varepsilon}$ from the structure function of wind velocity or from the Doppler spectrum width, 24,000 values of both $\hat{V}_r(R_k,t_l)$ and $\hat{\sigma}_S^2(R_k,t_l)$ were used.

Denote the estimates of the dissipation rate by the LSF and DSW methods as, respectively, $\hat{\varepsilon}_1$ and $\hat{\varepsilon}_2$. To calculate statistical characteristics of the dissipation rate estimates by these methods, 500 independent realizations of $\hat{\varepsilon}_i (i = 1, 2)$ were used. The following characteristics were calculated: the relative bias is

$$b_\varepsilon = (\langle\hat{\varepsilon}_i\rangle/\varepsilon - 1) \cdot 100\% \tag{4.53}$$

the relative error is

$$E_\varepsilon = [\langle(\hat{\varepsilon}_i/\varepsilon - 1)^2\rangle]^{1/2} \cdot 100\% \tag{4.54}$$

and the parameter of discrepancy between the dissipation rate estimates by LSF and DSW methods from the same simulated lidar data is

$$d_\varepsilon = 2[\langle(\hat{\varepsilon}_1 - \hat{\varepsilon}_2)^2/(\hat{\varepsilon}_1 + \hat{\varepsilon}_2)^{22}\rangle]^{1/2} \cdot 100\% \tag{4.55}$$

Figure 4.8(a) shows the calculated relative bias of dissipation rate estimates b_ε obtained from the simulated lidar data at the different turbulent intensities with the use of LSF and DSW. One can see that LSF yields the unbiased estimate of the dissipation rate ($\langle\hat{\varepsilon}\rangle = \varepsilon$) in the whole range of ε values specified in the simulation. The same is also true for DSW if we use the optimal value of n_{th} strictly corresponding to the SNR value, which depends on the distance R. However, as follows from the data shown as curves 1 and 2 in the figure, even small deviations of n_{th} from the optimal value lead to the significant bias of the estimate at weak turbulence ($\varepsilon \leq 10^{-4}$ m²/s³).

The relative errors E_ε of dissipation rate estimation by the LSF and DSW methods at different ε are shown in Figure 4.8(b) as, respectively, closed circles and squares. In the case of DSW, only optimal values of n_{th} were used (that is, the estimate $\hat{\varepsilon}_2$ is unbiased). It can be seen that the error of the estimate $\hat{\varepsilon}_1$ (LSF method) is nearly independent of the turbulence intensity for $\varepsilon \geq 10^{-5}$ m²/s³ and equal to about 20%. At $\varepsilon \geq 10^{-4}$ m²/s³, the error of estimation of the dissipation rate from measurements of the Doppler spectrum width is roughly halved compare to the case of the LSF method. In addition, Figure 4.8(b) shows the results of the d_ε calculation by (4.55).

Taking into account that in an actual experiment it is difficult to choose n_{th}, the DSW method can be practically efficient only for lidar measurements under conditions of moderate and strong turbulence, that is, at $\varepsilon > 10^{-4}$ m²/s³, when it (at SNR ≥ 1), according to Figure 4.8(b), can give a more accurate estimation compared to the LSF method. A disadvantage of the LSF method is that the $\hat{\varepsilon}$ estimation error by this method can be very high when the longitudinal dimension of the sensing

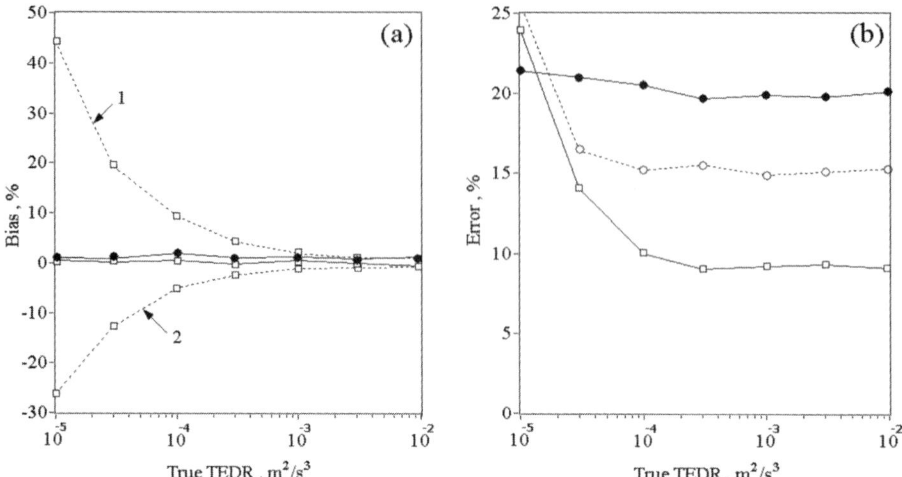

Figure 4.8   Relative bias b_ε (a), relative error E_ε, and parameter d_ε (b): (a) solid curves are for the results at the optimal noise threshold n_{th} in the case of LSF (circles) and DSW (squares); dashed curves 1 and 2 correspond to the case of DSW at $n_{th} = 1.13$ and $n_{th} = 1.14$, respectively; (b) solid curves are for the errors in the case of LSF (circles) and DSW (squares); the dashed curve is for the parameter d_ε. (© 2005 American Meteorological Society. Used with permission. From [8].)

volume Δz far exceeds the outer scale of turbulence L_V. In this case, as shown in [28], it is better to use the transverse structure function of wind velocity (TSF method) for estimation of the dissipation rate.

4.6   Experimental Studies of the Possibility of Turbulence Measurements by Pulsed CDLs in the Atmospheric Boundary Layer

Each year from 2001 to 2007, the Lidar Group (with participation of I.N. Smalikho) of the Institute of Atmospheric Physics of the German Aerospace Center in Oberpfaffenhofen took part in experimental campaigns aimed at the study of atmospheric effects on aircraft wake vortices. These experiments were conducted with a 2-μm pulsed CDL (lidar parameters are presented in Section 1.4). The geometry of lidar measurements is shown in Figure 4.7.

For vertical profiling of the atmospheric turbulence parameters by the LSF and DSW methods based, correspondingly, on the longitudinal spatial structure function and Doppler spectrum width, the data of lidar measurements from 27–28 August 2003 at the airfield of the Tarbes-Lourdes-Pyrenees International Airport in France were used [8]. The processing of these data was performed with the following parameters: $L' = 41$ (that is, the maximal sensing range $R_{\max} = 1{,}700$m); scan number $N = 20$ (corresponding to the measurement duration of 5 min at a scan duration of 15s); thickness of the averaging layer $\delta h = 50$m; maximal height of retrieval of wind and turbulence profiles $h_{\max} = 500$m (dashed line in Figure 4.7). The signal-to-noise ratio $\hat{SNR}(R_l, \varphi_i, n)$ took values from 1 to 5 (data of 27 August 2003) and from 1 to 20 (data of 28 August 2003). According to the results of numerical simulation and measured values of the SNR, the parameter n_{th} was taken in a range from 1.12 to 1.14.

Figure 4.9 exemplifies the vertical profiles of the dissipation rate $\varepsilon(h)$ and the outer scale of turbulence $L_V(h)$ retrieved from the data of lidar measurements for about 5 min with the use of the LSF and DSW methods. A total of 52 profiles of $\varepsilon(h)$ and $L_V(h)$ were obtained from the 27–28 August 2003, experimental data [8].

Figure 4.10 shows the temporal profile of the dissipation rate $\varepsilon(t)$ at a height $h = 200$m observed in the experiment on 27 August from 18:00 to 20:20 LT (Figure 4.10(a)) and 28 August from 8:00 to 10:00 (Figure 4.10(b)) [8]. One can see that the results on the vertical (Figure 4.9(a)) and temporal (Figure 4.10) profiles of the dissipation rate ε obtained by the LSF and DSW methods are close.

Figure 4.11 compares the turbulent energy dissipation rates estimated by the two methods with the use of all 52 vertical profiles (total measurement time of about 4.5 hours) [8]. It is seen that the dots are concentrated near the straight line corresponding to the coincidence of the estimates $\hat{\varepsilon}_1$ and $\hat{\varepsilon}_2$. The data of this figure were used in [8] for calculation of the discrepancy parameter (4.55), and it turned out that $d_\varepsilon = 41\%$.

According to the results of the numerical simulation shown in Figure 4.8(b), $d_\varepsilon \approx 25\%$ at $\varepsilon = 10^{-5}$ m^2/s^3 and $d_\varepsilon \approx 15\%$ at $\varepsilon > 10^{-4}$ m^2/s^3. However, in the field experiment $d_\varepsilon = 41\%$; that is, it is ~ 1.5 to 2.5 times over. Consequently, the error of lidar estimation of the turbulent energy dissipation rate by these methods in experiment exceeds the values of E_ε in Figure 4.8(b).

There are several possible reasons for the discrepancy between theory and experiment, in particular [8], (1) different sample sizes in the numerical and field experiments due to the different measurement geometry (it was assumed in the simulation that the probing beam was fixed in space and, consequently, the averaging volume in every realization is constant, while in the field experiment, as can be seen from Figure 4.7, the averaging area in a selected layer decreases with height starting from a height $h \sim 250$m); (2) anisotropy and inhomogeneity (vertical) of turbulence, regular

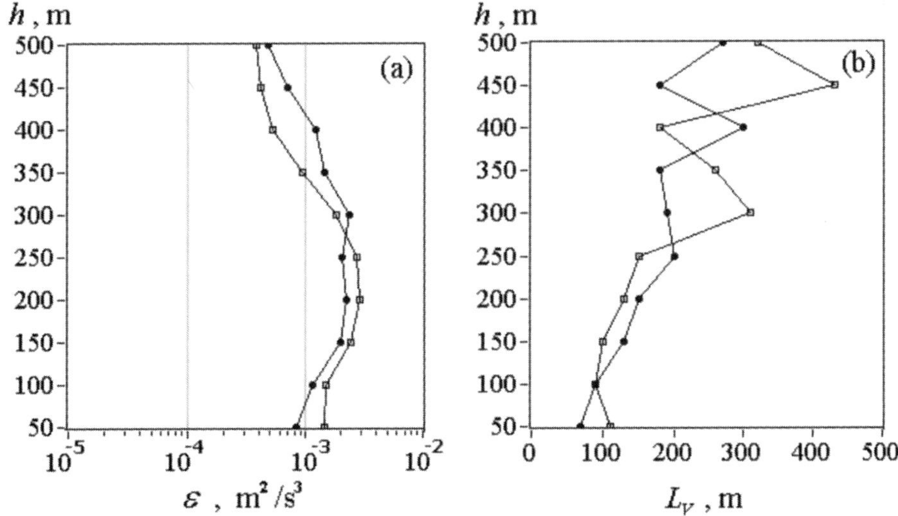

Figure 4.9 Vertical profiles of the (a) turbulent energy dissipation rate and (b) the integral scale of turbulence obtained with the use of the LSF (circles) and DSW (squares) methods. Measurements were performed on 27 August 2003 from 12:04 to 12:09 LT. (© 2005 American Meteorological Society. Used with permission. From [8].)

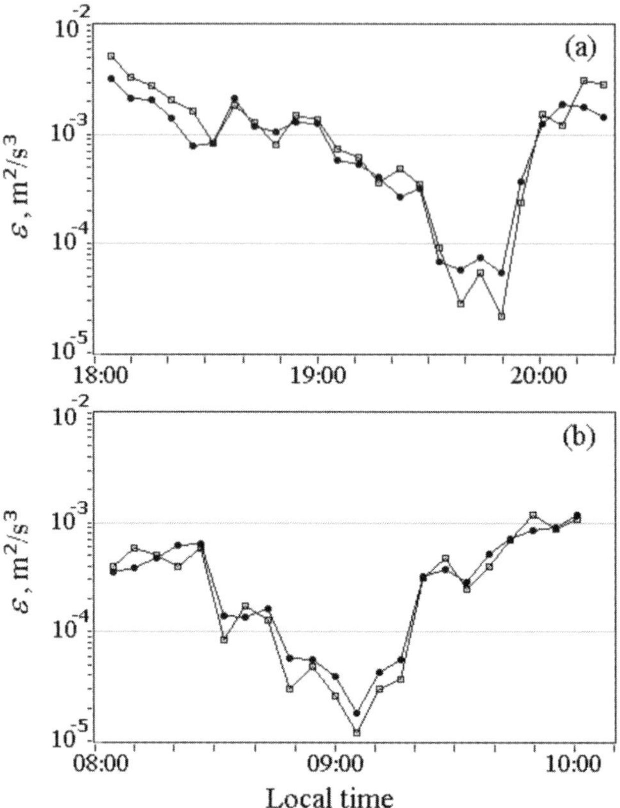

Figure 4.10   Temporal profiles of the dissipation rate at a height of 200m estimated by the LSF (circles) and DSW (squares) methods. Measurements were performed on (a) 27 August and (b) 28 August 2003. (© 2005 American Meteorological Society. Used with permission. From [8].)

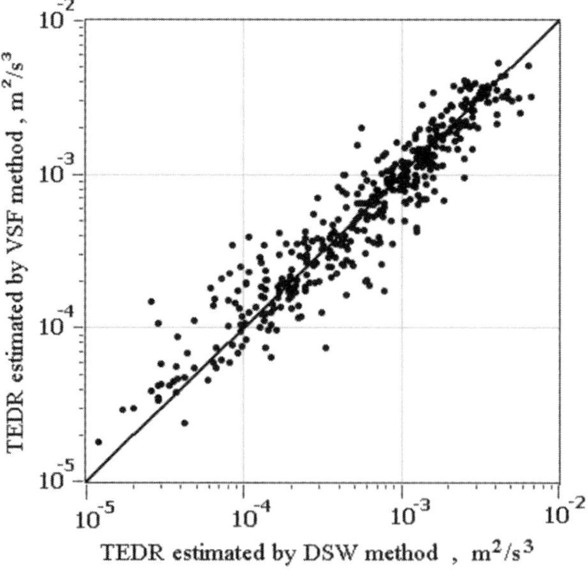

Figure 4.11   Comparison of individual estimates of the dissipation rate ε obtained from the longitudinal structure function of the radial velocity measured by lidar and the Doppler spectrum width. (© 2005 American Meteorological Society. Used with permission. From [8].)

inhomogeneity of wind and mesoscale processes in the real experiment [the structure function measured by lidar can be biased, that is, differ from (2.120)]; or (3) deviations (although small) of the probing pulse from the Gaussian distribution in time and regular (about several percent) deviation of system noise from the homogeneous distribution (white noise) in a selected bandwidth $B_F = 50$ MHz, which can ultimately influence the accuracy of the estimates $\hat{\varepsilon}_2$ (DSW method) under conditions of weak turbulence, and so on.

The results shown in Figures 4.9 through 4.11 were obtained with the 5-min averaging of lidar data and the maximal range $R_{\max} = 1,700$m. Let us increase 10-fold the averaging time ($N_s = 200$) using the raw data of lidar measurements for 27 August 2003 from 18:30 to 19:20 LT at relatively stationary atmospheric conditions and specify $R_{\max} = 2,900$m (L' = 81). The resultant retrieved vertical profiles of the wind velocity variance $\sigma_V^2(h)$, outer scale of turbulence $L_V(h)$, and the turbulent energy dissipation rate $\varepsilon(h)$ are depicted in Figure 4.12. The measured longitudinal structure function of wind velocity was used to determine ε and L_V (LSF method), and then the wind velocity variance σ_V^2 was calculated by (2.71). The values of these parameters were also obtained by the DSW method, that is, with the use of (4.47) and (4.49). One can see that these methods give close results, especially, for the dissipation rate. According to Figure 4.12(c), the deviation of the estimate $\hat{\varepsilon}_1$ (LSF method) from $\hat{\varepsilon}_2$ (DSW method) is, on average, ~ 10%.

To retrieve the vertical profiles of wind turbulence from the transverse spatial structure functions of the radial wind velocity $\hat{D}_{\bar{V}}(y_{m'}, h)$ (LSF method), the raw data of measurement by the 2-μm lidar on 26 and 27 May 2005 at the airfield of the Toulouse airport were used [28]. The duration of measurement was 20 min (from 12:52 to 13:12 LT on 26 May 2005) and 23 min (from 15:51 to 16:14 on 27 May 2005). The measurement geometry was similar to that shown in Figure 4.7. The following parameters were used for the data processing: $\Delta R = 30$m, $R_0 = 500$m, $R_{\max} = 2,000$m, and $\delta h = 40$m.

Figures 4.13(a) and 4.13(b) depict the vertical profiles of the projection (onto the plane of scanning by the probing beam) of the horizontal component of the

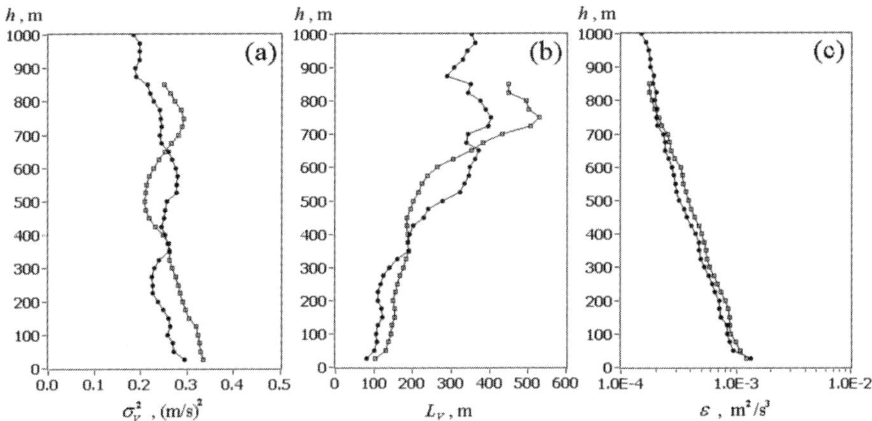

Figure 4.12 Vertical profiles of the (a) wind velocity variance, (b) outer scale of turbulence, and (c) turbulent energy dissipation rate retrieved from lidar measurements of the longitudinal structure function of wind velocity (circles) and by the DSW method (squares) on 27 August 2003 from 18:30 to 19:20 LT.

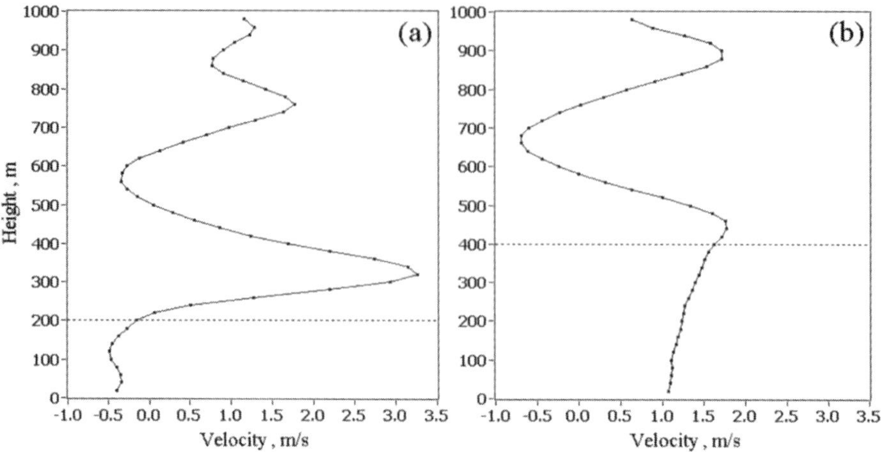

Figure 4.13 Vertical profiles of projection of mean wind velocity vector onto the plane of scanning by the probing beam as retrieved from lidar measurements on (a) 26 May and (b) 27 May 2005.

mean wind velocity vector $V_T(h)$ retrieved from the data of lidar measurements on, respectively, 26 and 27 May 2005 [28]. It can be seen that the mean wind field has a layered structure. The lower part of the atmosphere up to a height $h \approx 200$m (first day of measurements) and 400m (second day of measurements) is a layer of turbulent mixing of the wind flow, above which layers with high vertical gradients of wind velocity between different layers are observed. Attempts to apply the DSW, LSF, and TSF methods [8, 28, 42, 43] in order to obtain the estimates of the dissipation rate $\hat{\varepsilon}$ at heights beyond the turbulent mixing layer, where the strong regular inhomogeneity of wind along the vertical is observed, have led to results clearly contradicting the theory of turbulence of the atmospheric boundary layer (the value of $\hat{\varepsilon}$ could exceed the values of the dissipation rate in the mixing layer by an order of magnitude and even more). Thus, in the case of strong wind shears and jet flows, the considered methods do not work. Therefore, for the analysis of retrieval of vertical profiles of turbulence from lidar data, we restrict our consideration to only the mixing layer.

Figures 4.14 and 4.15 depict the obtained vertical profiles for the turbulent parameters [28]. The integral (outer) scale of turbulence L_V and the dissipation rate ε were determined from measurements of the transverse structure function (dashed curves) and Doppler spectrum width (solid curves). The estimates of wind velocity variance $\hat{\sigma}_V^2$ were calculated by (4.49), whose right-hand side is a sum of estimates of the variance of radial wind velocity $\hat{\sigma}_{\bar{V}}^2$ averaged over the sensing volume and the turbulent broadening of the Doppler spectrum $\hat{\sigma}_t^2$.

From the comparison of the contributions of $\hat{\sigma}_{\bar{V}}^2$ and $\hat{\sigma}_t^2$ to $\hat{\sigma}_V^2$, one can see that $\hat{\sigma}_t^2$ is about two times higher than $\hat{\sigma}_{\bar{V}}^2$ on the first measurement day (see Figure 4.14). This situation is possible only in the case for which the integral scale of turbulence L_V is far smaller than the longitudinal dimension of the sensing volume $\Delta z \approx 94$ m; that is, fluctuations of the radial velocity measured by lidar are averaged significantly and the turbulent energy for the most part comes for the broadening of the Doppler spectrum. The results of retrieval of the vertical profile of the integral

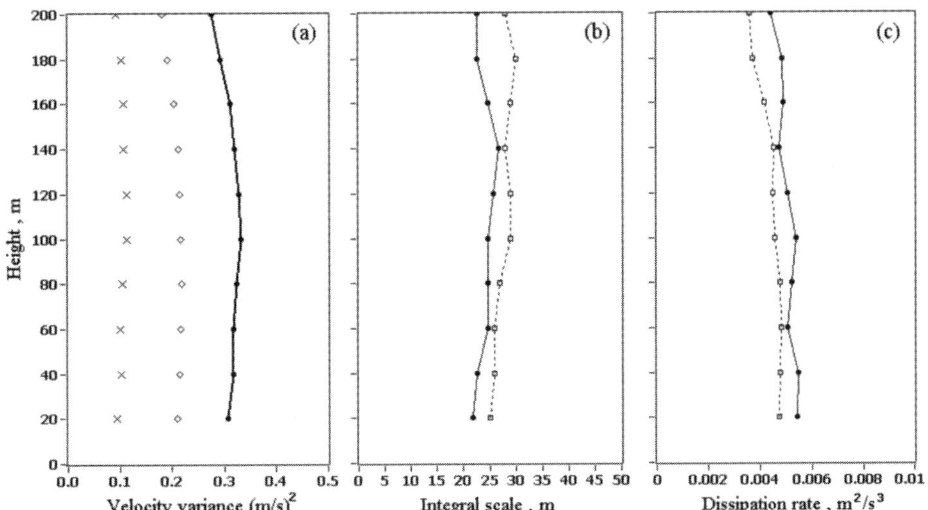

Figure 4.14 (a) Vertical profiles of turbulence parameters retrieved from data measured on 26 May 2005. Bold curve: variance of wind velocity $\hat{\sigma}_V^2$; crosses and rhombs: $\hat{\sigma}_{\tilde{V}}^2 - \hat{\sigma}_e^2$ and $\hat{\sigma}_t^2$, respectively. Estimates of (b) the integral scale of turbulence and (c) turbulent energy dissipation rate obtained from the Doppler spectrum width (dashed curves) and from the transverse structure function (thin solid curves).

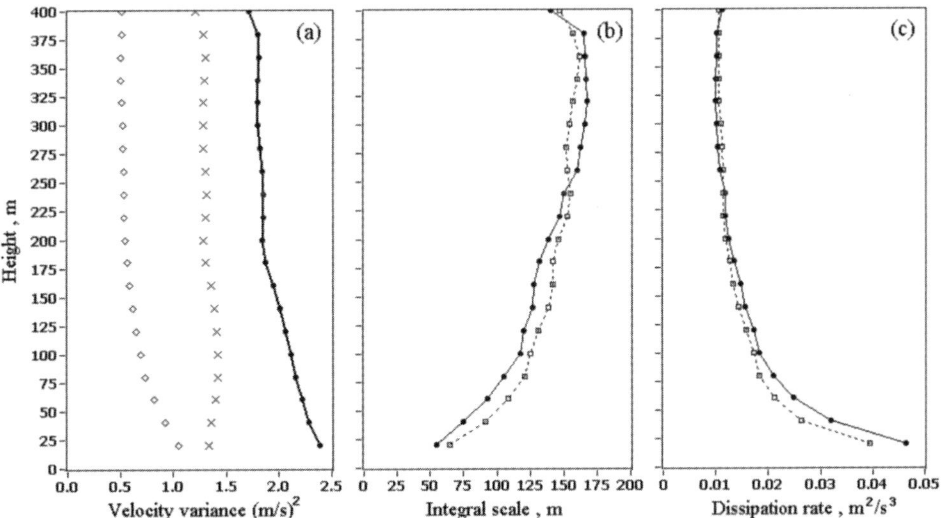

Figure 4.15 Vertical profiles of turbulence parameters retrieved from data measured on 27 May 2005. The remainder is the same as in Figure 4.14.

scale of turbulence are in agreement with this statement (see Figure 4.14(b)), where L_V is 20m to 30m. In this case, the methods considered for estimating L_V and ε give close results.

On the second measurement day, in contrast, the contribution from fluctuations of the radial wind velocity averaged over the sensing volume to the estimated variance $\hat{\sigma}_V^2$ is greater than the contribution from the turbulent broadening of the Doppler spectrum (Figure 4.15(a)). In this case, the ratio $\hat{\sigma}_{\tilde{V}}^2(h)/\hat{\sigma}_t^2(h)$ increases with height up to $h \approx 300$m. As expected, in this case, the estimated values of the integral scale of turbulence (Figure 4.15(b)) far exceed those shown in Figure 4.14(b). It can be seen from Figure 4.15 that the methods of L_V and ε estimation from the Doppler spectrum width (DSW method) and from the transverse structure function of wind velocity (TSF method) give close results, as on the first measurement day.

The same raw lidar measurement data, the results of whose processing are shown in Figures 4.14 and 4.15, were used for vertical profiling of the $\varepsilon(h)$ and $L_V(h)$ with estimation of these parameters from the longitudinal structure function (LSF method), and then these profiles were compared with the profiles obtained by the DSW and TSF methods. For the second measurement day (Figure 4.15), when the values of L_V exceed Δz, all three methods give relatively close results. However, for the first measurement day, when, according to Figure 4.14, L_V is far smaller than Δz, the estimates of the dissipation rate and the integral scale from the longitudinal structure function differ widely from the results obtained from the transverse structure function and the Doppler spectrum width methods. This is indicative of a large error in the estimation of turbulent parameters from the longitudinal structure function when $L_V < \Delta z$.

As mentioned earlier, in the case of strong inhomogeneity of the mean wind field (layered structure, jet flow, wind shears), none of the methods considered for estimating turbulent parameters from lidar data scanned in the vertical plane works efficiently despite the use of (4.43) through (4.46) in order to take into account variations of the mean wind with height in (4.47) through (4.51). If conical scanning by the probing beam is used during the measurements (see Figure 3.1) at small elevation angles (for example, $\varphi = 20°$), then the profiles $\varepsilon(h)$ and $L_V(h)$ can be retrieved from the transverse structure function of wind velocity even under conditions of abrupt changes of the mean wind with height. Thus, the authors of [14], having generalized the method of transverse structure function proposed in [13] for cw lidars (see Section 4.3), have obtained estimates of the turbulence profile with high vertical resolution from data of a conically scanning pulsed 2-μm CDL, which are in satisfactory agreement with the results of simultaneous turbulence measurements by wind sensors installed on a vertically movable platform.

The representativeness of estimates of turbulence parameters and capabilities of the methods described above can be judged only after the comparative analysis of results of simultaneous measurements by a coherent lidar (scanning in the vertical plane) and another device, whose measurement accuracy is rather high and well known.

4.6.1 Comparison of Lidar Estimates of the Turbulent Energy Dissipation Rate with Data from Sonic Anemometers

Experiments [60] conducted in the fall of 2004 at the airfield of the Frankfurt airport in Germany involved not only a 2-μm lidar, but also other measurement devices, in particular, a sonic anemometer set at a height of 20m. The anemometer was used to determine the diurnal profile of ε calculated from temporal spectra of the wind velocity.

Figure 4.16 shows four examples of comparison of the results of $\varepsilon(t)$ measurements with the lidar estimates from the longitudinal structure function (at a thickness for the average layer of $\delta h = 20$m and a measurement time of ~16 min) and with a sonic anemometer [46]. In general, these two devices demonstrate similar tendencies of variation in measured ε with time. Wide discrepancies in the measurement results corresponding to the early hours of November 8 and 12 are likely caused by the very weak turbulence ($\varepsilon < 10^{-5}$ m²/s³), which is a restriction for lidar measurements.

During the experiment in September 2003 in southeast Colorado (USA) within the Lamar Low-Level Jet Project, both conical scanning of the probing beam (see Section 3.4) and scanning in the vertical plane (Figure 4.7) were used. As already mentioned in Section 3.4, this experiment involved four sonic anemometers installed on a 120m meteorological tower (at heights of 54m, 67m, 85m, and 116m) and a 2-μm pulsed CDL. This lidar had parameters similar to those presented in Section 1.4 with the only difference being that the probing pulse duration was $\tau_P = 200$ ns and the pulse repetition frequency was 200 Hz.

For the comparative analysis of the results of simultaneous measurements of the dissipation rate by the lidar and sonic anemometers, raw experimental data obtained on 15 September 2003 were selected [32], because on this day the wind direction was such that the shadow effects of the meteorological tower on the data from the sonic anemometers can be neglected.

The estimates of the spectral density function of the longitudinal component of wind velocity $\hat{S}_V(f_k)$ were obtained from the data of sonic anemometers with the use of (4.12) at $M = 20000$ and $\Delta t = 0.05$s; that is, the time for measurement of one spectrum was $T = M\Delta t = 1{,}000$s ≈ 16.7 min. The mean wind velocity $\hat{U}$ was measured for the same time. An example of the spectral density function $\hat{S}_V(f_k)$ obtained with the use of 20 degrees of freedom for averaging is shown in Figure 4.17. The dissipation rate

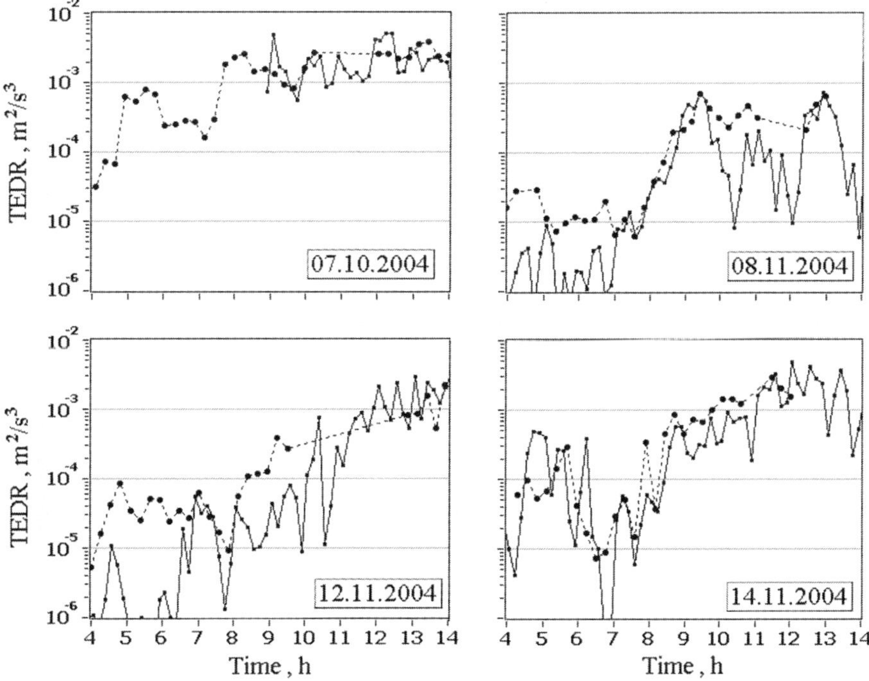

Figure 4.16 Comparison of the results of measurement of the turbulent energy dissipation rate by a sonic anemometer (solid curves) and by a lidar (dots connected by dashed lines).

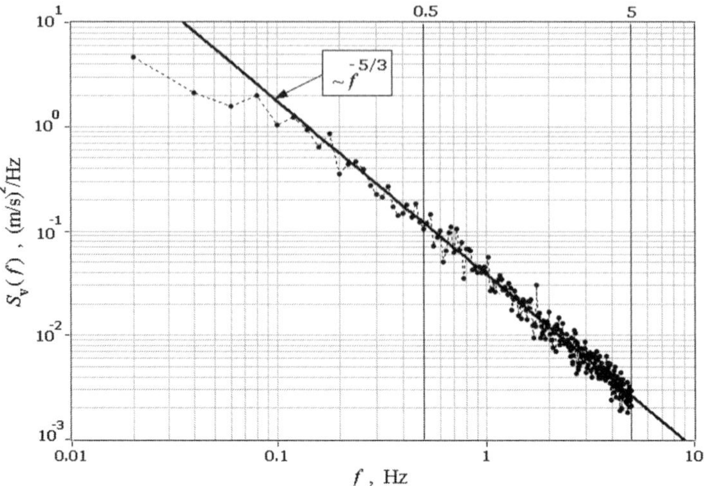

Figure 4.17 Typical example of the spectral density function of wind velocity fluctuations measured by sonic anemometers on 15 September 2003 with the use of 20 degrees of freedom for smoothing (dots); Kolmogorov-Obukhov spectrum $S_V(f) = 0.073 C_K \varepsilon^{2/3} U^{2/3} f^{-5/3}$ (solid line).

was calculated by (4.16), where Q_i was taken in the form $Q_i = 0.073 C_K \hat{U}^{2/3} f_{k_0+i}^{-5/3}$ on the assumption of a fulfilled frozen turbulence hypothesis and according to (2.94). In this case, $\hat{S}_V(f_k)$ data in the frequency range of 0.5 Hz $\leq f_{k0+i} \leq$ 5 Hz ($n = 4,500$) were used. These frequencies correspond quite well to the inertial interval of turbulence even at the strong wind and the small outer scale of turbulence L_V (for example, at $U = 15$ m/s and $L_V = 30$m). The calculations by (4.17) at $n = 4,500$, $\alpha = 1$, $\sigma_V/U = 0.1$, $\tau_V = 20$s, and $T = 1,000$s have shown that the relative error E_ε of estimation of the turbulent energy dissipation rate from sonic anemometer data is 3%.

The raw lidar data were used (with the rectangular temporal window at $T_W = 200$ ns and averaging over $L = 50$ shots for every Doppler spectrum) to obtain the arrays of estimates of radial wind velocities $\hat{V}(R_l, \varphi_i, n')$, where $R_l = R_0 + l\Delta R$, $R_0 = 190$m, $\Delta R = 30$m; $l = 0, 1, 2, ..., L' - 1$; $L' = 40$; $\varphi_i = i\Delta\varphi$; $i = 0, 1, 2, ..., 80$; $\Delta\varphi = 0.25°$; and $n' = 1, 2, 3, ...$ is the scan number. According to (2.34), at $T_W = 200$ ns and $\sigma_P = 120$ ns, the longitudinal dimension of the sensing volume is $\Delta z \approx 40$m. These arrays, in their turn, were used to calculate the longitudinal structure functions of velocity $\hat{D}_{\bar{V}}(r_{k'}, h)$ by (4.50) for different heights h. An example of this function is shown in Figure 4.18 [32]. The thickness of the averaging layer δh was taken equal to 20m, while the measurement time for each structure function was about 20 min. The estimates $\hat{\varepsilon}$ and $\hat{L}_V$ were obtained through the fitting aimed at minimization of the functional (4.52).

It is shown in Section 4.5 with the use of numerical simulation that at a 5-min measurement by a coherent lidar the error of estimation of the dissipation rate is $E_\varepsilon \approx 20\%$. In our case, the measurement time is four times longer. The rough estimates (with regard to the mean wind velocity, area, and time of data averaging) show that in the field experiment under consideration the accuracy of lidar measurements of the dissipation rate should be ~10% to 15%.

The vertical profiles of the turbulent energy dissipation rate estimated from lidar data are exemplified as two curves in Figure 4.19 [32], which also shows the results of ε determination from the data of sonic anemometers at four heights. It can be seen that the dissipation rate calculated from both the lidar data and the sonic anemometer data at 01:30 UTC is about an order of magnitude lower than that at 09:00 UTC (local time is UTC − 6h). The explanation is that in the time interval between these measurements the wind strength and turbulence increased. At the same time, ε at a height of 116m often was an order of magnitude smaller than that at lower heights of anemometer installation. This vertical inhomogeneity of the dissipation rate is determined by the features of the nocturnal boundary layer, whose thickness is much smaller than in the daytime.

The time dependences of the dissipation rate $\varepsilon(t)$ determined from the data of sonic anemometers at four heights of 54m, 67m, 85m, and 116m are shown as curves in Figure 4.20. The symbols in this figure are for the lidar estimates of ε at the same heights obtained from 10 retrieved vertical profiles of the dissipation rate [32]. The estimates of the outer scale of turbulence L_V from the same lidar data show that it usually varied slightly inside the layer 54m $\leq h \leq$ 116m and took values of, on average, about 50m. In the rough approximations the variances of random errors of lidar estimates ε at these heights should be nearly identical. However, for height $h = 116$m, near which a sharp change of the vertical profile of the dissipation rate is often observed, the lidar estimate $\hat{\varepsilon}$ obtained from the 20-m-thick layer can be biased. It can be seen in Figure 4.20 that the difference between the results of lidar and anemometer estimates of ε at a height of 116m is greater than that at other heights (54m, 67m, and 85m).

The results of simultaneous measurements of the dissipation rate by coherent lidar scanning in the vertical plane and the sonic anemometers are compared in Figure 4.21 with the use of the data of Figure 4.20. The theoretical estimates of the error of dissipation rate measurement by a sonic anemometer ($E_S = 3\%$) and lidar ($E_L = 10\%$ to 15%) are presented above. It follows from these estimates that the

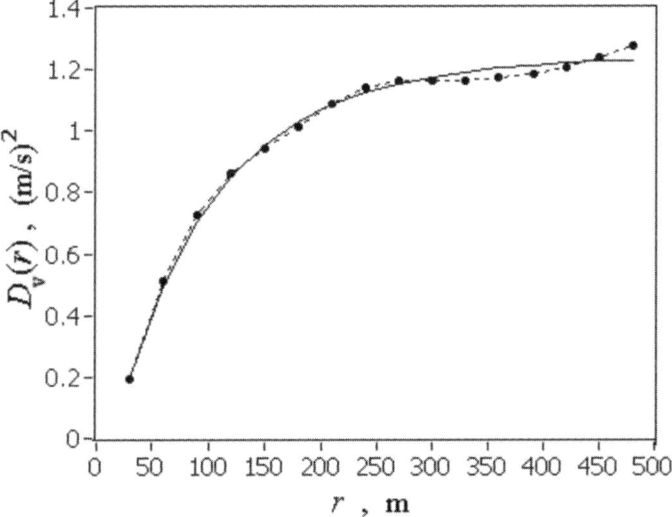

Figure 4.18 Typical example of the structure function of radial wind velocity obtained from lidar measurements on 15 September 2003 (dots). The structure function was calculated by (2.120), (2.121), and (2.118), where the parameters ε and L_V are determined through minimization of the functional (4.52) (solid curve).

measurements by sonic anemometers have very high accuracy and can be taken as reference ones for comparisons with lidar data. We introduce the following characteristics: normalized bias $b_\varepsilon = \langle (\hat{\varepsilon}_L - \hat{\varepsilon}_S) / \hat{\varepsilon}_S \rangle_a \times 100\%$ and normalized standard deviation of the lidar estimate $\sigma_\varepsilon^{(E)} = \sqrt{\langle [(\hat{\varepsilon}_L - \hat{\varepsilon}_S)/\hat{\varepsilon}_S]^2 \rangle_a - b_\varepsilon^2}$, where $\hat{\varepsilon}_L$ and $\hat{\varepsilon}_S$ are the dissipation rate estimates obtained, respectively, from the lidar and sonic anemometer data, and $\langle x \rangle_a = \sum_{i=1}^{40} x_i/40$. Using the data shown in Figure 4.21, we obtain $b_\varepsilon = -7\%$ and $\sigma_\varepsilon^{(E)} = 25\%$ [32]. Thus, random deviations of $\hat{\varepsilon}_L$ from $\hat{\varepsilon}_S$ far exceed the regular discrepancies of these estimates. The nonzero value of the b_ε

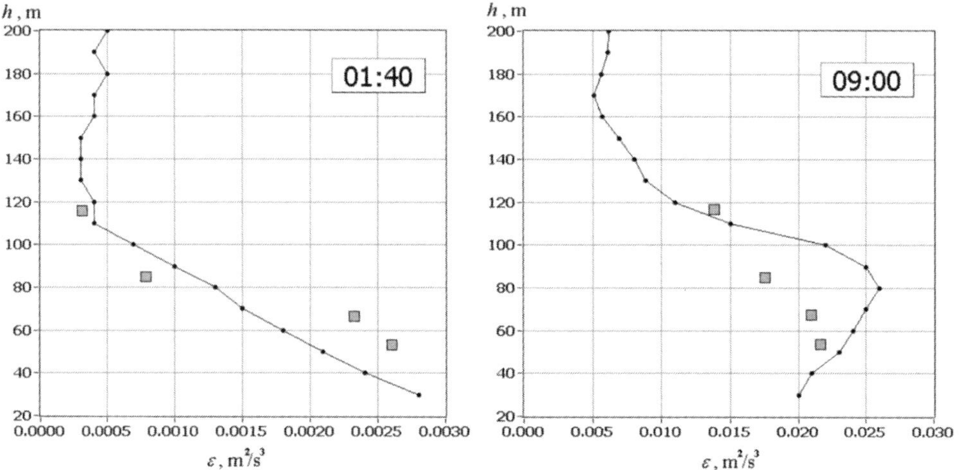

Figure 4.19 Vertical profiles of the turbulent energy dissipation rate retrieved from CDL data measured on 15 September 2003 at 01:40 and 09:00 UTC and estimates of the dissipation rate obtained from the simultaneous measurements by sonic anemometers at heights of 54m, 67m, 85m, and 116m (squares).

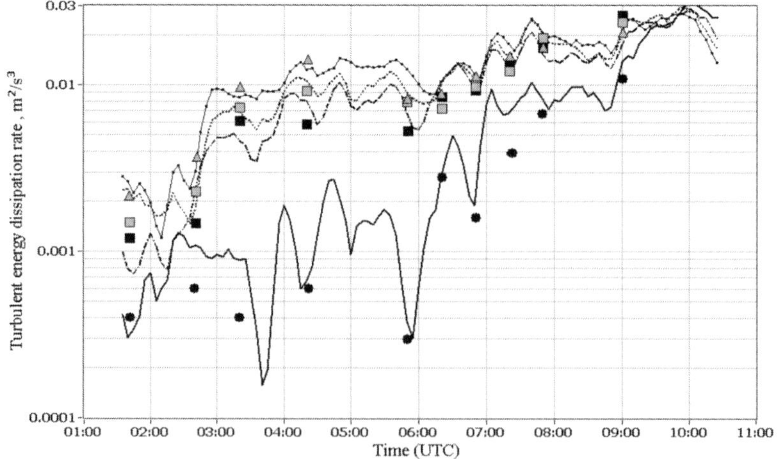

Figure 4.20 Temporal profile of the turbulent energy dissipation rate measured by sonic anemometers at heights $h = 54$m (dots connected by thin lines), 67m (dashed curve), 85m (dot-and-dash curve), and 116m (bold solid curve) and lidar estimates $\hat{\varepsilon}$ at heights $h = 54$m (triangles), 67m (gray squares), 85m (black squares), and 116m (circles).

parameter is likely caused by the error of lidar estimate $\hat{\varepsilon}_L$ due to the regular vertical inhomogeneity of the dissipation rate inside the 20m layer.

The errors of estimates of the dissipation rate obtained from raw lidar and sonic anemometer data can be considered to be statistically independent. Then, for the theoretical estimation of the normalized standard deviation $\sigma_\varepsilon^{(T)} = \sqrt{\langle [(\hat{\varepsilon}_L - \hat{\varepsilon}_S)/\hat{\varepsilon}_S]^2 \rangle}$ at $|\hat{\varepsilon}_S - \varepsilon| \ll \varepsilon$, we can use the equation $\sigma_\varepsilon^{(T)} = \sqrt{E_S^2 + E_L^2}$. Given $E_L = 10\%$ and $E_S = 3\%$, we obtain $\sigma_\varepsilon^{(T)} = 10.4\%$, that is, the value of $\sigma_\varepsilon^{(T)}$ is almost completely determined by the accuracy of the lidar estimate of the dissipation rate ($\sigma_\varepsilon^{(T)} \approx E_L$). The experimental estimate of the normalized standard deviation of $\sigma_\varepsilon^{(E)} = 25\%$ exceeds by ~1.5 to 2.5 times the theoretical estimate $\sigma_\varepsilon^{(T)}$. The discrepancy between theory and experiment can be explained by various reasons including those mentioned earlier in Section 4.6. At the same time, we can conclude that the error of the lidar measurements of the turbulent energy dissipation rate does not exceed 25%. This accuracy is quite acceptable for analysis of the influence of turbulence on the wake vortex lifetime based on the data of lidar measurements, whose results are presented in Chapter 5.

Thus, three methods of retrieval of the vertical profiles of wind turbulence parameters from data measured by a pulsed lidar scanning in the vertical plane (DSW, LSF, and TSF methods) have been considered above. Each of these methods has advantages and disadvantages. The LSF method is the simplest one (in the sense of computer processing of the raw data). However, for the conditions, when the outer scale of turbulence L_V is much smaller than the longitudinal dimension of the sensing volume Δz, we prefer the other two methods. In the case of moderate and strong turbulence when SNR > 1, the DSW method is more accurate than the TSF method. In the case of weak turbulence ($\varepsilon < 10^{-4}$ m^2/s^3) and a small outer scale of turbulence ($L_V \ll \Delta z$ and $L_V \sim \Delta R$), the TSF method can provide the best result (but

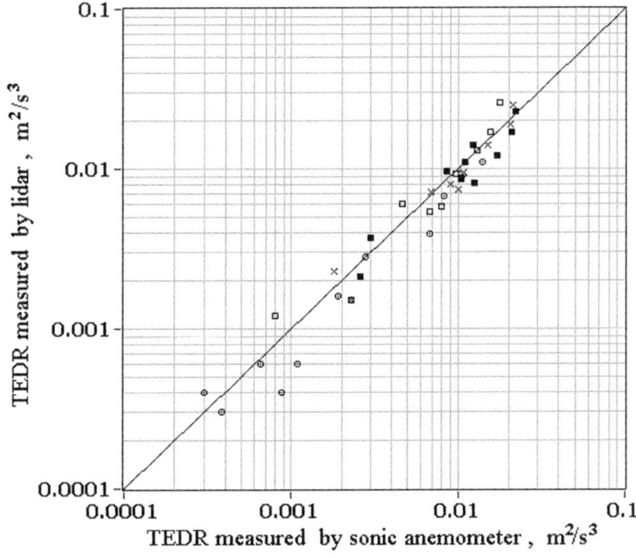

Figure 4.21 Comparison of estimates of the turbulent energy dissipation rate obtained from simultaneous measurements by coherent lidar and sonic anemometers at heights of 54m (closed squares), 67m (crosses), 85m (open squares), and 116m (circles).

for this method, as well as for applicability of the LSF method, the SNR should be no lower than −5 dB). Taking into account that SNR, ε, and L_V vary with height, a combination of the DSW, LSF, and TSF methods can be used for the retrieval of vertical turbulence profiles. A general feature of the considered methods is that neither of them can provide the required accuracy in the case of lidar measurements under conditions of strong vertical inhomogeneity of the mean wind (strong wind shears and narrow jet flows). In addition, field experiments [8, 28, 42, 43; see also Figures 4.9–4.12, 4.14–4.16, 4.19, and 4.20) show that the lidar estimation of the dissipation rate is possible only for the range bounded from below $\varepsilon \geq 10^{-5}$ m²/s³. At lower values of ε, turbulent distortions of the lidar echo signal are too weak to estimate ε from them.

4.7 Estimation of the Turbulence Energy Dissipation Rate from Data Measured with a Conically Scanning Pulsed CDL

In [14], based on the results of [7] (see also Section 4.3), a method is proposed for estimating the dissipation rate ε, outer (integral) scale of turbulence L_V, and variance of the wind velocity σ_V^2 from the azimuth (transverse) structure function of the radial wind velocity obtained from measurements by a pulsed CDL that uses sector conical scanning by its probing beam. The possibilities of estimating the dissipation rate from the data measured by a pulsed CDL with the use of the full conical scanning were studied in [15, 16]. Below we describe the procedure for the processing of lidar data obtained with a conical-scanning probing beam in order to estimate the turbulence energy dissipation rate by the transverse structure function method of fluctuations of radial velocity. We then present the results of testing of this approach in numerical and field experiments.

Let the elevation angle φ be fixed while the 2-μm CDL takes measurements and the conical scanning by the probing beam be carried out at a constant angular rate ω_0. In this case, the azimuth angle $\theta = \omega_0 t$ varies with time t from 0° to 360° (full rotation around the vertical axis passing through the point of lidar location). For different distances $R_i = R_0 + i\Delta R$ from the lidar to the center of the sensing volume and azimuth angles $\theta_m = \theta_0 + m\Delta\theta$, Doppler spectra are obtained from the raw lidar data with the use of the rectangular temporal window T_W and accumulation of spectral estimates over L laser shots, where $R_0 \gg \Delta p$, $\Delta p = \sigma_P c/2$, $i = 0, 1, 2, ..., I - 1$, $\Delta R = cT_W/2$, $m = 1, 2, 3, ..., M$, $\Delta\theta = \omega_0 L T_P$, and T_P is the time interval between shots.

According to (2.35), the unbiased estimate of the radial velocity $\hat{V}_r(R_i, \theta_m)$ can be represented, with allowance for the averaging of the wind velocity over the azimuth angle, in the form [15]

$$\hat{V}_r(R_i, \theta_m) = \overline{V}_r(R_i, \theta_m) + V_e(R_i, \theta_m) \tag{4.56}$$

where

$$\overline{V}_r(R_i, \theta_m) = L^{-1} \sum_{l=1}^{L} \int_{-\infty}^{+\infty} dz' Q_s(z') V_r(R_i + z', \theta_{m-1} + l\omega_0 T_P) \tag{4.57}$$

and $Q_s(z')$ is given by (2.33).

By applying the sine wave fitting procedure (see Chapter 3) to the obtained array $\hat{V}_r(R_i, \theta_m)$, we can obtain estimates of the mean wind velocity vector $\hat{V} = \{\hat{V}_z, \hat{V}_x, \hat{V}_y\}$ at heights $h_i + h_L + R_i \sin\varphi$ (h_L is the height from which the probing pulse is launched into the atmosphere). Then fluctuations of estimates of the radial velocity are calculated as

$$V'_R(R_i, \theta_m) = \hat{V}_r(R_i, \theta_m) - \hat{V}(h_i) \cdot S(\theta_m) \tag{4.58}$$

From the array of fluctuations of lidar estimates of the radial velocity, we obtain the estimate of the azimuth structure function $\hat{D}_L(n\Delta\theta; R_i)$ as

$$\hat{D}_L(n\Delta\theta; R_i) = (M - n)^{-1} \sum_{m=1}^{M-n} [\hat{V}'_r(R_i, \theta_m + n\Delta\theta) - \hat{V}'_r(R_i, \theta_m)]^2 \tag{4.59}$$

where $n = 1, 2, \ldots, N \ll M$. This azimuth structure function estimate carries the information about turbulent parameters, in particular, about the dissipation rate ε. To obtain the estimate of the dissipation rate $\hat{\varepsilon}$ from $\hat{D}_L(n\Delta\theta; R_i)$, the dissipation rate should be represented by the azimuth structure function of the lidar measured radial velocity $D_L(n\Delta\theta; R_i) = \langle \hat{D}_L(n\Delta\theta; R_i) \rangle$ in an explicit form.

Taking the conditions $N_a \gg 1$, $R_i \gg L_V$, and $N\Delta\theta \ll \pi/2$ to be true, we assume the following. Turbulence is stationary, homogeneous, isotropic, and described by the von Karman model [51]. The speed of displacement of the sensing volume along the boundary of the scanning cone base $\omega_0 R_i \cos\varphi$ far exceeds the mean wind velocity U. This allows us to disregard the mean wind transport of turbulent inhomogeneities in the derivation of the equation for $D_L(n\Delta\theta; R_i)$.

Based on the aforesaid, in (4.57) we pass the radial velocity V_r from the polar coordinate system $\{z', \theta\}$ to the rectangular coordinate system on a plane $\{z', y'\}$ (z' is the longitudinal coordinate axis and y' is the transverse one) and, upon replacement of the summation with integration in (4.57), we obtain the following equation:

$$\overline{V}_r(R_i, \theta_m) = \frac{1}{\Delta y_i} \int_{-\Delta y_i/2}^{\Delta y_i/2} dy' \int_{-\infty}^{+\infty} dz' Q_s(z') V_r(R_i + z', R_i\theta_m \cos\varphi + y') \tag{4.60}$$

where $\Delta y_i = R_i\Delta\theta \cos\varphi$ is the transverse (on the scanning surface) dimension of the sensing volume. Then, replacing $\hat{V}$ with $\langle V \rangle$ in (4.58), from (4.56) and (4.58) through (4.60), after the corresponding manipulations, we obtain the equation

$$D_L(n\Delta\theta; R_i) = \overline{D}_L(n\Delta\theta; R_i) + 2(1 - \delta_n)\sigma_e^2 \tag{4.61}$$

where

$$\overline{D}_L(n\Delta\theta; R_i) = 8\int_0^\infty d\kappa_z \int_0^\infty d\kappa_y S_V^{(2)}(\kappa_z, \kappa_y) H_P(\kappa_z) H_\perp(\kappa_y)[1 - \cos(2\pi n\Delta y_i\kappa_y)] \tag{4.62}$$

is the transverse structure function of the radial velocity averaged over the sensing volume, $S_V^{(2)}(\kappa_z, \kappa_y)$ is the two-dimensional spatial spectrum of turbulent fluctuations of wind velocity [see (2.86) and (2.71)], $H_p(\kappa_z)$ is the transfer function of the low-frequency filter along the axis z' given by (2.118), and

$$H_\perp(\kappa_y) = \sin c^2(\pi \Delta y \kappa_y) \tag{4.63}$$

is the transfer function of the low-frequency filter along the y' axis and $\mathrm{sinc}(x) = \sin(x)/x$.

According to (4.61), the difference of the structure functions of the radial velocity measured by lidar $\Delta D_L(n\Delta\theta; R_i) = D_L(n\Delta\theta; R_i) - D_L(\Delta\theta; R_i)$ is independent of the error of lidar estimation of the radial velocity σ_e. Therefore, the wind turbulence parameters should be estimated from the difference of measured structure functions $\Delta \hat{D}_L(n\Delta\theta; R_i) = \hat{D}_L(n\Delta\theta; R_i) - \hat{D}_L(\Delta\theta; R_i)$ through the fitting of the difference of the structure functions $\Delta \bar{D}_L(n\Delta\theta; R_i) = \bar{D}_L(n\Delta\theta; R_i) - \bar{D}_L(\Delta\theta; R_i)$ [calculated by (4.62), (2.71), (2.86), (2.118), and (4.63)] to $\Delta \hat{D}_L(n\Delta\theta; R_i)$. In this case, it is sufficient to calculate the function $F_L(n\Delta\theta; R_i, L_V) = \Delta D_L(n\Delta\theta; R_i)/\varepsilon^{2/3}$ depending on only one parameter, L_V.

To obtain the estimates of the dissipation rate $\hat{\varepsilon}$ and the outer scale of turbulence $\hat{L}_V$, we use the following algorithm [15]:

$$\min\{\rho(L_V)\} = \rho(\hat{L}_V) \tag{4.64}$$

and

$$\hat{\varepsilon} = [\mu(\hat{L}_V)]^{3/2} \tag{4.65}$$

where

$$\rho(L_V) = \sum_{n=2}^{N} \left[\frac{\Delta \hat{D}(n\Delta\theta; R_i)}{F_L(n\Delta\theta; R_i, L_V)} - \mu(L_V) \right]^2 \tag{4.66}$$

$$\mu(L_V) = \frac{1}{N-1} \sum_{n=2}^{N} \frac{\Delta \hat{D}(n\Delta\theta; R_i)}{F_L(n\Delta\theta; R_i, L_V)} \tag{4.67}$$

With the obtained estimates $\hat{\varepsilon}$ and $\hat{L}_V$, the variance of wind velocity $\hat{\sigma}_V^2$ is calculated by (2.71).

The practical implementation of the algorithm of (4.64) through (4.67) includes the calculation of F_L at different L_V. According to the known experimental data [56], in the atmospheric boundary layer at heights from 25m to 300m the integral scale of fluctuations of the longitudinal wind velocity component varies from 30m to

450m depending on the temperature stratification. That is why in [15, 16] F_L was calculated for L_V ranging from 20m to 500m with a 10m step.

To decrease fluctuations of lidar estimates of the azimuth structure function of the radial velocity, we can use additional averaging over the $I' + 1$ layers along the z' axis; that is, replace $\Delta\hat{D}_L(n\Delta\theta; R_i)$ with $(I' + 1)^{-1} \sum\limits_{i'=-I'/2}^{I'/2} \Delta\hat{D}_L(n\Delta\theta; R_i + i'\Delta R)$ in (4.66) and (4.67), where I' is an even integer. Correspondingly, $F_L(n\Delta\theta; R_i, L_V)$ should be replaced with $(I' + 1)^{-1} \sum\limits_{i'=-I'/2}^{I'/2} F_L(n\Delta\theta; R_i + i'\Delta R, L_V)$. If several complete conical scans are carried out during the experiment, then estimates of structure functions obtained from the data of every scan can be averaged.

In [15], the applicability of the method of transverse structure function [see (4.62) through (4.67)] for estimating the turbulent energy dissipation rate from conically scanning pulsed CDL data with maximal separation of the azimuth angles $N\Delta\theta = \pi/6$ is justified numerically. It is shown that the procedure of numerical simulation of random realizations of lidar estimates of the radial velocity can be significantly simplified through the transition from the polar coordinate system $\{R, \theta\}$ to the Cartesian one $\{x, y\}$ by taking $x = R$ and $y = R\theta\cos\varphi$ (where $0 \le y \le 2\pi R\cos\varphi$). This simplification allows the statistical analysis of lidar estimates of the turbulence parameters with allowance for the averaging (over the sensing volume) and the error of estimation of the radial wind velocity.

The simulation was carried out with the following values for the parameters: $T_P = 5$ ms, $\sigma_P = 120$ ns ($\Delta p = 18$m), $2\pi/\omega_0 = 60$s, $T_W = 200$ ns ($\Delta R = 30$m), and $L = 50$. To calculate one value of the error of lidar estimation of the radial velocity σ_e, we used 10^5 independent realizations of $\hat{V}_r - \bar{V}_r$. The numerical experiment has shown that for any intensity of wind turbulence under the condition SNR ≥ 0.2 (or -7 dB) the lidar estimate of the radial velocity is 100% unbiased. At SNR = -8 dB, the fraction of bad estimates of $\hat{V}_r$ uniformly distributed in the estimation range from -25 m/s to $+25$ m/s [17, 61, 62] is about 10^{-4}. Consequently, for this SNR the estimate also can be considered unbiased, so much more that if 1 of 240 estimates per one scan at a fixed range appears to be bad, it can be easily rejected when obtaining the azimuth structure function of the radial velocity from this array of estimates.

Figure 4.22 shows the calculated error of the lidar estimate of the radial velocity σ_e as a function of the SNR at $L_V = 100$m and the dissipation rate $\varepsilon = 0$, 10^{-3} ($\sigma_V = 0.524$ m/s), 10^{-2} m^2/s^3 ($\sigma_V = 1.128$ m/s). One can see that with no turbulence or at very weak turbulence (the result calculated at $\varepsilon = 10^{-4}$ m^2/s^3 almost completely coincides with curve 1), σ_e varies from 0.1 m/s (at SNR = 10 dB) to 0.4 m/s (at SNR = -8 dB). In the case of strong turbulence ($\varepsilon = 10^{-2}$ m^2/s^3) and SNR ~ 10, the error σ_e is roughly twice as high as in the absence of turbulence. As SNR decreases, σ_e increases from 0.2 to 0.46 m/s (see curve 3).

Figure 4.23 shows the calculated relative error of the lidar estimate of the dissipation rate E_ε [determined by (4.54)] as a function of the ratio σ_e/σ_V at $N_{\text{scan}} = 1$ and $N_{\text{scan}} = 4$. As expected, the error E_ε increases with an increase in the error of the lidar estimate of the radial velocity σ_e at fixed σ_V or with an increase of the wind turbulence intensity at fixed σ_e. Under the condition $\sigma_e/\sigma_V < 0.3$, the influence of the error of the radial velocity estimate on the accuracy of estimating the dissipation

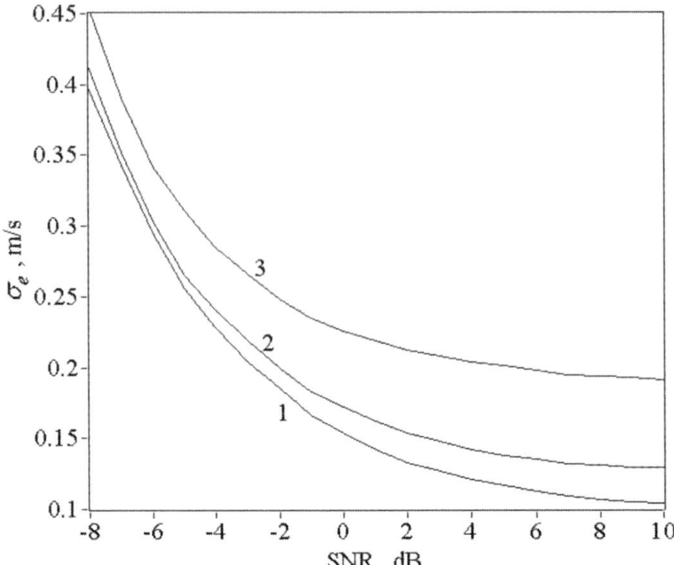

Figure 4.22 Error of lidar estimate of the radial velocity σ_e as a function of the SNR at $\varepsilon = 0$ (curve 1), 10^{-3} (curve 2), and 10^{-2} m^2/s^3 (curve 3). (© 2013 Pleiades Publishing, Ltd. Used with permission. From [16].)

rate can be neglected. When σ_e and σ_V are comparable, the error E_ε can reach 40% at $N_{\text{scan}} = 1$. A fourfold increase of the scan number N_{scan} allows the error to be nearly halved.

Taking into account (2.71), from the data shown in Figures 4.22 and 4.23 for the case of $L_V = 100$m and $R \approx 458$m, we can find the smallest value of the dissipation rate ε_{min} determined from the conically scanning lidar data with an error of less than 25% (for $N_{\text{scan}} = 4$, the error $E_\varepsilon = 25\%$ at $\sigma_e/\sigma_V = 1.16$). Thus, $\varepsilon_{\text{min}} = 6 \cdot 10^{-3}$ m^2/s^3 at $N_{\text{scan}} = 1$ when SNR $= -8$ dB, and $\varepsilon_{\text{min}} = 2.7 \cdot 10^{-4}$ m^2/s^3 when SNR $= 0$ dB, and at $N_{\text{scan}} = 4$ $\varepsilon_{\text{min}} = 2.8 \cdot 10^{-4}$ m^2/s^3 (SNR $= -8$ dB) and $\varepsilon_{\text{min}} = 1.5 \cdot 10^{-5}$ m^2/s^3(SNR $= 0$ dB).

In addition to the numerical analysis of the accuracy of lidar estimates of the dissipation rate, in [16] the considered technique was tested with the data of the Lamar Low-Level Jet Project field experiment (see Section 3.4). The data obtained from the conical scanning by the probing beam around the vertical axis at elevation angle $\varphi = 10°$ were used. The duration of one full scan was 1 min. The scanning was conducted every hour from 00:00 to 3:00 LT. The data were processed with the algorithm of (4.62) through (4.67). For heights $h > 120$m, when SNR could take values smaller than -8 dB (but not smaller than -12 dB), the procedure of filtering of good estimates of the radial velocity was used: The smaller the SNR compared to -8 dB, the higher the probability of getting a bad estimate of the radial velocity [63]. The vertical profiles of the turbulent energy dissipation rate retrieved from these data are depicted in Figure 4.24.

The curves in Figure 4.24 show the vertical profiles of ε determined from lidar data, while symbols show the results of measurement by sonic anemometers. Figure 4.25 compares the measurements of the dissipation rate by sonic anemometers and lidar.

In the experiment, the SNRs at the heights where the sonic anemometers were located ranged from -8 to -6 dB. Based on the information about turbulent parameters obtained from the sonic anemometer data and on the results shown in

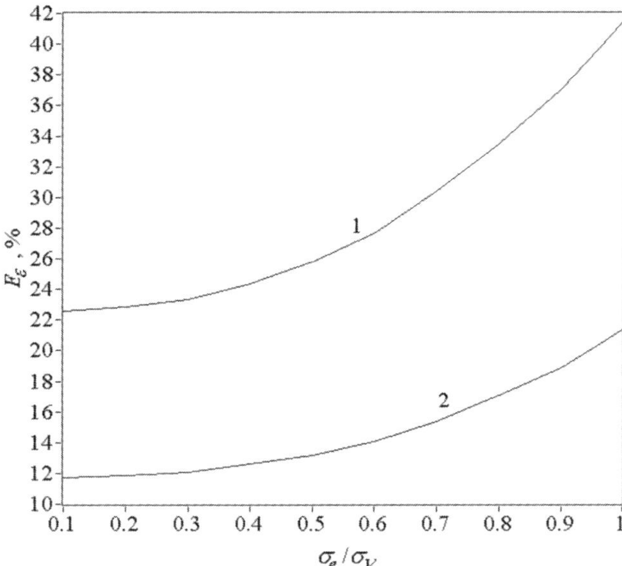

Figure 4.23 Related error of the lidar estimate of the dissipation rate of turbulent energy E_ε as a function of the ratio σ_e/σ_V at $N_{\text{scan}} = 1$ (curve 1) and 4 (curve 2). (© 2013 Pleiades Publishing, Ltd. Used with permission. From [16].)

Figures 4.22 and 4.23, we have calculated the relative error of the lidar estimate of the dissipation rate E_ε. According to the results obtained for the conditions of this experiment, the value of E_ε should vary from 24% to 40% (on average, 32%).

Taking into account the highly accurate measurements of the dissipation rate by the sonic anemometers, the relative error of the lidar estimate of the dissipation rate can be calculated from field experiment data as follows:

$$\hat{E}_\varepsilon = \sqrt{\frac{1}{K} \sum_{k=1}^{K} \left[\hat{\varepsilon}_L(k)/\hat{\varepsilon}_S(k) - 1 \right]^2} \times 100\% \tag{4.68}$$

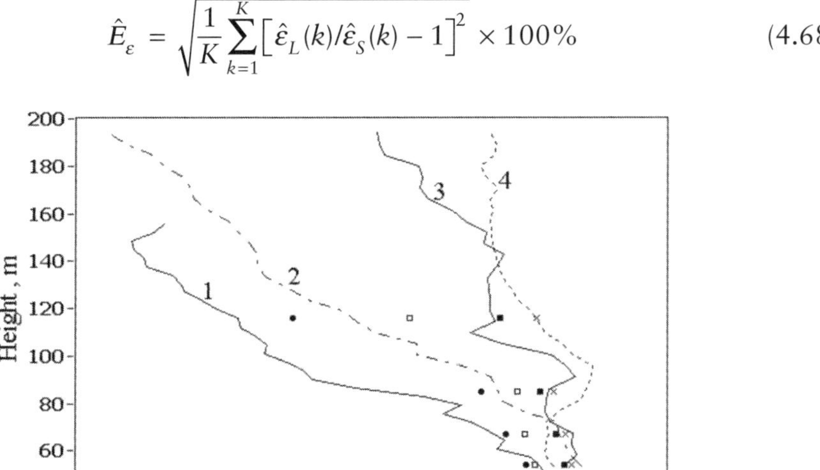

Figure 4.24 Vertical profiles of the turbulent energy dissipation rate retrieved from measurements by lidar (curves 1–4) and sonic anemometers (symbols) on 15 September 2003 at 00:00 (1, circles), 01:00 (2, triangles), 02:00 (3, squares), and 03:00 (4, crosses). (© 2013 Pleiades Publishing, Ltd. Used with permission. From [16].)

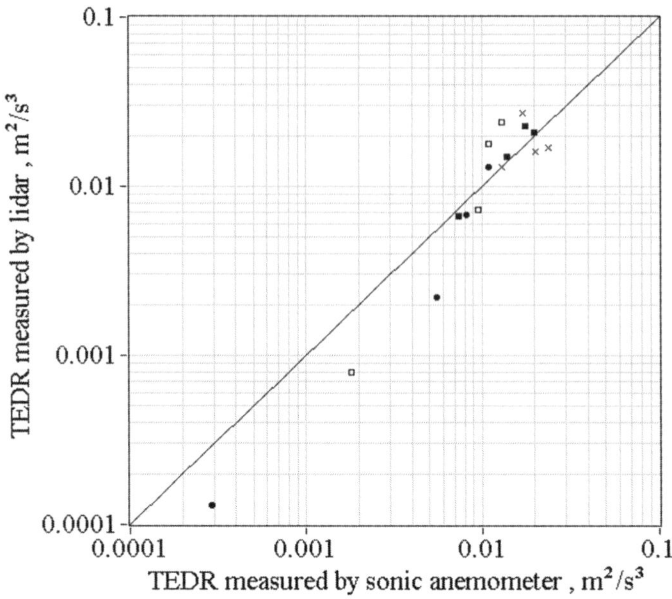

Figure 4.25 Comparison of the turbulent energy dissipation rate estimated from data measured with sonic anemometers and lidars at heights of sonic anemometer locations. (© 2013 Pleiades Publishing, Ltd. Used with permission. From [16].)

The use of all estimates of the dissipation rate from the lidar data $(\hat{\varepsilon}_L)$ and the sonic anemometer data $(\hat{\varepsilon}_S)$ is depicted in Figure 4.25. According to (4.68), $\hat{E}_\varepsilon \approx 42\%$. This estimate is 10% higher than the theoretical value of E_ε presented earlier. Certainly, at $K = 16$ (only 16 estimates are shown in Figure 4.25) the estimate $\hat{E}_\varepsilon$ may be too rough. Nevertheless, if the error of the lidar estimate of the dissipation rate determined from measurement data of one full scan ($N_{\text{scan}} = 1$) is about 40%, then, as the computer simulation shows (Figure 4.23), the use of lidar data obtained for four full independent scans allows the error E_ε to be decreased to 20%.

4.8   Simulation of Clear Air Turbulence Detection by Coherent Doppler Lidars

The main causes of clear air turbulence in the free atmosphere are temperature and wind velocity contrasts, which are formed due to the meeting and interaction of air masses with different characteristics near atmospheric fronts and high-altitude frontal zones, loss of stability by waves formed in inversion layers at the tropopause and near other atmospheric interfaces, deformation of airflows by mountains, and the appearance of wave perturbations in the lee of mountain obstructions [48]. According to the data of airborne measurements [48], in the upper troposphere (at altitudes of 6 to 10 km) and in the lower stratosphere (up to a height of ~25 km), the turbulence with occurrence up to 25% concentrates in discontinued (with characteristic horizontal dimensions of parts ~10 km) and continuous (with horizontal dimensions ~100 km) zones, which are transported by the mean flow. In this case, the vertical dimension of the turbulence zone is ~1 km.

Clear air turbulence (CAT) zones constitute a danger for aircraft, and the early detection of these zones in the aircraft flight path is necessary to increase aviation safety. The use of Doppler lidars seems to be promising for this purpose. Next, a numerical simulation of the echo signals of pulsed airborne CDLs is used to study the possibility of determining the clear air turbulence intensity level from estimates of the Doppler spectrum width.

The measurements of spectral densities of fluctuations of the horizontal (longitudinal) wind velocity component

$$S_u(\kappa) = (2\pi)^{-1} \int\limits_{-\infty}^{+\infty} dr B_u(r) \exp(-j\kappa r) \tag{4.69}$$

where $B_u(r) = \langle u'(r_0 + r)u'(r_0)\rangle$ is the correlation function, and the u' are wind velocity fluctuations, have shown [48] that in the free atmosphere at frequencies $\kappa \geq 10^{-3}$ m^{-1} (up to the boundary of the turbulence viscous interval) the spectrum obeys the Kolmogorov-Obukhov law $[S_u(\kappa) \sim \kappa^{-5/3}]$, whereas at low frequencies (beyond the boundary of turbulent energy income) the exponent v in the power dependence of the spectrum on the spatial frequency $[S_u(\kappa) \sim \kappa^v]$ differs from the $-5/3$ law and, by various sources, varies from -2.2 to -3.1 (on average, $v \approx -2.7$). The presence of two parts with different power dependencies in the turbulent spectrum is determined by the fact that in the upper troposphere and stratosphere the conditions of stable temperature stratification usually take place. Therefore, the process of wind perturbation energy transfer from larger scales to smaller ones is first accompanied by the expenditure of kinetic energy of motion for the work against the Archimedes buoyant force (significant part of the turbulence energy transforms into the potential energy of stratification) and only then, in the scales corresponding to the inertial interval, can the process of energy transfer in the spectrum begin to proceed without losses, with the constant rate equal to the rate of transformation of the turbulence energy into the thermal one [48].

It is commonly accepted that the effect of clear air turbulence on aircraft flight stability is evaluated using a four-point scale: weak (b_1), moderate (b_2), strong (b_3), and very strong or storm (b_4). Based on the data for the spectrum $S_u(\kappa)$ at $\kappa = 2 \cdot 10^{-3}$ m^{-1} presented in [48] and taking into account that for $\kappa \geq 10^{-3}$ m^{-1}

$$S_u(\kappa) = 0.25\varepsilon^{2/3}\kappa^{-5/3} \tag{4.70}$$

we have calculated the values of the turbulence energy dissipation rates ε corresponding to the levels of this four-point scale. The results are summarized in Table 4.2.

Aircraft do not react to any turbulent inhomogeneities of a wind flow, but only to perturbations with the scales corresponding to the relatively narrow part of the spectrum of turbulence. In particular, the flight of an aircraft at subsonic speeds can be affected by vortical structures with dimensions from 10m to 20m to 3 to 4 km (for heavy aircraft, 6 to 7 km) [48]. The estimates of the turbulent energy $\int_{\kappa_1}^{\kappa_2} d\kappa S_u(\kappa)$ in the frequency range from $\kappa_1 = 3 \cdot 10^{-4}$ m^{-1} to $\kappa_2 = 3 \cdot 10^{-1}$ m^{-1} on the assumption that the spectrum is described by (4.70) for $\kappa \geq 10^{-3}$ m^{-1} and $S_u(\kappa) = A\kappa^{-2.7}$ for $\kappa < 10^{-3}$ m^{-1}, where $A \leq 10^{-6}$ (m^3/s)m$^{-2.7}$ according to [48], and with the use of only (4.70) for the entire interval $[\kappa_1, \kappa_2]$ have shown that both calculations give close results for the turbulent energy for nearly all values of ε given in Table 4.2. That is why the spectral model of (4.70) is used for the whole frequency spectrum in this section.

Let CDL be installed aboard an aircraft and turbulence be measured in the direction of aircraft flight. The lidar receiving system reacts to echo signals $Z(mT_s, l)$, where L is the shot number, from which the estimates of Doppler spectra $\hat{S}(f_k)$ are obtained [see (2.42)]. The numerical simulation of lidar echo signals and their following virtual processing allows for the detailed study of possibilities of lidar detection of zones with intense CAT. The algorithms for numerical simulation of random realizations of the wind velocity and lidar echo signals are described in Sections 2.3 and 2.6. The needed values of the SNR for different values of the backscattering coefficient β_π [64] taking place in the atmosphere at heights of 5 to 20 km, wavelength $\lambda = 2$ μm, and different energies of the probing pulse E_P (from 1 to 1,000 mJ) were calculated in [31]. The SNR values as functions of β_π and E_P can vary by 5 orders of magnitude—from ~10^{-4} to ~10.

An important aspect of lidar detection in CAT zones at a certain distance in the aircraft flight direction is the promptness of obtaining the required information. In this respect, the best method is the estimation of turbulence intensity from the width of the power spectrum of the Doppler echo signal determined by the random (due to wind turbulence) velocity spread of scattering particles in the sensing volume.

From (2.22) and (2.32) under the condition that the width of the temporal window far exceeds the probing pulse duration ($T_W \gg \sigma_P$), for the second frequency moment $\overline{\sigma}_f^2$ we obtain the equation

$$\overline{\sigma}_f^2 = \frac{1}{8\pi\sigma_P^2} + \left(\frac{2}{\lambda}\right)^2 \int_{-\infty}^{+\infty} dz' Q_s(z')[V_r(R + z') - \overline{V}_r(R)]^2 \tag{4.71}$$

where R is the distance from the lidar to the center of the sensing volume, and $\overline{V}_r(R)$ is the radial velocity averaged over the sensing volume along the beam axis [see (2.29)]. Since the rectangular temporal window is used when obtaining the Doppler spectra, the function $Q_s(z')$ is described by (2.33). For the squared width of the Doppler spectrum (in velocity units) $\sigma_S^2 = (\lambda/2)^2 \langle \overline{\sigma}_f^2 \rangle$ from (4.71), upon averaging

Table 4.2 Doppler Spectrum Widths

Aircraft Ride Quality	Turbulence Energy Dissipation Rate ε (m^2/s^3)	Doppler Spectrum Width (m/s)
b_1	$10^{-3} - 5 \cdot 10^{-3}$	$0.97 - 1.65$
b_2	$5 \times 10^{-3} - 1.5 \cdot 10^{-2}$	$1.65 - 2.39$
b_3	$1.5 \times 10^{-2} - 4.5 \cdot 10^{-2}$	$2.39 - 3.45$
b_4	$\geq 4.5 \cdot 10^{-2}$	≥ 3.45

with the use of the turbulence model of (4.70) and taking into account the condition $T_W \gg \sigma_P$, we obtain a simple equation [31]:

$$\sigma_S^2 = (\lambda/2)^2 \ / \ (8\pi\sigma_P^2) + 0.45\varepsilon^{2/3}(T_W c/2)^{2/3} \tag{4.72}$$

Table 4.2 summarizes the widths of the Doppler spectrum σ_S calculated by (4.72) corresponding to the different levels of turbulence. The calculations have been performed for the following values of the parameters: $\lambda = 2$ μm, $\sigma_P = 1$ μs, and $T_W = 20.48$ μs ($T_W = T_s M$, where $T_s = 20$ ns and $M = 1{,}024$), at which the longitudinal dimension of the sensing volume Δz is approximately 3 km. In the absence of turbulence ($\varepsilon = 0$), the width of the Doppler spectrum is determined by the probing pulse duration ($\sigma_S \sim \lambda/\sigma_P$) and $\sigma_S \approx 0.1$ m/s, which is comparable to the spectral velocity resolution $\Delta V = \lambda/(2T_W) \approx 0.05$ m/s. With an increase of the turbulence intensity, the spectrum width increases up to 3.45 m/s and even higher. According to the data of Table 4.2, for the chosen parameters σ_P and T_W at $\sigma_S < 0.97$ m/s, the turbulence does not markedly affect an aircraft.

The longitudinal dimension of the sensing volume of 3 km at $T_W \approx 20$ μs is comparable with the upper boundary of scales of turbulent wind perturbations influencing an aircraft at subsonic speed. It is obvious that even at statistical homogeneity of wind turbulence the estimated spectral width $\hat{\sigma}_S$ fluctuates significantly at the transition from one sensing volume to another, since $V_r(z)$ is a random function and the largest scale of wind perturbations is much larger than the dimension of the sensing volume. To obtain the stable estimate $\hat{\sigma}_S$, it is necessary to perform the averaging over an ensemble of realizations corresponding to nonoverlapping sensing volumes. Consequently, the total longitudinal dimension of the sensing volume $N\Delta z$, where N is the number of realizations, can exceed significantly the length of the sensing path R. However, if the lidar installed aboard an aircraft is intended for the advance warning of the pilot about the presence of intense turbulence at a certain distance from the aircraft, then the averaging of single estimates of the spectral width is meaningless. In this case, the following is proposed. The spectral width $\hat{\sigma}_S$ (from the sensing volume lying, for example, at a distance $R = 10$ km ahead of the aircraft) is measured. Then, depending on the range of Table 4.2 within which the Doppler spectrum width $\hat{\sigma}_S$ falls, the turbulence energy dissipation rate is determined along with the ride quality experienced (if the estimate $\hat{\sigma}_S$ is true) by the aircraft after the distance R. Thus, Table 4.2 can be used for determination of the "current" level of turbulence from the value of $\hat{\sigma}_S$ measured in the routine regime.

The results of numerical simulation at SNR < 1 have shown that algorithm (2.49) understates the spectral width. Therefore, to obtain the estimate $\hat{\sigma}_S$, the procedure of least-squares fitting of the model spectrum is applied:

$$S_M(V_k) = \frac{\hat{\text{SNR}} \cdot B_V}{\sqrt{2\pi}\hat{\sigma}_S} \exp\left[-\frac{(V_k - \hat{\bar{V}}_r)^2}{2\hat{\sigma}_S^2} \right] + 1 \tag{4.73}$$

to the measured normalized Doppler spectrum $\hat{S}(V_k)$.

In (4.73), $\hat{SNR}$ and $\hat{\bar{V}}_r$ are the estimates of the SNR and the radial wind velocity averaged over the sensing volume; $B_V = M\Delta V$; $V_k = (k - M/2)\Delta V$, and $k = 0, 1, \ldots, M - 1$. The procedure of determination of $\hat{\sigma}_S$, $\hat{SNR}$, and $\hat{\bar{V}}_r$ is described thoroughly in [31].

The calculation of $\bar{\sigma}_S = (\lambda/2)\bar{\sigma}_f$ by (4.71) and the estimation of $\hat{\sigma}_S$ [through the fitting of the Gaussian model of the spectrum given by (4.73) to the measured spectrum] at SNR $\gg$ 1 or at the number of probed pulses used for accumulation of spectra $L \gg 1$ give close results (on average, the difference within 10%).

In the numerical simulation, the values of the dissipation rate ε were chosen corresponding to the boundaries of the turbulence levels b_i (see Table 4.2); that is, $\varepsilon = 10^{-3}, 5 \cdot 10^{-3}, 1.5 \cdot 10^{-2}$, and $4.5 \cdot 10^{-2}$ m^2/s^3. Figure 4.26 exemplifies the numerical simulation of Doppler spectra at $L = 1$ and SNR = 0.1, as well as the results of least-squares fitting to the Gaussian model of the spectrum given by (4.73) to simulated spectra for different levels of turbulence intensity b_i.

To estimate the reliability of detection of CAT zones from the routinely measured "instantaneous" (without averaging over turbulent variations of wind) Doppler spectrum width $\bar{\sigma}_S = (\lambda/2)\bar{\sigma}_f$ [see (4.71)], we have calculated the root mean square deviation of $\hat{\sigma}_S$ from $\bar{\sigma}_S$: $\sqrt{\langle \varepsilon_w^2 \rangle} = \sqrt{\langle (\bar{\sigma}_S - \hat{\sigma}_S)^2 \rangle}$ at different SNRs and values for L. For the calculation of the deviation $\sqrt{\langle \varepsilon_w^2 \rangle}$, we used a sample of 1,000 independent random realizations. The results of our calculations (in meters per second) are tabulated in Table 4.3. According to the data of Tables 4.2 and 4.3, the value of $\sqrt{\langle \varepsilon_w^2 \rangle}/\sigma_S$ characterizing the relative error of determination of the intensity of CAT zones that are potentially dangerous for an aircraft is 25% at $L = 1$ and SNR = 1; 50% at $L = 1$ and SNR = 0.1, 37% at $L = 100$ and SNR = 0.01 for practically any values of the turbulence energy dissipation rate ε presented in Table 4.3.

Figure 4.27 depicts the histograms of distribution of the random deviation $\varepsilon_w = \bar{\sigma}_S - \hat{\sigma}_S$ calculated from simulated data in intervals 0.2 m/s wide, and N_e is the number of events (of 1,000) that ε_w falls within some or other interval. It follows from the data presented in Table 4.3 and Figure 4.27 that the absolute error of estimation of the Doppler spectrum width increases with an increase in the turbulence intensity (ε). A decrease in the SNR also leads to an increase in the error (both absolute and relative). The results of numerical experiments show that, at SNR < 0.01, to determine the intensity of CAT with acceptable accuracy, a large number of shots ($L < 100$) must be used for accumulation of spectra, which is not always possible at typical pulse repetition rates and aircraft speeds.

4.9   Conclusions

1. The accumulation of Doppler spectra centered about the first spectral moment allows us to determine, under the conditions of statistical homogeneity of wind flow, the mean value of the turbulent spectral broadening with accuracy sufficient for retrieval of the information about the parameters of wind turbulence of any intensity from lidar data.

2. The methods of estimation of the dissipation rate of kinetic energy of turbulence from the Doppler spectrum width of ground-based continuous-wave CO_2 CDLs and from the temporal spectrum of the radial velocity measured

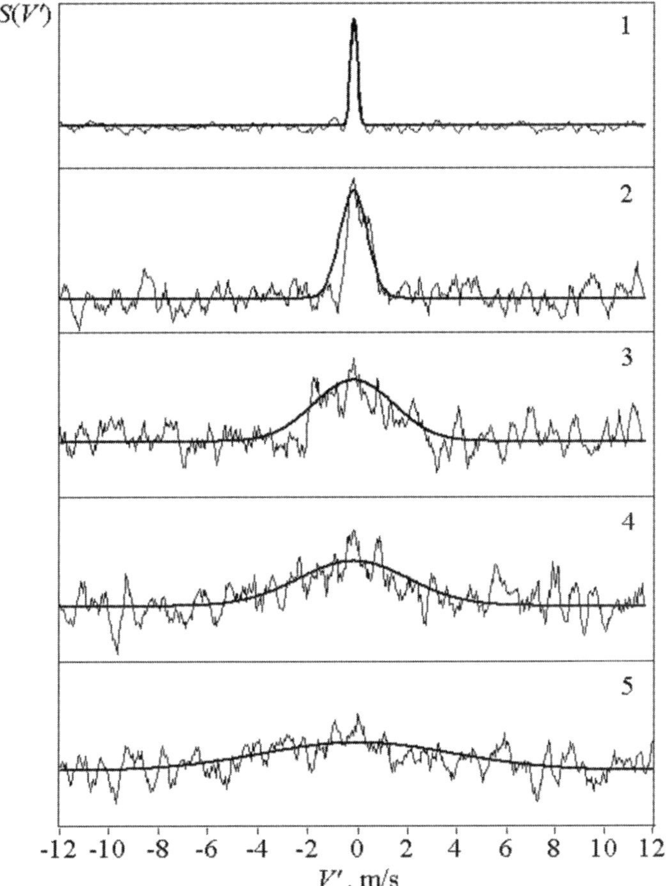

Figure 4.26 Examples of simulation of Doppler spectra (solid curves) and results of fitting of the Gaussian dependence on velocity (bold curves) at SNR = 0.1 and $L = 1$, $\varepsilon = 0$ (1), 10^{-3} (2), $5 \cdot 10^{-3}$ (3), $1.5 \cdot 10^{2}$ (4), and $4.5 \cdot 10^{-2}$ m^2/s^3 (5) to simulated spectra.

Table 4.3 Root Mean Square Deviation of the Doppler Spectrum Width (in m/s)

Dissipation Rate ε (m^2/s^3)	$L = 1, SNR = 1$	$L = 1, SNR = 0.1$	$L = 100, SNR = 0.01$
$\varepsilon = 10^{-3}$	0.25	0.48	0.36
$\varepsilon = 5 \cdot 10^{-3}$	0.39	0.86	0.61
$\varepsilon = 1.5 \cdot 10^{-2}$	0.60	1.24	0.92
$\varepsilon = 4.5 \cdot 10^{-3}$	0.76	1.51	1.21

by these CDLs allow the vertical profiles of the dissipation rate to be retrieved up to a height of ~200m. The information about wind turbulence at higher altitudes of the atmospheric boundary layer can be obtained from measurements of the Doppler spectrum width only with regard to the outer scale of turbulence and if the mean wind does not change within the sensing volume.

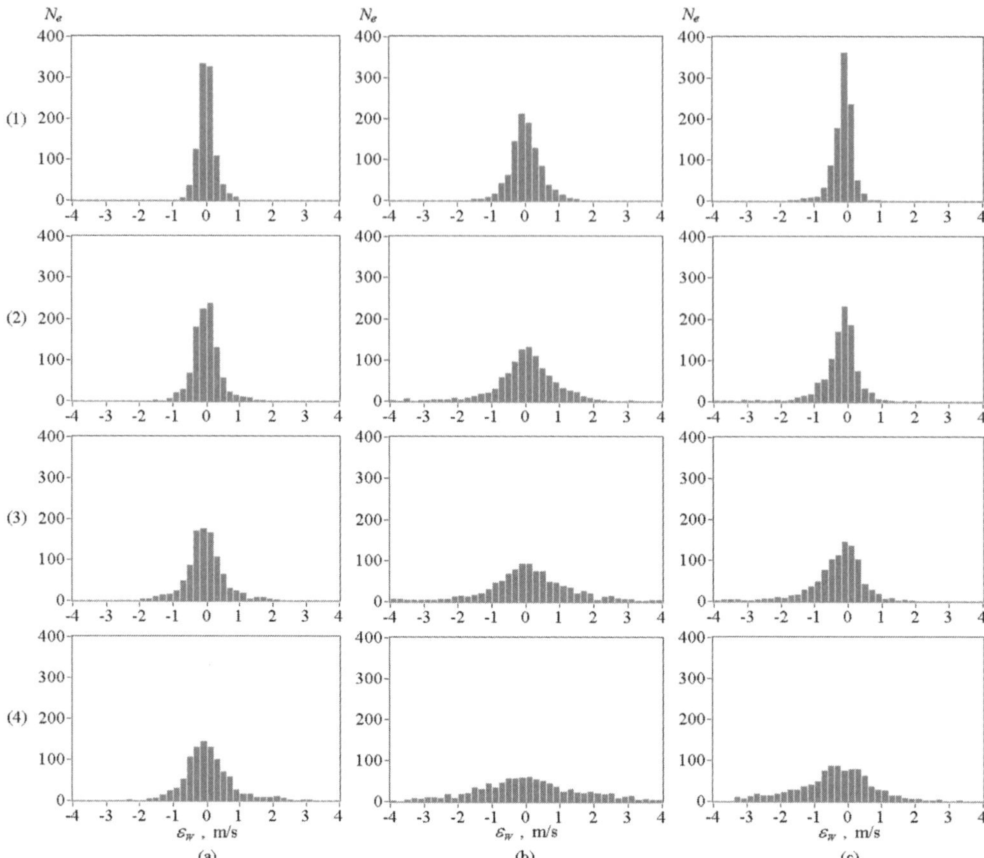

Figure 4.27 Histograms of errors of estimation of the Doppler spectrum width ε_w at $\varepsilon = 10^{-3}$ (1), $5 \cdot 10^{-3}$ (2), $1.5 \cdot 10^{-2}$ (3), and $4.5 \cdot 10^{-2}$ m^2/s^3 (4): (a) $L = 1$ and SNR $= 1$; (b) $L = 1$ and SNR $= 0.1$; (c) $L = 100$ and SNR $= 0.01$.

3. The method for estimating the dissipation rate of the kinetic energy of turbulence from the transverse structure function of wind velocity measured by cw CDLs with conical scanning by their probing beam allows for the retrieval of the vertical profiles of the dissipation rate up to heights of ~700m for any intensity of wind and refractive turbulence.

4. The wind turbulence parameters can be estimated from the data of pulsed CDLs obtained with scanning by their probing beams in the vertical plane (1) from the Doppler spectrum width, (2) from the longitudinal structure function, and (3) from the transverse structure function of the radial velocity measured by lidar in the absence of strong wind shears and narrow jet flows. The range of application of these methods for the estimation of the turbulence energy dissipation rate is limited from below by the values $\varepsilon \approx 10^{-5}$ m^2/s^3.

5. The DSW method is applicable under the condition that the SNR exceeds unity. To obtain an unbiased estimate of the turbulence energy dissipation rate ε by this method, it is necessary to accumulate Doppler spectra centered about the first spectral moment. In the case of strong inhomogeneity of the SNR in the averaging layer, when determining the second spectral moments,

one should use a threshold that is different from the noise component of the spectrum and calculated in accordance with the experimental values of SNR. If SNR $\gg 1$, then in the case of strong and moderate turbulence, the estimates ε obtained by the DSW method are more accurate than the estimates obtained by the LSF and TSF methods.

6. The LSF and TSF methods provide unbiased estimates of the dissipation rate if SNR ≥ -5 dB. The accuracy of the LSF method of estimating ε decreases drastically if the outer scale of turbulence is significantly (more than once) smaller than the longitudinal dimension of the sensing volume. The TSF is free of this disadvantage.

7. The relative error of estimation of the dissipation rate ε by the DSW, LSF, and TSF methods with the use of vertically scanning pulsed CDL data does not exceed 25% if the measurement time is ~15 min. The estimates of ε obtained simultaneously from lidar and sonic anemometer data are in satisfactory agreement if the conditions of applicability of the lidar methods are fulfilled. The vertical profiles of wind turbulence parameters retrieved by these methods from lidar data do not contradict the theory of developed turbulence of the atmospheric boundary layer and the available results of direct sensor measurements.

8. If the estimate of radial velocity is unbiased and the ratio σ_e/σ_V does not exceed unity, then the TSF method applied to data of a conically scanning pulsed CDL provides a relative error of estimation of the turbulent energy dissipation rate of less than 40% in the case where data from one full scan is used. The use of data from four full scans allows the dissipation rate to be estimated by this method with a relative error of less than 20%.

9. Clear air turbulence zones can be detected from aboard an aircraft from the broadening of the Doppler spectrum of the echo signal of a pulsed CDL. The accuracy of estimation of the turbulence level ahead of the aircraft from measurements of the Doppler spectrum width depends on the turbulence intensity (dissipation rate ε), the SNR, and the order of spectrum accumulation (number of shots). In the case of only one shot, the turbulence intensity can be determined only at SNR ≥ 0.1. In the case of SNR $= 0.01$, for this purpose, Doppler spectra must be accumulated from at least 100 probing pulses.

References

[1]   Eberhard, W.L., Cupp, R.E., and Healy, K.R., "Doppler lidar measurements of profiles of turbulence and momentum flux," *Journal of Atmospheric and Oceanic Technology*, Vol. 6, 1989, pp. 809–819.

[2]   Gal-Chen, T., Xu, M., and Eberhard, W.L., "Estimations of atmospheric boundary layer fluxes and other turbulence parameters from Doppler lidar data," *Journal of Geophysical Research*, Vol. 97, No. D17, 1992, P. 18,409–18,423.

[3]   Banta, R.M., Pichugina, Y.L., and Brewer, W.A., "Turbulent velocity-variance profiles in the stable boundary layer generated by a nocturnal jet," *Journal of Atmospheric Science*, Vol. 62, 2006, pp. 2700–2719.

[4]   Gordienko, V.M., et al., "Coherent CO_2 lidars for measuring wind velocity and atmospheric turbulence," *Optical Engineering*, Vol. 33, No. 10, 1994, pp. 3206–3213.

[5]   Byzova, N.L., et al., "Joint measurements of wind velocity by Doppler lidar and high-tower anemometers," *Meteorol. Gidrol.*, No. 3, 1991, pp. 114–117.

[6]   Banakh, V.A., and Smalikho, I.N., "Determination of the turbulent energy dissipation rate from lidar sensing data," *Atmos. Oceanic Opt.*, Vol. 10, No. 4–5, 1997, pp. 295–302.

[7]   Banakh, V.A., et al., "Measurements of turbulent energy dissipation rate with a cw Doppler lidar in the atmospheric boundary layer," *Journal of Atmospheric and Oceanic Technology*, Vol. 16, No. 8, 1999, pp. 1044–1061.

[8]   Smalikho, I.N., Köpp, F., and Rahm, S., "Measurement of atmospheric turbulence by 2-mkm Doppler lidar," *Journal of Atmospheric and Oceanic Technology*, Vol. 22, No. 11, 2005, pp. 1733–1747.

[9]   Smalikho, I.N. "On measurement of the dissipation rate of the turbulent energy with a cw Doppler lidar," *Atmos. Oceanic Opt.*, Vol. 8, No. 10, 1995, pp. 788–793.

[10] Banakh, V.A., et al., "Turbulence measurements with a cw Doppler lidar in the atmospheric boundary layer," *Atmos. Oceanic Opt.*, Vol. 8, No. 12, 1995, pp. 955–959.

[11] Banakh, V.A., et al., "Fluctuation spectra of wind velocity measured with a Doppler lidar," *Atmos. Oceanic Opt.*, Vol. 10, No. 3, 1997, pp. 202–208.

[12] Banakh, V.A., Werner, Ch., and Smalikho, I.N., "The effect of aerosol microstructure on the error in estimating wind velocity with a Doppler lidar," *Atmos. Oceanic Opt.*, Vol. 13, No. 8, 2000, pp. 685–691.

[13] Banakh, V.A., et al., "Measurement of the turbulent energy dissipation rate with a scanning Doppler lidar," *Atmos. Oceanic Opt.*, Vol. 9, No. 10, 1996, pp. 849–853.

[14] Frehlich, R.G., et al., "Measurements of boundary layer profiles in urban environment," *Journal of Applied Meteorology and Climatology*, Vol. 45, No. 6, 2006, pp. 821–837.

[15] Smalikho, I. N., and Banakh, V. A., "Accuracy of estimation of the turbulent energy dissipation rate from wind measurements by a conically scanning pulsed coherent Doppler lidar. Part I. Algorithm of data processing," *Atmos. Oceanic Opt.*, Vol. 26, No. 3, 2013, pp. 213–219.

[16] Smalikho, I. N., et al., "Accuracy of estimation of the turbulent energy dissipation rate from wind measurements by a conically scanning pulsed coherent Doppler lidar. Part II. Numerical and atmospheric experiment," *Atmos. Oceanic Opt.*, Vol. 26, No. 3, 2013, pp. 220–225.

[17] Banakh, V.A., and Smalikho, I.N., "Estimation of the turbulence energy dissipation rate from the pulsed Doppler lidar data," *Atmos. Oceanic Opt.*, Vol. 10, No. 12, 1997, pp. 957–965.

[18] Frehlich, R.G., "Effect of wind turbulence on coherent Doppler lidar performance," *Journal of Atmospheric and Oceanic Technology*, Vol. 14, No. 2, 1997, pp. 54–75.

[19] Hannon, S.M., et al., "Windshear, turbulence and wake vortex characterization using pulsed solid-state coherent lidar," *SPIE Proc. Air Traffic Control Technology*, Orlando, FL, USA, 18–19 April 1995, Vol. 2464, pp. 94–102.

[20] Hannon, S.M., and Thomson, J.A., "Aircraft wake vortex detection and measurement with pulsed solid-state coherent laser radar," *Journal of Modern Optics.*, Vol. 41, 1994, pp. 2175–2196.

[21] Constant, G., et al., "Coherent laser radar and the problem of aircraft wakes," *Journal of Modern Optics*, Vol. 41, No. 11, 1994, pp. 2153–2173.

[22] Köpp, F., "Doppler lidar investigation of wake vortex transport between closely spaced runways," *AIAA Journal*, Vol. 32, No. 4, 1994, pp. 805–810.

[23] Holzäpfel, F., et al., "Strategies for circulation evaluation of aircraft wake vortices measured by lidar," *Journal of Atmospheric and Oceanic Technology*, Vol. 20. No. 8, 2003, pp. 1183–1195.

[24] Brockman, P. B., et al., "Coherent pulsed lidar sensing of wake vortex position and strength, winds and turbulence in the terminal area," *Proc. 10th Coherent Laser Radar Conference*, Mount Hood, OR, USA, 28 June–2 July 1999, pp. 12–15.

[25] Köpp, F., et al., "Comparison of wake-vortex parameters measured by pulsed and continuous-wave lidars," *Journal of Aircraft*, Vol. 42, No. 4, 2005, pp. 916–923.

[26] Rahm, S., Smalikho, I.N., and Köpp, F., "Characterization of aircraft wake vortices by airborne coherent Doppler lidar," *Journal of Aircraft*, Vol. 44, No. 3, 2007, pp. 799–805.

[27] Banakh, V.A., et al., "Estimates of turbulence parameters from measurements of wind velocity with a pulsed coherent Doppler lidar," *Atmos. Oceanic Opt.*, Vol. 18, No. 12, 2005, pp. 955–957.

[28] Banakh, V.A., et al., "Measurement of atmospheric turbulence parameters by vertically-scanning pulsed coherent lidar," *Atmos. Oceanic Opt.*, Vol. 20, No. 12, 2007, pp. 1019–1023.

[29] Smalikho, I.N. "Accuracy of the turbulent energy dissipation rate estimation from the temporal spectrum of wind velocity fluctuations," *Atmos. Oceanic Opt.*, Vol. 10, No. 8, 1997, pp. 559–563.

[30] Banakh, V.A., et al., "Computer simulation of cw Doppler wind lidar operation in the turbulent atmosphere," *Atmos. Oceanic Opt.*, Vol. 12, No. 10, 1999, pp. 905–911.

[31] Banakh, V.A., Werner, Ch., and Smalikho, I.N., "Remote sensing of clear sky turbulence using Doppler lidar: Numerical simulation," *Atmos. Oceanic Opt.*, Vol. 14, No. 10, 2001, pp. 856–863.

[32] Banakh, V.A., et al., "Representativeness of measurements of the turbulence energy dissipation rate by a scanning coherent Doppler lidar," *Atmos. Oceanic Opt.*, Vol. 22, No. 10, 2009, pp. 966–972.

[33] Banakh, V.A., et al., "Turbulent energy dissipation rate measurements by coherent lidar," *SPIE Proc. Lidar Techniques for Remote Sensing II*, Paris, France, 25–26 September 1995, Vol. 2581, pp. 243–253.

[34] Banakh, V.A., et al., "Estimations of turbulent energy dissipation rate from Doppler lidar data," *Proc. 8th Coherent Laser Radar Conference*, Keystone, CO, USA, July 1995, pp. 116–111.

[35] Banakh, V.A., et al., "Turbulent characteristics of wind velocity measured by Doppler lidar," *Proc. 9th Coherent Laser Radar Conference*, Linköping, Sweden, 23–27 June 1997, pp. 166–169.

[36] Banakh, V.A., et al., "Simulation of scanning cw Doppler lidar run," *Proc. 19th International Laser Radar Conference*, Annapolis, MD, USA, 6–10 July 1998, pp. 731–734.

[37] Banakh, V.A., et al., "Measurement of turbulence parameters in the atmospheric boundary layer with a cw Doppler lidar," *Proc. 19th International Laser Radar Conference*, Annapolis, MD, USA, 6–10 July 1998, pp. 727–729.

[38] Banakh, V.A., et al., "Simulation of scanning cw Doppler lidar run," *SPIE Proc. "V International Symposium on Atmospheric and Ocean Optics,"* Tomsk, Russia, 15–18 June 1998, Vol. 3583, pp. 360–365.

[39] Banakh, V.A., et al., "Development of methods for lidar sounding of the turbulent field of wind velocity in the atmospheric boundary layer," *Fifth International Scientific–Technical Conference on Radar Detection and Ranging, Navigation, and Communication*, Vol. 3, Voronezh, Russia, 1999, pp. 1900–1904.

[40] Banakh, V.A., Smalikho, I.N., and Werner, Ch., "Numerical simulation of Doppler lidar detection of clear air turbulence," *SPIE Proc. Laser Radar: Ranging and Atmospheric Lidar Techniques III*, Toulouse, France, 17–18 September 2001, Vol. 4546, pp. 67–78.

[41] Banakh, V.A., Werner, Ch., and Smalikho, I.N., "Numerical simulation of Doppler lidar detection of clear air turbulence," *Digest of the VIII Joint International Symposium on Atmospheric and Ocean Optics, Atmospheric Physics*, Irkutsk, Russia, 25–29 June 2001, p. 169.

[42] Smalikho, I.N., Köpp, F., and Rahm, S., "Measurement of atmospheric turbulence by 2-μm Doppler lidar," *Proc. Workshop on Aircraft Vortices and Atmospheric Turbulence*, Institute of Atmospheric Physics, DLR, Oberpfaffenhofen, Germany, September 2004, pp. 20–31.

[43] Smalikho, I.N., Köpp, F., and Rahm, S., "Measurement of atmospheric turbulence by 2-μm Doppler lidar," Report No. 200, DLR, Oberpfaffenhofen, August 2004, p. 37.

[44] Falits, A.V., et al., "Retrieval of atmospheric turbulence parameters from the coherent 2-μm wind Doppler lidar data," *SPIE Proc. Europe International Symposium on Remote Sensing*, Bruges, Belgium, 19–22 September 2005, Vol. 5984, pp. 43–49.

[45] Smalikho, I.N., and Rahm, S., "Doppler lidar measurements of parameters of aircraft wake vortices and atmospheric turbulence," *XV International Symposium on Atmospheric and Ocean Optics, Atmospheric Physics*, Krasnoyarsk, Russia, 22–28 June 2008, CI-07, pp. 97.

[46] Banakh, V.A., and Smalikho, I.N., "Measurements of atmospheric turbulence parameters by coherent Doppler lidar," *Proc. XVI International Symposium on Atmospheric and Ocean Optics and Atmospheric Physics*, Tomsk, Russia, 12–15 October 2009, IAO SB RAS, pp. 529–533.

[47] Frehlich, R.G., and Conman, L.B., "Coherent Doppler lidar signal spectrum with wind turbulence," *Applied Optics*, Vol. 38, No. 36, 1999, pp. 7456–7466.

[48] Vinnichenko, N.K., et al., *Turbulence in the Free Atmosphere*, 2nd ed., Consultants Bureau, New York,1980, p. 310.

[49] Bendat, J.S. and Piersol, A.C., *Random Data: Analysis and Measurement Procedures*, Wiley, New York, 1971.

[50] Van Trees, H.L., *Detection, Estimation, and Modulation Theory*, Wiley, New York, 1968.

[51] Monin, A.S., and Yaglom, A.M., *Statistical Fluid Mechanics, Volume II: Mechanics of Turbulence*, MIT Press, Cambridge, MA, 1971.

[52] Tatarskii, V.I., *Wave Propagation in a Turbulent Medium*, McGraw-Hill, New York, 1961.

[53] Patrushev, G.Ya., Rostov, A.P., and Ivanov, A.P., "Automated ultrasonic anemometer-thermometer for measuring the turbulent characteristics in the ground atmospheric layer," *Atmos. Oceanic Opt.*, Vol. 7, No. 11–12, 1994, pp. 890–891.

[54] Lumley, J.L., and Panofsky, H.A., *The Structure of Atmospheric Turbulence*, Interscience Publishers, New York, 1964.

[55] Zilitinkevich, S.S., *Dynamics of Atmospheric Boundary Layer*, Gidrometeoizdat, Leningrad,1970, p. 292.

[56] Byzova, N.L., Ivanov, V.N., and Garger, E.K., *Turbulence in Atmospheric Boundary Layer*, Gidrometeoizdat, Leningrad, 1989, p. 263.

[57] Volkovitskaya, Z.I., and Ivanov, V.N., "Dissipation of turbulence energy in the atmospheric boundary layer," *Izv. AN SSSR. Fiz. Atmos. Okeana*, Vol. 6, No. 5, 1970, pp. 435–444.

[58] Pichugina, Y.L., et al., "Nocturnal boundary layer height estimate from Doppler lidar measurements," *Proc. 18th Symposium on Boundary Layer and Turbulence*, Stockholm, Sweden, June 2008, 7B.6.

[59] Frehlich, R.G., and Cornman, L.B., "Estimating spatial velocity statistics with coherent Doppler lidar," *Journal of Atmospheric and Oceanic Technology*, Vol. 19, No. 3, 2002, pp. 355–366.

[60] Smalikho, I.N., and Rahm, S., "Measurements of aircraft wake vortex parameters with a coherent Doppler lidar," *Atmos. Oceanic Opt.*, Vol. 21, No. 11, 2008, pp. 854–868.

[61] Frehlich, R.G., and Yadlowsky, M.J., "Performance of mean-frequency estimators for Doppler radar and lidar," *Journal of Atmospheric and Oceanic Technology*, Vol. 11, No. 5, 1994, pp. 1217–1230.

[62] Ray, B.J., and Hardesty, R.M., "Discrete spectral peak estimation in incoherent backscatter heterodyne lidar. I: Spectral accumulation and Cramer-Rao lower bound," *IEEE Trans. on Geoscience and Remote Sensing*, Vol. 31, No. 1, 1993, pp. 16–27.

[63] Banakh, V.A., et al., "Wind velocity and direction measurement with a coherent Doppler lidar under conditions of weak echo signal," *Atmos. Oceanic Opt.*, Vol. 23, No. 5, 2010, pp. 333–340.

[64] Zuev, V.E., and Krekov, G.M., *Optical Models of the Atmosphere.* Gidrometeoizdat, Leningrad, 1986, p. 256.

CHAPTER 5

Lidar Investigations of Aircraft Wake Vortices

5.1 Introduction

Investigations of turbulent wind fields have shown that coherent vortical structures can be present in the atmosphere along with the random mixing of air masses [1–5]. These structures can form in a turbulent flow at the expense of the flow energy at flow reversal and then collapse, making the air motion chaotic. They also can be dissipative and, in contrast, consume the energy of the chaotic motion in a flow [5].

The formation of an aerodynamic lift force as a result of aircraft motion in airspace is always accompanied by the formation of a pair of vortices. Being examples of coherent structures of technogenic origin, these vortices transform into stable vortex plates. An aircraft wake can extend to many kilometers and constitute a danger for other aircraft. Current increases in air transportation and heavy operation of international airports make the problem of wake vortex safety especially important [6]. For aircraft safety, it is necessary to know the minimum acceptable distance between consecutive aircraft. This distance depends on the initial strength of the wake vortex (aircraft type, mass and wingspan, flight altitude and speed of aircraft), flight configuration (takeoff mode, land approach, cruise mode, and so on), and wake location and its "age" [7]. Theoretical investigations [6–22] have shown that the wake vortex axis trajectory and wake vortex lifetime in a fixed plane across the wake depend significantly on the state of the atmosphere: wind velocity and direction, wind turbulence intensity, presence of wind shear, and so on.

Along with measurements of wind and atmospheric turbulence, coherent Doppler lidars can also be used for detection and investigation of aircraft wake vortices. In face, CDLs were initially developed with parameters such that the information about aircraft wake vortices could be retrieved from the array of measured lidar echo signals [23–26]—and these efforts were not made in vain. Now CDLs, along with the developed measuring strategies and methods for processing of raw lidar data, are the most efficient tools for experimental investigation of atmospheric effects on aircraft wake vortices.

Attempts to measure wake vortices with cw CDLs were undertaken soon after the advent of the first systems of this type. The experiments involved scanning by the probing beam across the wake. The results of these measurement using a cw lidar are published in, for example, [24, 26–31]. However, the accuracy of wake vortex measurements by one cw CDL was often low. The point is that during the scanning by the beam through the wake vortex axis, the vortex core should be within the sensing volume. Otherwise, the wake vortex becomes invisible to the lidar. Because the distance to the vortex core is often unknown, especially sometime after the wake vortex has formed, it is not always possible to correctly focus the probing beam. In addition, as the focal length increases, the spatial resolution of a cw lidar worsens so that it may appear to be insufficient for wake vortex detection. Therefore, at least two cw lidars are necessary for the measurement of aircraft wake vortex parameters.

The problems considered above do not apply to pulsed CDLs. The first results of wake vortex measurements by a 2–μm pulsed CDL were reported in [32] along with a description of the measurement strategy and the procedure for processing the raw lidar data. The processing of data in [32] consists of estimating the radial wind velocity from measured Doppler spectra. The estimates thus obtained are then used to determine vortex axis coordinates and circulation with the aid of some model for the tangential velocity of vortices. However, this approach does not allow a real profile of the tangential velocity to be obtained from lidar data, and the real profile can differ widely from the model profile due to vortex deformation under the effect of wind shear and atmospheric turbulence.

In [33], profiles of the wake vortex tangential velocity were obtained from Doppler spectra measured by pulsed CDLs in the form of velocity envelopes by specifying some threshold value of the spectrum. (Velocity envelopes are determined at the points of intersection of the measured spectral curve with the threshold.) Then these envelopes were used to calculate the wake vortex parameters. However, if in the case of cw CDLs this threshold can be found empirically, then in this case it depends significantly on the signal-to-noise ratio (SNR) and the vortex circulation, which is an unknown, sought-after parameter. Thus, to obtain wake vortex measurements of acceptable accuracy using pulsed CDLs, other approaches are required.

From 2001 until 2007, the Lidar Group of the Institute of Atmospheric Physics of the German Aerospace Center (DLR, Oberpfaffenhofen) conducted wake vortex measurements using two types of DLR coherent Doppler lidars: cw CO_2 CDLs (2001 and 2003) and pulsed 2-μm CDLs (2002–2007). These experiments involved also two research groups from France (ONERA, 2001–2003) and the United Kingdom (QinetiQ, 2001 and 2002), each with its own cw CO_2 CDL. Before 2005, only ground-based measurements were conducted. The raw data (of which 85% was obtained with 2-μm CDLs) obtained during this period have allowed the retrieval of 1,500 profiles of parameters of a pair of wake vortices (vortex axis coordinates and circulation as functions of time) generated by different types of aircraft. The data from both ground-based and airborne lidar measurements were used. The measurement results and methods used for their processing are reported in [34–50].

The first experiments were conducted by three groups from DLR, ONERA, and QinetiQ in the spring of 2001 near Oberpfaffenhofen (Germany). Parameters of wake vortices generated by the Advanced Technologies Testing Aircraft System (ATTAS) operated by DLR Braunschweig [34] were measured by three cw CDLs. In the following experiments conducted in France at the airfield of Tarbes airport in summer 2002, a DLR 2-μm pulsed CDL was used in addition to two ONERA and QinetiQ cw lidars. These experiments were designed to study wake vortices behind a large transport aircraft (LTA). Independent simultaneous measurements by the pulsed lidar, on one hand, and by two cw lidars, on the other hand, have allowed the accuracy of wake vortex parameter estimates to be assessed [36]. A year later, analogous measurements were carried out at the same place, but a DLR cw lidar was used in place of the QinetiQ lidar.

In 2004 to 2007, experiments were carried out by the DLR Lidar Group using only one pulsed CDL. To study the influence of the Earth's surface on wake vortex behavior, lidar measurements of wake vortices generated by aircraft landing on two runways at the Frankfurt am Main Airport were conducted in September–December 2004. Ground-based lidar measurements for LTAs of various types and various flight configurations were carried out in France (Toulouse, May–June 2005; Istres, July–December of 2005; Tarbes, April 2007) and in Germany (Oberpfaffenhofen, March and July–August 2006).

Lidar measurements of aircraft wake vortices in the free atmosphere are possible only from aboard another aircraft. A pulsed 2-μm CDL integrated into the DLR research aircraft Falcon-20 was used in April 2005 to measure wake vortices generated by the ATTAS [37]. In March, June, and November 2006, similar experiments were conducted for the case of wake generation by LTAs of different types. Because the concentration of background aerosol in the free atmosphere is insufficient for such lidar measurements, smoke generators were installed on the wings of the aircraft and their wake vortices studied. In the experiments conducted in March and June 2006, smoke generators were not used, because weather conditions favored moisture condensation of the exhaust gases from the aircraft nozzle into the atmosphere.

For the 2001–2007 period, along with modernization of technical facilities and gaining experience in lidar studies of wake vortices, algorithms for processing of experimental data were improved. This allowed for a permanent increase in the accuracy of wake vortex parameter estimates and for extension of the range of variation of atmospheric conditions in lidar experiments.

This chapter describes the strategies used to measure wake vortex parameters using coherent Doppler lidars and the methods for processing the experimental data used in wake vortex studies. We present the most characteristic results of the experiments mentioned above. The influence of wind and atmospheric turbulence on wake vortices is analyzed.

Another example of a coherent structure of technogenic origin is a wake generated downstream of a wind turbine. For this reason we have included in this chapter the results of recent lidar investigations of wind turbine wakes that were carried out in cooperation with our colleagues at the NOAA Earth System Research Laboratory, Boulder, Colorado, USA [51–54].

This chapter is based on the results of [34–43, 45–55].

5.2   Influence of Aircraft Wake Vortices on the Form of Doppler Spectra: Velocity Envelopes and Integration Method

Figure 5.1 illustrates the process of wake formation behind an aircraft. For a pair of formed wake vortices, the velocity vector can be presented in complex form as a superposition of the fields of two isolated vortices [56]:

$$V^{(2)}(r) = V_1^{(1)}(r - r_1) + V_2^{(1)}(r - r_2) \tag{5.1}$$

where $V_i^{(1)}(r) = (-1)^i \dfrac{\Gamma_i}{2\pi} \cdot \dfrac{\mu(|r|/r_C)}{|r|^2} jr$ is the velocity vector of the ith isolated vortex in the coordinate system centered at the point r_i (the points r_i are coordinates of the axis of the ith vortex; subscript $i = 1$ for the left vortex and $i = 2$ for the right one), $r = y + jz$ is the radius vector, $j = \sqrt{-1}$, z is the vertical coordinate, y is the horizontal coordinate in the plane perpendicular to the flight direction (see Figure 5.1), $V = V_y + jV_z$, V_z and V_y are, respectively, the vertical and horizontal components of the velocity vector, Γ_i is circulation of the ith vortex, and r_C is the vortex core radius. For the Lamb-Oseen model [56], the function $\mu(x)$ is $\mu(x) = 1 - \exp(-1.256x^2)$.

Figure 5.2 exemplifies the vertical velocity component $V_z(0, y)$ calculated by (5.1) with the use of the Lamb-Oseen model at $r_1 = -b/2$ and $r_2 = b/2$, where $b = |r_2 - r_1|$ is the separation between the vortex axes. For arbitrary values of the coordinates r_1 and r_2, the tilt angle of the vortex pair with respect to the horizontal axis y is determined as $\theta_p = \arg(r_2 - r_1)$.

In the formed pair of vortices, air masses rotate inward. Therefore, being in the field of each other, the cores of these vortices move downward (in the absence of significant upward wind flow). As follows from (5.1), the vertical velocity of motion of the vortex core $w = V_z(b/2)$ in the unperturbed atmosphere at $\Gamma_1 = \Gamma_2 = \Gamma$, $r_1 = -b/2$, and $r_2 = b/2$ can be determined as $w = -\Gamma/(2\pi b)$. According to theory [7], the initial separation between the vortex axes is $b_0 = (\pi/4)B_a$, where B_a is the wingspan. Correspondingly, for the initial velocity of vertical motion of the vortex cores, we have $w_0 = -\Gamma_0/(2\pi b_0)$, where Γ_0 is the initial vortex circulation, which can be calculated as [7]

$$\Gamma_0 = M_a g/(\rho_a b_0 V_a) \qquad (5.2)$$

In (5.2), g is the freefall acceleration, ρ_a is the air density at the flight height, M_a is the aircraft mass, and V_a is the true speed of the aircraft. Given b_0 and Γ_0, we

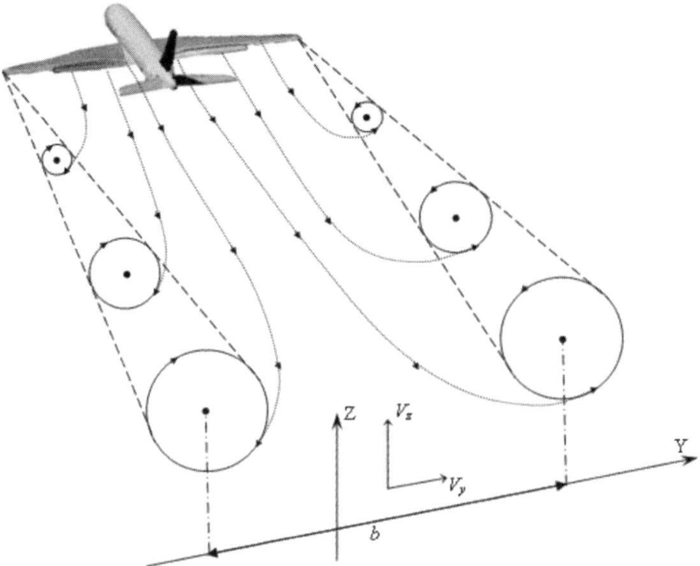

Figure 5.1 Formation of wake behind an aircraft.

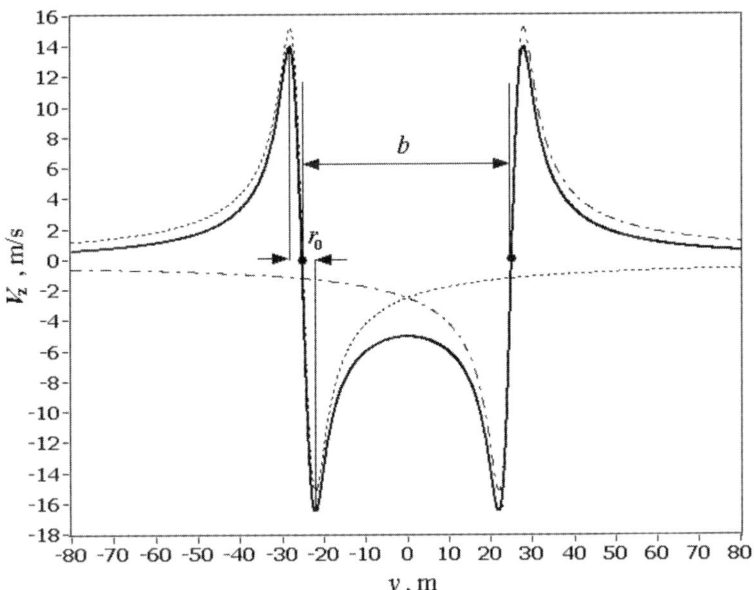

Figure 5.2 Vertical velocity component $V_z(0, y)$ calculated by (5.1) at $b = 50$m, $r_c = 3$m, and $\Gamma_1 = \Gamma_2 = 400$ m²/s (solid curve); $\Gamma_1 = 400$ m²/s, $\Gamma_2 = 0$ (dashed curve); and $\Gamma_1 = 0$, $\Gamma_2 = 400$ m²/s (dot-and-dash curve).

can calculate the time for which the vortices move downward to a distance equal to the initial distance between the vortex axes $t_0 = -b_0/w_0$ as follows:

$$t_0 = 2\pi b_0^2 \,/\, \Gamma_0 \tag{5.3}$$

The calculated values of b_0, Γ_0, and t_0 are often used for normalization of the results of wake vortex measurements.

The Doppler spectrum measured by lidar carries information about the distribution of radial velocities inside the sensing volume. Assume that the sensing volume moves along a straight line perpendicular to the optical axis in the plane ZY (see Figure 5.1) so that the center of the sensing volume at some instant intersects the axis of one of the vortices. Then, neglecting the influence of vortices on each other and the background wind, the maximum (in the absolute value) radial velocity of aerosol particles being in the sensing volume is the tangential velocity of the wake $\left|V_i^{(1)}\right|$ as a function of the distance between the axis of the corresponding vortex and the center of the sensing volume.

Consider Doppler spectra (normalized spectra of lidar signal power) $S_D(V) = (2B_V/\lambda)\overline{S}_s(2V/\lambda) + 1$ for two CDL types. In the case of cw lidar, from (2.14), (2.17), and (2.32), we have

$$S_D(V) = \text{SNR} \frac{B_V}{\Delta V} \int\limits_0^\infty dz Q_s(z) \,\text{sinc}^2\left[\frac{\pi}{\Delta V}(V - V_r(z))\right] + 1 \tag{5.4}$$

where $V = (k - M/2)\Delta V$, $\Delta V = \lambda/(2T_W)$ is the velocity resolution, $B_V = M\Delta V$, and $Q_s(z)$ is the function described by (2.55). In the case of a pulsed lidar, with the use of the Gaussian temporal window, according to (2.24), the Doppler spectrum $S_D(V)$ can be represented in the form

$$S_D(V) = \mathrm{SNR}\,\frac{B_V}{\sqrt{2\pi}\sigma_{VI}} \int\limits_{-\infty}^{+\infty} dz'\, Q_s(z') \exp\left\{\frac{[V - V_r(R + z')]^2}{2\sigma_{VI}^2}\right\} + 1 \qquad (5.5)$$

where $\sigma_{VI} = (\lambda/2)\sigma_{fl}$ and $Q_s(z')$ are described by (2.28) and (2.25), respectively. Let the velocity values $V = (k' - M'/2)\delta V \in [-B_V/2,\, B_V/2]$ $(B_V = M'\delta V)$ in (5.5) be set with a step $\delta V = 0.1$ m/s.

The Doppler spectra were calculated by (5.4) for the following parameters: SNR $= 1$, $B_V = 40$ m/s, $\Delta V = 0.1$ m/s, $\lambda = 10.6$ μm, $a_0 = 7.5$ cm, and $F = 200$m ($\Delta z \approx 37$m). Taking $V_r(z) = V_z(z - F, y)$, where V_z is the velocity component described by (5.1), and specifying the vortex parameters, the solid curve in Figure 5.2 is calculated.

Figure 5.3(a) shows the Doppler spectra calculated at different y with a step $\Delta y = 0.1$m. Let us consider the spectra to the left and to the right from the vortex core at $y = -35$m and $y = -15$m shown, respectively, by lines 1 and 2 in Figure 5.3(a). These spectra are depicted in Figures 5.3(b) and 5.3(c), where the corresponding values of

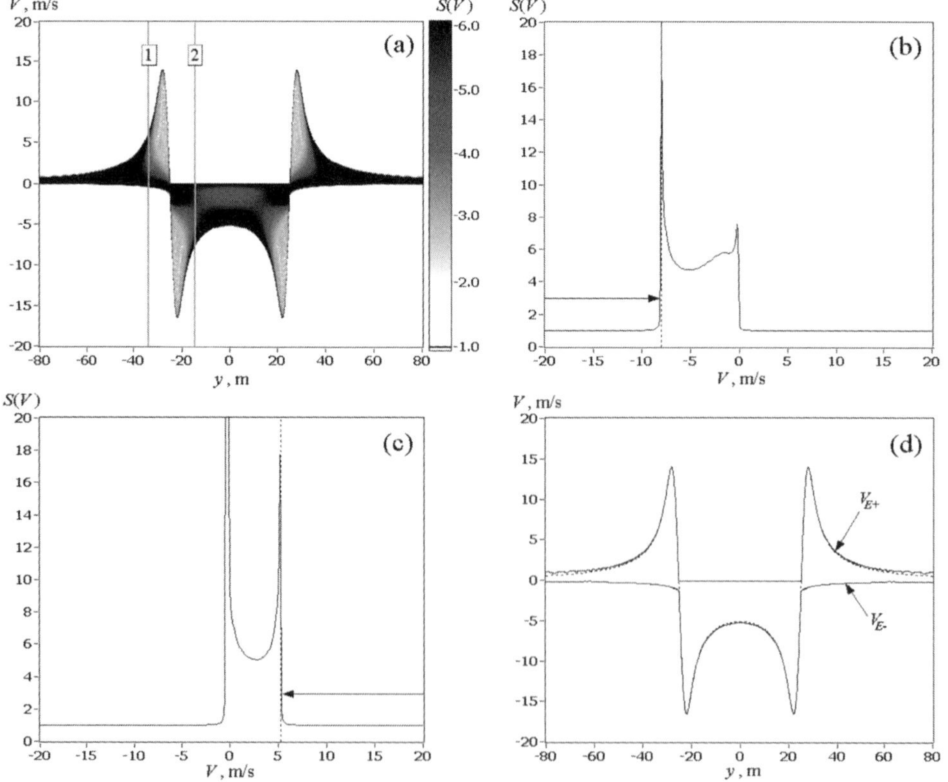

Figure 5.3 (a–c) Simulated Doppler spectra in the case of a cw CDL for the sensing volume lying in the region where the vortex axes are located and (d) velocity envelopes obtained from these spectra.

$V_z(0, y)$ are shown at the points of intersection of the dashed lines with axis V. If we specify some threshold value q_{th} and go from the left end of the spectral range of velocities to the first point of intersection with the spectral curve, $q_{th} = S_D(V_{E-})$, then we can find the value of the negative velocity envelope V_{E-}. If an analogous procedure is performed from the right, we can obtain the positive velocity envelope V_{E+} [$S_D(V_{E+}) = q_{th}$]. The points of intersection of the spectral curves with the given threshold are shown by arrows in Figures 5.3(b) and 5.3(c). One can see that the obtained values V_{E-} and V_{E+} are very close to, respectively, $V_z(0, y_1)$ ($y_1 = -35\mathrm{m}$) and $V_z(0, y_2)$ ($y_2 = -15\mathrm{m}$). In this case, as follows from the figure, the threshold q_{th} is not very critical for determination of V_{E-} and V_{E+} and can be specified in a wide range of values. The velocity envelopes $V_{E-}(y)$ and $V_{E+}(y)$ calculated at $q_{th} = 3$ are shown in Figure 5.3(d) along with $V_z(0, y)$, which is shown as a dashed curve. The joining of the velocity envelopes $V_{E+}(y)$ at $y < -25\mathrm{m}$ with $V_{E-}(y)$ at $-25\mathrm{m} < y < 25\mathrm{m}$ and with $V_{E+}(y)$ at $y > 25\mathrm{m}$ can yield the velocity profile $V_E(y)$, which, according to Figure 5.3(d), is very close to $V_z(0, y)$. Thus, the velocity envelopes measured by a cw CDL can be used to find the profile of air mass rotation velocity in the wake cross section, which, in its turn, can be used (if the vortex cores are within the sensing volume) to determine vortex axis coordinates and vortex circulation.

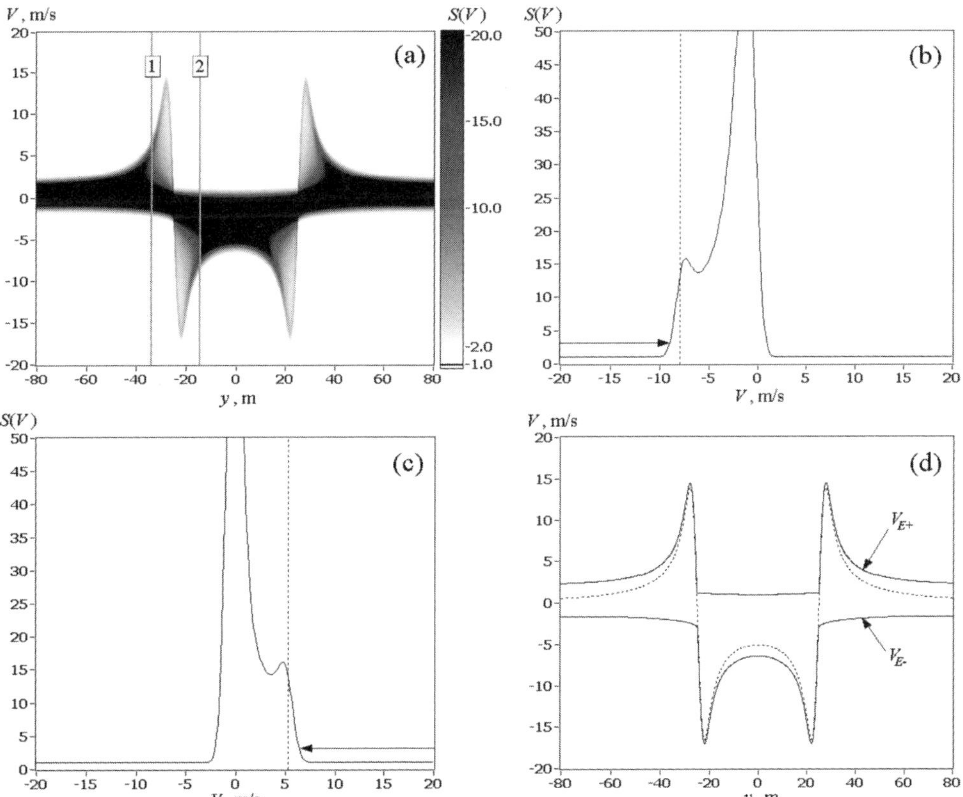

Figure 5.4   (a–c) Simulated Doppler spectra in the case of a pulsed CDL for the sensing volume lying in the region where the vortex axes are located and (d) velocity envelopes obtained from these spectra.

For the case of a 2-μm pulsed CDL, the spectra $S_D(V; y)$ were calculated by (5.5), where $V_r(R + z') = V_z(z', y)$, at SNR = 5 and $\sigma_W = \sigma_P = 0.25$ μs ($\Delta z = 94$m, $\sigma_S = 0.64$ m/s). The calculation results are shown in Figure 5.4(a). Figures 5.4(b) and 5.4(c) illustrate more clearly the difference between Doppler spectra in the cases of cw and pulsed CDLs (compare with Figures 5.3(b) and 5.3(c)). The difference is connected to the fact that the power spectrum of a cw lidar's echo signal is mostly determined by the distribution of radial velocities in the sensing volume, whereas in the case of pulsed lidar, the probing pulse duration and the width of the temporal window affect significantly the shape of the spectrum. As a result, as follows from Figures 5.4(b) through 5.4(d), the velocity envelopes $V_{E-}(y)$, $V_{E+}(y)$, and $V_E(y)$ obtained at the same spectral threshold of $q_{th} = 3$ can markedly, by 1 to 2 m/s, exceed (in the absolute value) the velocity $V_z(0, y)$. As a consequence, estimates of the wake vortex circulation from measured velocity envelopes can exceed the true value of Γ by a few tens of percent. Nevertheless, the velocity envelopes shown in Figure 5.4(d) can be used to determine the y coordinates of the axes of the right and left vortices at the corresponding points between the positions of the maximum of $V_{E+}(y)$ and minimum of $V_{E-}(y)$.

For CDL measurement of the combined velocity envelope $V_E(y)$, scanning by the probing beam in the vertical plane transverse to the wake is used. In this case, the radial velocity V_r as a function of the elevation angle (varying during the scanning) φ and the distance R from the lidar to the observation point can be written as

$$V_r(R,\varphi) = \mathrm{Re}\left\{\varsigma^* \mathbf{V}^{(2)}(R\varsigma)\right\} \tag{5.6}$$

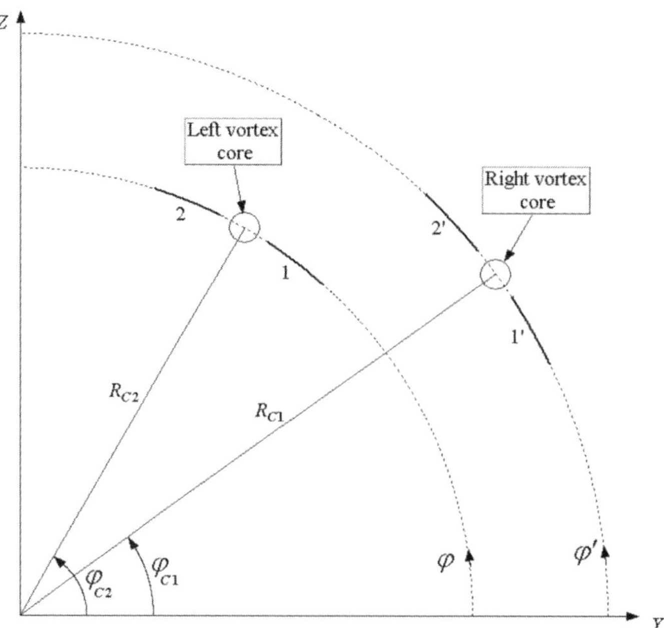

Figure 5.5 Geometry of wake vortex measurement by a scanning lidar.

where $\varsigma = \cos\varphi + j\sin\varphi$ is the unit vector directed along the axis of the probing beam. Assume that the coordinates of the axes of the right $\{R_{C1}, \varphi_{C1}\}$ and left $\{R_{C2}, \varphi_{C2}\}$ vortices (see Figure 5.5) and the velocity envelopes $V_{E1}(\varphi)$ and $V_{E2}(\varphi)$ are determined from measured data.

When the observation points $\{R_{C1}, \varphi\}$ and $\{R_{C2}, \varphi'\}$ lie beyond vortex cores and the conditions $R_{C1}^2|\varsigma(\varphi) - \varsigma(\varphi_{C1})|^2 \gg r_C^2$ and $R_{C2}^2|\varsigma(\varphi') - \varsigma(\varphi_{C2})|^2 \gg r_C^2$ are fulfilled, we can take $\mu(|r/r_C|) = 1$ in (5.1). Reference [35] shows that $V_{E1}(\varphi) = V_r(\tilde{R}_1, \varphi)$ and $V_{E2}(\varphi) = V_r(\tilde{R}_2, \varphi)$, where $\tilde{R}_i = R_{Ci}\cos(\varphi - \varphi_{Ci})$. Then, from (5.6) and (5.1), we obtain [35]

$$V_{E1}(\varphi) = -\frac{\Gamma_1}{2\pi}\frac{1}{r_1(\varphi)} + \frac{\Gamma_2}{2\pi}\frac{r_2(\varphi)}{r_2^2(\varphi) + d^2(\varphi)}$$

$$V_{E2}(\varphi') = -\frac{\Gamma_1}{2\pi}\frac{r_1(\varphi)}{r^2_1(\varphi) + d^2(\varphi)} + \frac{\Gamma_2}{2\pi}\frac{1}{r_2(\varphi)}$$

$$(5.7)$$

where

$$r_1(\varphi) = R_{C1}\sin(\varphi - \varphi_{C1}) \tag{5.8}$$

$$r_2(\varphi) = R_{C2}\sin(\varphi - \varphi_{C2}) \tag{5.9}$$

and

$$d(\varphi) = R_{C1}\cos(\varphi - \varphi_{C1}) - R_{C2}\cos(\varphi - \varphi_{C2}) \tag{5.10}$$

Equations (5.7) through (5.10) demonstrate the relationship between the measured parameters $\{R_{C1}, \varphi_{C1}\}$, $\{R_{C2}, \varphi_{C2}\}$, $V_{E1}(\varphi)$, and $V_{E1}(\varphi')$ with the circulation of the right Γ_1 and left Γ_2 vortices. As a result of solution of the system of linear equations (5.7), we obtain two estimates of the circulation $\tilde{\Gamma}_1(\varphi, \varphi')$ and $\tilde{\Gamma}_2(\varphi, \varphi')$. In practice, these values are measured with some error. To decrease the error of estimation of the vortex circulation, an integration method is applied [57]; that is, the circulation is estimated as

$$\hat{\Gamma}_i = \frac{1}{N}\sum_{n=1}^{N}\frac{1}{N'}\sum_{n'=1}^{N'}\tilde{\Gamma}_i(\varphi_n, \varphi_{n'}) \tag{5.11}$$

where $i = 1, 2$. The summation in (5.11) is performed over values of individual estimates obtained at the angles φ_n and $\varphi_{n'}$ satisfying the conditions $r_{min} \leq |r_1(\varphi_n)| \leq r_{max}$ and $r_{min} \leq |r_2(\varphi_{n'})| \leq r_{max}$, where r_{min} and r_{max} are, respectively, the minimum and maximum distances from the axis of the ith vortex. Figure 5.5 shows the integration domains 1 ($\varphi_n \in [\varphi_1, \varphi_{N/2}]$) and 2 ($\varphi_n \in [\varphi_{N/2+1}, \varphi_N]$) for the left vortex and 1' ($\varphi_{n'} \in [\varphi_{1'}, \varphi_{N'/2}]$) and 2' ($\varphi_{n'} \in [\varphi_{N'/2+1}, \varphi_{N'}]$) for the right vortex. If the internal integration domains 1 and 2' overlap, that is, $\varphi_1 < \varphi_{N'}$, then in (5.11) one can use only the external domains 2 and 1' for the summation [35]. The integration ranges

$r_{min} = 3m$, $r_{max} = 8m$ for small aircraft and $r_{min} = 5m$, $r_{max} = 15m$ for large aircraft are quite acceptable [57].

If the wake vortex cores are so far from each other that the vortices are almost independent (isolated), the estimates of the circulation $\hat{\Gamma}_i$ can be obtained separately for each vortex by the equations

$$\hat{\Gamma}_1 = \frac{2\pi}{N} \sum_{n=1}^{N} |V_{E1}(\varphi_n)| |r_1(\varphi_n)| \tag{5.12}$$

$$\hat{\Gamma}_2 = \frac{2\pi}{N'} \sum_{n'=1}^{N'} |V_{E2}(\varphi_{n'})| |r_2(\varphi_{n'})| \tag{5.13}$$

Equations (5.12) and (5.13) can also be used at arbitrary distances between vortex axes b, if the integration domains 1 and 2' (see Figure 5.5) do not overlap (that is, $\varphi_1 > \varphi_{N'}$) in the case of a cw CDL or if a special procedure described in Section 5.4 is used to obtain the envelopes $V_{E1}(\varphi)$ and $V_{E2}(\varphi)$ from a pulsed CDL data.

5.3   Measurement of Wake Vortex Parameters Using a Continuous-Wave CDL

A necessary requirement for taking wake vortex measurements with a cw CDL is the focusing of the probing beam to the distance F, at which the distances between the lidar and the vortex axes fall within the range $[F - \Delta z/2, F + \Delta z/2]$, where, according to (1.58), the longitudinal dimension of the sensing volume is $\Delta z \sim F^2$. Otherwise, it is practically impossible to derive information about wake vortices from the Doppler spectra $S_D(V)$ measured at different angles φ.

Figure 5.6(a) exemplifies the Doppler spectra measured by the DLR cw homodyne CO_2 CDL [39]. The values of the Doppler spectra S_D at every point (V, t) are shown in the gray scale, where t is time (abscissa) passed after the wake vortices' appearance (once the aircraft has crossed the plane of scanning by the probing beam). This example incorporates 375 spectra $S_D(V)$ measured every 13.3 ms during scanning by the probing beam (i.e., a scanning angular rate of 8 deg/s). The plot for the elevation (scan) angle $\varphi(t)$ for the considered period of time is shown by the dot-and-dash curve in Figure 5.6(d). Because the measurements were carried out by the homodyne scheme (intermediate frequency $f_I = 0$), from one Doppler spectrum we can derive the information about only one absolute value of the radial velocity $|V_r|$. That is why the velocity range is $V \in [0, B_V]$, where in the demonstrated example $B_V = 14$ m/s, and only one velocity envelope $|V_E(\varphi)| \equiv |V_E[\varphi(t)]|$ can be obtained from the spectra in Figure 5.6(a).

The Doppler spectra shown in Figure 5.6(a) were measured at the probing beam focused to the distance F close to the distances between the lidar and the axes of the right (R_{C1}) and left (R_{C2}) vortices. The spectra demonstrated by lines 1 and 2 in Figure 5.6(a) are depicted, respectively, in Figures 5.6(b) and 5.6(c). With the threshold $q_{th} = 4$, the velocity envelope $|V_E[\varphi(t)]|$ was obtained from the spectra shown in Figure 5.6(a). It is shown as a solid curve in Figure 5.6(d). From this envelope, one

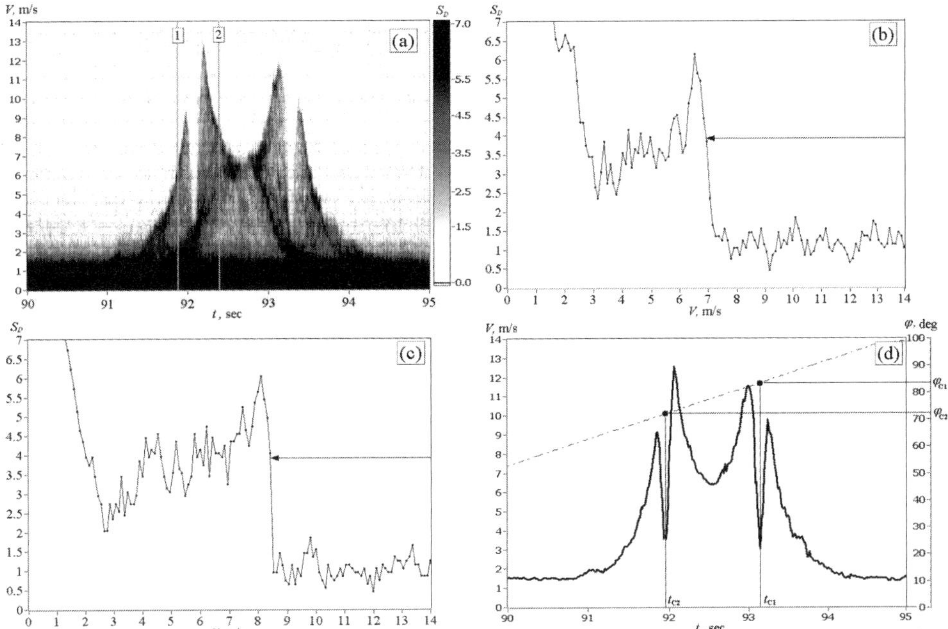

Figure 5.6 (a–c) Doppler spectra and (d) velocity envelope measured by a DLR cw lidar for one scan by the probing beam in the vicinity of the wake vortex pair.

can readily determine the angles φ_{C1}, φ_{C2} and the instants of time t_{C1}, t_{C2}, at which the probing beam intersects, respectively, the axes of the right and left vortices (see Figure 5.6(d)). To obtain an estimate of the vortex circulation by (5.12), (5.13), (5.8), and (5.9), it is necessary to know the distances R_{C1} and R_{C2}. If the values of $|F/R_{C1} - 1| \times 100\%$ and $|F/R_{C2} - 1| \times 100\%$ do not exceed a few percent, then in (5.8) and (5.9) we can use the known focal length F in place of the distances R_{C1} and R_{C2} and calculate the wake circulation with relatively high accuracy. [Using (5.7) through (5.13), it is easy to show that $|\hat{\Gamma}_i/\Gamma_i - 1| \approx |F/R_{Ci} - 1|$, if R_{Ci} is replaced with F at the negligibly small error of determination of the velocity envelope $V_E(\varphi)$].

Because the coordinates of the vortex axes vary with time, it is necessary to change the focal length F during measurements by one cw CDL for the vortex cores to stay within the sensing volume. Having the information about the initial coordinates of the vortex axes and the projection of the wind velocity vector onto the plane of scanning by the probing beam, one can predict the vortex core trajectory. The focal length F is changed during the measurements in accordance with this prediction. However, the true values of $R_{C1}(t)$ and $R_{C2}(t)$ can differ widely from $F(t)$ due to the random inhomogeneity of the wind flow. In this case, although the condition $F(t) - \Delta z/2 \leq R_{Ci}(t) \leq F(t) + \Delta z/2$ $(i = 1, 2)$, which allows the angular coordinates φ_{C1} and φ_{C2} to be determined with good accuracy, is fulfilled, the error of estimation of the vortex circulation with replacement of the radial coordinates R_{C1} and R_{C2} by F in (5.12) and (5.13) can be rather large.

The problems discussed above do not arise if the wake vortex parameters are measured by two cw CDLs. The geometry of measurement by two cw CDLs scanning in the same plane and separated by some distance ΔY is shown in Figure 5.7.

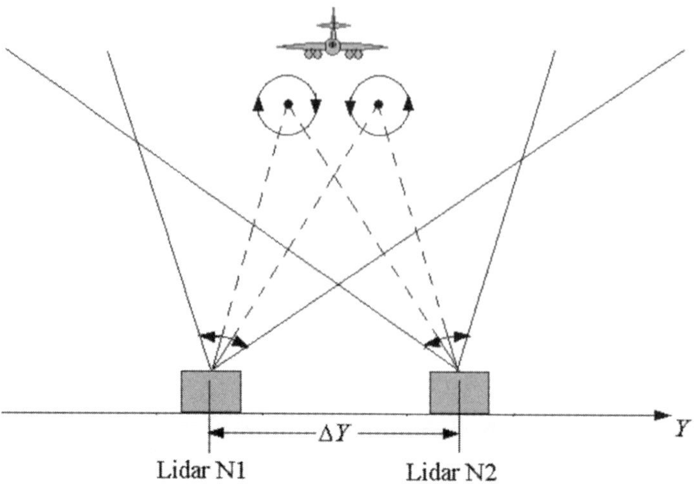

Figure 5.7 Geometry of taking measurements with two scanning cw CDLs.

The measured angular coordinates of the vortex axes $\varphi_{Ci}^{(1)}(t_{in})$ (lidar 1) and $\varphi_{Ci}^{(2)}(t'_{in})$ (lidar 2), where the subscript $i = 1$ is for the right vortex and $i = 2$ is for the left vortex, and $n = 1, 2, 3, \ldots$ is the scan number, are used to calculate the distances from lidar 1 (or 2) to the axis of the ith vortex; that is, $R_{Ci}^{(1)}(t_{in})$ [or $R_{Ci}^{(2)}(t'_{in})$] through triangulation (ΔY is known). As a rule, $t_{in} \neq t'_{in}$. The use of interpolation (for example, linear one) for the transition from $\varphi_{Ci}^{(2)}(t'_{in})$ to $\varphi_{Ci}^{(2)}(t_{in})$ [or from $\varphi_{Ci}^{(1)}(t_{in})$ to $\varphi_{Ci}^{(1)}(t'_{in})$] for the calculation of the radial coordinate of the vortex axis by the equation $R_{Ci}^{(1)} = [\sin\varphi_{Ci}^{(2)}/\sin(\varphi_{Ci}^{(2)} - \varphi_{Ci}^{(1)})]\Delta Y$ [or $R_{Ci}^{(2)} = \sin\varphi_{Ci}^{(1)}/\sin(\varphi_{Ci}^{(2)} - \varphi_{Ci}^{(1)})]\Delta Y$] can often lead to the significant random variation of the estimates $R_{Ci}^{(1)}$ [$R_{Ci}^{(2)}$] due to errors (even small) in estimation of the angles $\varphi_{Ci}^{(1)}(t_{in})$ and $\varphi_{Ci}^{(2)}(t'_{in})$ and instants of time t_{in} and t'_{in}. The accuracy of estimation of the radial coordinates of vortex axes can be increased significantly through the use of the smoothing Kalman filter [34, 58]. Thus, it is shown in [34] that the use of this filter allows one to estimate the vortex axis coordinates with accuracy of approximately ±4m (the core radius of a vortex generated by LTA is ~3m to 4m), which is sufficient for the estimation of vortex circulation by (5.12), (5.13), (5.8), and (5.9).

The requirement that the distance from the lidar to the vortex core should fall within the range $[F - \Delta z/2, F + \Delta z/2]$ cannot always be met in practice by specifying the focal length of the probing beam based on the *a priori* information about the initial vortex axis coordinates and the wind speed and direction. Another significant disadvantage of the considered method is the restriction to the measurement time connected with the wake stay time in overlapping sectors of scanning by probing beams (see Figure 5.7). If the lateral wind quickly transports the wake vortices, then the measurement time can be only a few tens of seconds (two to three scans). These problems do not arise if pulsed CDL is used for wake vortex investigation. At the not very strong lateral wind, the pulsed lidar allows for observation of the wake vortex evolution over long periods, from the moment of wake generation to the progressed state of vortex decay.

5.4   Measurement of Wake Vortex Parameters Using a Pulsed CDL

The geometry needed for measuring wake vortex parameters with a ground-based pulsed CDL is shown in Figure 5.8. As in the case of the cw lidar, scanning by the probing beam in the vertical plane (alternatively up and down) across the wake is also used in this case. The lidar is assumed to be far enough from the runway, along which the aircraft flies at a certain altitude, for the motion and evolution of wake vortices generated by this aircraft to be fully traced. At a flight altitude of no higher than 400m, the distance from the lidar to the runway (~700m) is quite sufficient. We can see from Figure 5.7 that the probing beams of the cw lidars scan the wake from below, whereas the pulsed lidar's scanning is carried out at much smaller angles φ (that is, practically from the side).

The results presented here were obtained with the use of a 2-μm pulsed CDL. The parameters of this lidar are given in Section 1.4. In the case of ground-based measurements, scanning by the probing beam with an angular rate of 2 deg/s was applied. The elevation angle φ varied from 0° to 20°–30° (duration of one scan is ~10s to 15s).

The procedure for processing the experimental data consisted of the following. The echo signal (series of samples of the backscatter signal) measured at one shot was used to estimate the power spectra of the signal at different distances from the lidar to the center of the sensing volume $R_l = R_0 + l\Delta z$, where $l = 0, 1, 2, ..., L_R, L_R = 100, \Delta R = 12$m, and $R_0 = 360$m, through the application of the Gaussian temporal window with $\sigma_W = \sigma_P$ and the fast Fourier transform. In this case, the longitudinal spatial resolution (longitudinal dimension of the sensing volume) is $\Delta R \approx 94$m. Then, estimates of the spectra obtained at other (neighboring) shots are averaged at fixed R_l. For this purpose (in ground-based measurements), 25 shots are used. Because the scan rate and the pulse repetition frequency are, respectively, 2 deg/s and 500 Hz, the scan angle resolution at this spectral averaging is $\Delta\varphi = 0.1$ deg. Consequently, the transverse dimension of the sensing volume in the scan plane $R_l\Delta\varphi$ varies from 0.63m at $R_l = 360$m to 2.7m at $R_l = 1560$m. After normalization of the measured spectra to the mean level of the noise spectral component, we obtain an array of normalized signal power spectra (Doppler spectra) $\hat{S}_D(V, R_l, \varphi_{i'}; n)$, where $\varphi_{i'} = \varphi_0 + i'\Delta\varphi$ is the elevation (scan) angle, $i' = 0, 1, 2, ..., I, I = 300, n = 0, 1, 2, 3, ...$ is the scan number, and $V \in [-25$ m/s, $+25$ m/s].

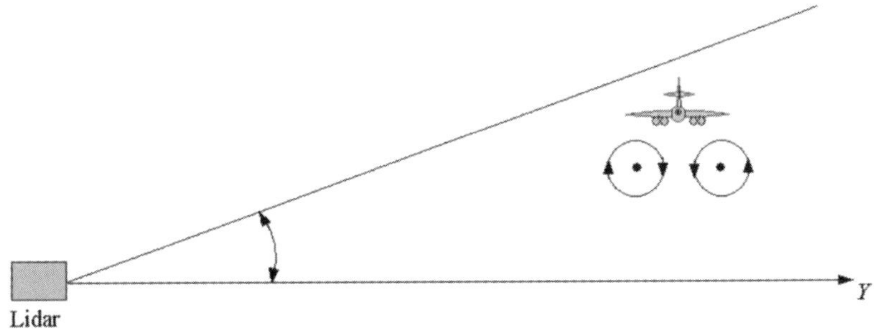

Figure 5.8   Geometry for measurements made by a scanning pulsed CDL.

Figures 5.9(b) through 5.9(d) exemplify the Doppler spectra measured by pulsed CDL (ground-based measurements, Frankfurt, 2004) [39]. The positions of the cores of the right and left vortices are shown in Figure 5.9(a). The spectrum measured between these cores (point of intersection of the horizontal and vertical lines in the figure) is shown in Figure 5.9(b). The spectra measured at different ranges R and the fixed angle φ, as well as at different φ and fixed R are depicted in Figures 5.9(c) and 5.9(d), respectively.

Given a certain threshold q_{th}, from the measured Doppler spectrum we can find the values of the positive V_{E+} and negative V_{E-} velocity envelopes (see Figure 5.9(b)). The value of the threshold q_{th} is determined by the level of fluctuations of the noise spectrum component. This level depends on the averaging number L. The probability that peak values of the normalized noise $\hat{S}_N(V) \equiv \xi$ ($\langle\xi\rangle = 1$) exceed q_{th} should be very low. On the other hand, the threshold q_{th} should not exceed significantly the mean noise level. The particular value of q_{th} can be obtained from the following considerations. According to (2.50), the probability density of the normalized noise is described by the equation

$$p(\xi) = (L^L/(L-1)!)\xi^{L-1}\exp(-L\xi) \tag{5.14}$$

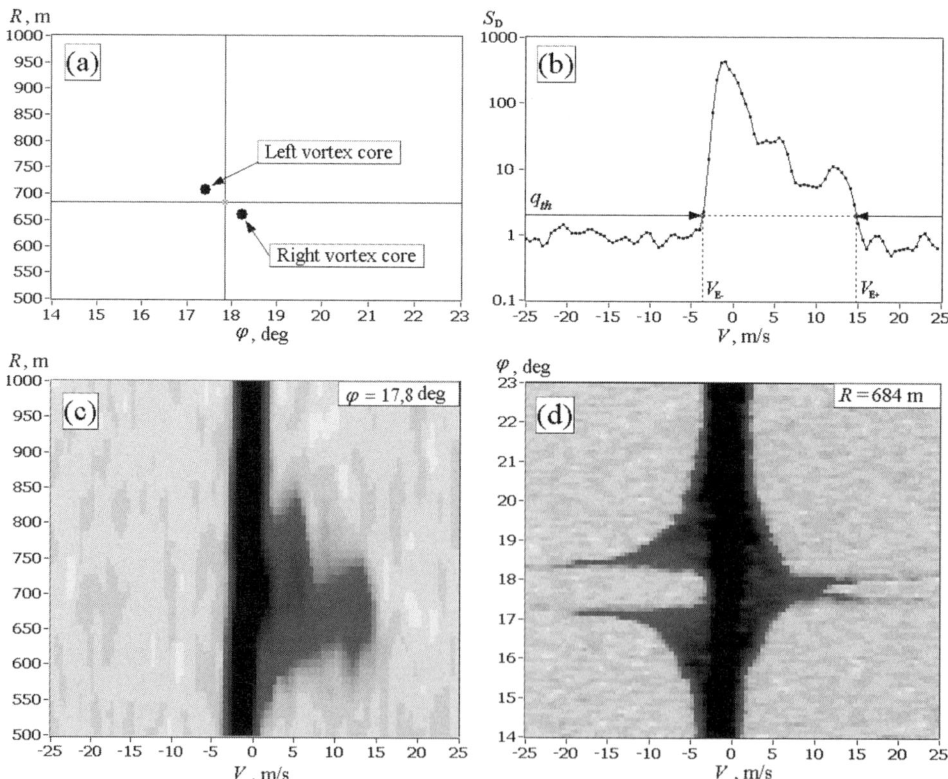

Figure 5.9 Example of Doppler spectra measured by 2-μm pulsed CDL (b–d): (b) single spectrum at the point of intersection of the vertical and horizontal lines in part (a); (c) spectra measured at different distances R from the lidar and at the fixed scanning angle φ; (d) spectra measured at different φ and fixed R.

In our case, $L = 25$. The probability that for the measurement time at least one value of ξ exceeds the threshold q_{th} can be written as

$$p(\xi > q_{th}) = 1 - \left[\int_0^{q_{th}} d\xi\, p(\xi) \right]^{N^*} \tag{5.15}$$

where $N^* = N_{ch} \times N_{sp} \times N_{sc}$, $N_{ch} = [B_V \sqrt{2\pi}\sigma_W 2/\lambda]$ is the number of independent frequency channels in one spectrum [59], $N_{sp} = I \cdot [\Delta RL_R/\Delta z]$ is the number of independent spectra obtained for one scan, and N_{sc} is the number of scans required to observe the wake evolution. For $B_V = 50$ m/s, $\sigma_W = 0.24\ \mu$s, $I = 300$, $\Delta RL_R = 1{,}200$m, $\Delta z = 94$m, and $N_{sc} = 10$, the value of N^* is $N^* \sim 10^6$. It follows from (5.14) and (5.15) that at $L = 25$ and $N^* = 10^6$ the probability $P(\xi > q_{th}) = 0.01$ when $q_{th} \approx 2.5$.

To determine the coordinates of the vortex axis, it is sufficient to use the value of the threshold q_{th} identical for any R_l and $\varphi_{i'}$. An example of the obtained velocity envelopes $V_{E+}(R_l, \varphi_{i'})$ and $V_{E-}(R_l, \varphi_{i'})$ (measurements in Tarbes airfield in 2003) at $q_{th} = 2.5$ is shown in Figure 5.10 [39]. In the figure, one can see the zone with a couple of vortices. The coordinates of the vortex axes $\{R_{C1}, \varphi_{C1}\}$ and $\{R_{C2}, \varphi_{C2}\}$ are estimated separately for the right and for the left vortex as a position of the point equidistant from the points of the maximum of the positive envelope and the minimum of the negative envelope. In Figure 5.10, the positions of these cores are shown by circles: white (left vortex) and gray (right vortex).

The coordinates of the wake vortex cores $\{R_{C1}, \varphi_{C1}\}$ and $\{R_{C2}, \varphi_{C2}\}$ and the spectra $\hat{S}_D(V - V_B, R_{C1}, \varphi_{i'}; n)$ and $\hat{S}_D(V - V_B, R_{C2}, \varphi_{i'}; n)$, where V_B is the radial velocity of the background wind determined from lidar measurements [35], can be used from the estimation of the vortex circulation at the nth scan by the probing beam. Figures 5.11(a) and (b) exemplify these spectra (ground-based measurements, Frankfurt, 2004) [39].

If we use the threshold $q_{th} = 2.5$ identical for any angles $\varphi_{i'}$, then from the Doppler spectra (Figures 5.11(a) and 5.11(b)) we obtain the positive $V_{E+}(R_{Ci}, \varphi_{i'})$ and negative $V_{E-}(R_{Ci}, \varphi_{i'})$ ($i = 1, 2$) velocity envelopes shown by dashed curves in Figures 5.11(c) and 5.11(d). The vortex circulation estimates obtained by the integration method of (5.7) through (5.13), using the parts in these velocity envelopes (shown by the gray colors in Figures 5.11(c) and 5.11(d) at $r_{min} = 5$m, $r_{max} = 15$m), have a large

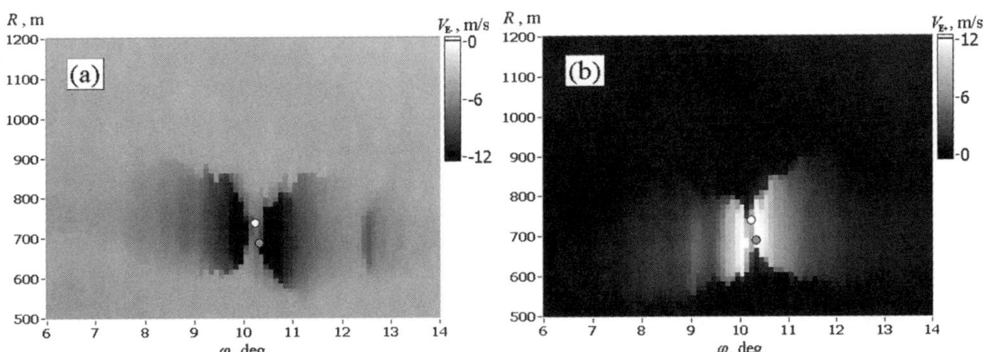

Figure 5.10   (a) Negative and (b) positive velocity envelopes.

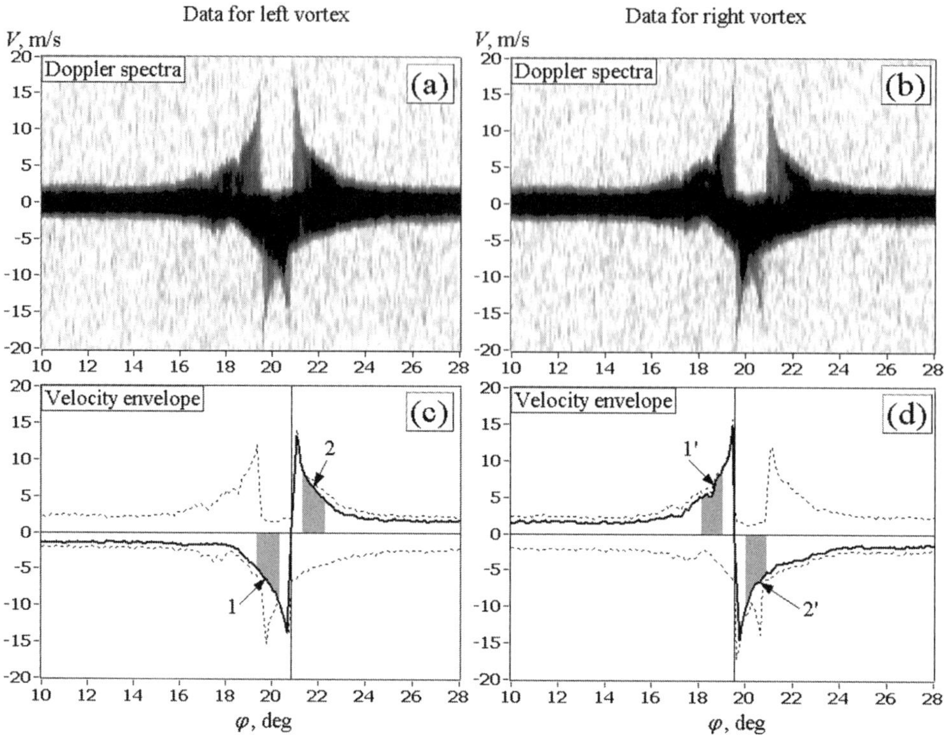

Figure 5.11 Example of Doppler spectra used for estimation of circulation of the (a) left and (b) right vortex. Velocity envelopes obtained from these spectra (c and d) at the fixed and floating threshold are shown, respectively, by dashed and solid curves.

error. Actually, in this example, when the angle between the vortex axes is $\varphi_{C2} - \varphi_{C1}$ ~ 1° and the distance between them is b ≈ 50m, which is almost half the longitudinal dimension of the sensing volume Δ_z = 94m, in the inner integration domains 1 and 2′ (see Figures 5.11(c) and 5.11(d)), the absolute value of the velocity envelope $|V_{E-}(R_{Ci}, \varphi_{i'})|$ obtained separately for the right (left) vortex in the range 2′ (1) is the distribution of the tangential velocity of the left (right) vortex around its maximum. In this case, to avoid large errors, we can estimate the vortex circulation $\hat{\Gamma}_1$ and $\hat{\Gamma}_2$ by using only the outer integration domains 1′ and 2 (see Figures 5.11(c) and 5.11(d)).

According to the data of Figure 5.4, where the same threshold q_{th} = 3 for any spectrum was used to obtain the velocity envelopes, in the integration domains the absolute values of the velocity envelopes exceed the velocity $|V_z(0, y)|$. The higher the SNR, the larger this excess. Even if the outer integration domains 1′ and 2 from the data shown as the dashed curve in Figures 5.11(c) and 5.11(d) are used, the estimates of the vortex circulations $\hat{\Gamma}_1$ and $\hat{\Gamma}_2$, according to the theoretical calculations, exceed the true circulation values by about 10%.

The calculations of $q_{th} = S_D(V_{E1}(\varphi))$ and $q_{th} = S_D(V_{E2}(\varphi))$ with the use of (5.5), where $V_{E1}(\varphi) = V_r(\tilde{R}_1, \varphi)$, $V_{E2}(\varphi) = V_r(\tilde{R}_2, \varphi)$, $\tilde{R}_i = R_{Ci} \cos(\varphi - \varphi_{Ci})$, and $V_r(R, \varphi)$ is described by (5.6) and (5.1), have shown that the threshold q_{th} is a function of the elevation angle $q_{th} = q_{th}(\varphi)$. Thus, to obtain the velocity envelopes closest to the tangential velocities of the vortices in the integration intervals (including the overlapping intervals 1 and 2′), it is necessary first to calculate the floating threshold $q_{th}(\varphi_{i'})$ for each measured Doppler spectrum $\hat{S}_D(V - V_B, R_{C1}, \varphi_i; n)$ and $\hat{S}_D(V - V_B, R_{C2}, \varphi_i; n)$ using the spectral model described by (5.5), (5.6), and (5.1). All of the parameters in these equations (including the vortex core radius r_C determined by the aircraft type) are known, except for the vortex circulations Γ_1 and Γ_2. The following iterative procedure was proposed for the estimation of the vortex circulations from the measured Doppler spectra [38, 39].

The estimates $\hat{\Gamma}_1$ and $\hat{\Gamma}_2$ obtained from the velocity envelopes at the constant threshold $q_{th} = 2.5$ (see dashed curves in Figures 5.11(c) and 5.11(d)) are used as a first iteration. Then, after the calculation of the floating thresholds $q_{th}^{(i)}(\varphi_{i'})$, new velocity envelopes $V_{E-}(R_{Ci}, \varphi)$ and $V_{E+}(R_{Ci}, \varphi)$ are obtained for the right ($i = 1$) and left ($i = 2$) vortices, and the procedure repeats. The iterative process converges fast. As a rule, no more than three iterations are sufficient. The resultant velocity envelopes $V_E(R_{C1}, \varphi)$ and $V_E(R_{C2}, \varphi)$ are shown in Figures 5.11(c) and 5.11(d) as solid curves.

5.5   Comparative Analysis of the Results of Simultaneous Measurements of Wake Vortex Parameters Using Pulsed and Continuous-Wave Lidars

In June 2002 at the Tarbes airfield (France), three lidar teams from Germany (DLR, Oberpfaffenhofen), France (ONERA, Palaiseau), and the United Kingdom (QinetiQ, Malvern) conducted a field experiment on measurement of wake vortices [36] with the use of the DLR pulsed 2-μm CDL and two ONERA and QinetiQ cw CO_2 coherent lidars. The experiment allowed a comparative analysis to be conducted on the results of wake vortex parameters measured simultaneously by two cw CDLs and one pulsed CDL.

Figure 5.12 shows the geometry of lidar measurements of wake vortices. The separation between the ONERA (L3) and QinetiQ (L2) cw CDLs was 122m. The pulsed CDL (L1) was separated by 824m from L3 at a small angle to the Y axis (Figure 5.12, top view). The experiment on measurement of wake vortices generated by a LTA included scanning by the probing beams in the zones shown in Figure 5.12 (side view) with angular rates of 2 deg/s (L1), 10 deg/s (L2), and 12 deg/s (L3). The aircraft circled in above the measurement site, flying along the line parallel to the X axis with a period of intersection of the $\{Z, Y\}$ plane equal to about 8 min. The flight altitude varied from 180m to 430m. The data measured by the cw lidars L2 and L3 allow us to estimate only the parameters of the wake vortices falling within the zone of the overlapping scanning sectors (Figure 5.12). As noted earlier, during a strong lateral wind these measurements cannot be long. In the case of the pulsed lidar (L1), the wake vortex's evolution can be observed, as a rule, until the complete decay of the wake vortex.

The measurements were conducted over 4 days. During this time, the LTA crossed the measurement plane (ZY plane in Figure 5.12) a total of 65 times. The duration of one scan by the probing beam of the pulsed lidar was 11s. This periodicity was quite sufficient under typical atmospheric conditions for the detailed observation of the behavior (vortex core motion) and evolution (decay) of wake vortices.

Figure 5.13 shows two examples of the trajectories of the wake vortex axes measured by the pulsed lidar (dashed curves) and the results of triangulation (solid

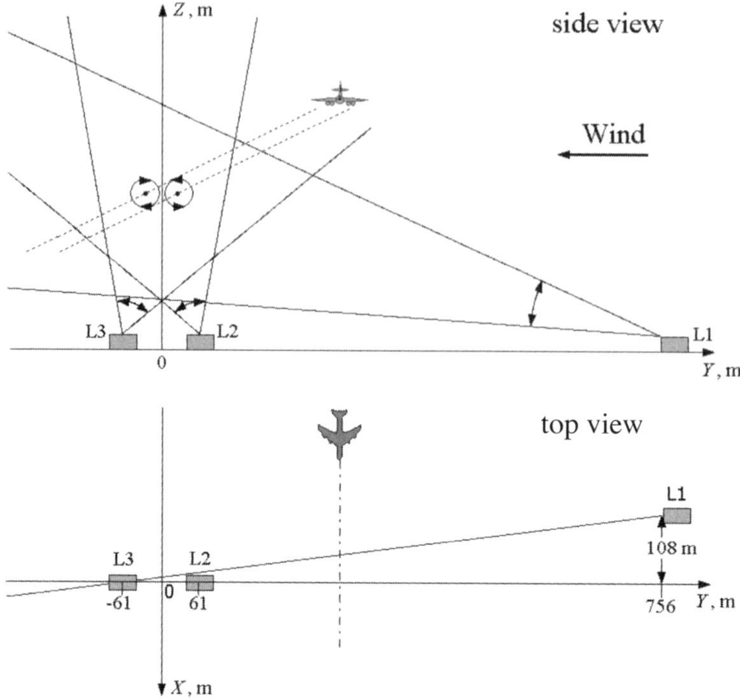

Figure 5.12 Geometry of lidar measurements at the Tarbes airfield for one pulsed CDL (L1) and two cw CDLs (L2 and L3). The trajectories of the vortex cores transported by wind are shown by the dashed curves. (© 2005 American Institute of Aeronautics and Astronautics. From [36].)

curves) obtained from the measurement data of the two cw lidars. These measurements were conducted on 13 and 14 June 2002, when the wind velocity component transverse to the X axis was, respectively, 4.5 m/s (Figure 5.13(a)) and 2.5 m/s (Figure 5.13(b)). For the examples shown in Figure 5.13, the aircraft coordinates (at $X = 0$; see Figure 5.12) were as follows: $Z = 334$m, $Y = 424$m (Figure 5.13(a)) and $Z = 347$m, $Y = 297$m (Figure 5.13(b)). The vortex core trajectories were measured by the pulsed lidar for the periods of 52s to 138s (Figure 5.13(a)) and 70s to 224s (Figure 5.13(b)) after overflight, that is, after intersection of the ZY plane by the aircraft at $X = 0$. The results of triangulation were obtained for the corresponding intervals of 39s to 96s (Figure 5.13(a)) and 57s to 182s (Figure 5.13(b)). One can see from Figure 5.13 that with time the vortices move down and to the left. The velocities of horizontal motion are close to the mentioned velocities of the lateral wind. In these two examples, the time of wake vortex observation by the pulsed lidar is, on average, 78s longer than that of the cw lidars.

The scanning beams of the pulsed and cw lidars intersect the vortex axes at different moments in time. Therefore, linear interpolation was used to compare the results of simultaneous measurements. For comparison, Figure 5.14 shows the results of triangulation Z_{TR} and Y_{TR} along with the vertical Z_{L1} (Figure 5.14(a)) and horizontal Y_{L1} (Figure 5.14(b)) coordinate results of the vortex axis measured by the pulsed lidar. The data shown in this figure were used in [36] to calculate two statistical parameters, $\hat{\sigma}_Z$ and $\hat{\sigma}_Y$, which characterize the level of discrepancy in

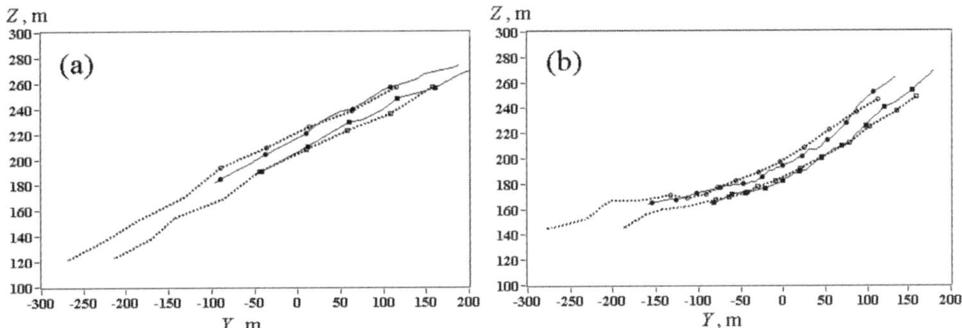

Figure 5.13 Results of lidar measurements of wake vortex core trajectories in Tarbes on (a) 13 June and (b) 14 June 2002. Local time of the overflight was (a) 21:34:03 and (b) 20:13:11. Trajectories measured by one pulsed CDL (dashed lines), and results of triangulation obtained from the data of two cw CDLs (solid curves). (© 2005 American Institute of Aeronautics and Astronautics. From [36].)

the vertical and horizontal coordinates of the vortex axis obtained from the data of the pulsed lidar and cw lidars (with the use of triangulation) by the equations

$$\hat{\sigma}_Z = \sqrt{\frac{1}{K}\sum_{k=1}^{K}[Z_{L1}(t_k) - Z_{TR}(t_k)]^2} \tag{5.16}$$

$$\hat{\sigma}_Y = \sqrt{\frac{1}{K}\sum_{k=1}^{K}[Y_{L1}(t_k) - Y_{TR}(t_k)]^2} \tag{5.17}$$

where $K = 284$ is the number of points in every plot drawn in Figure 5.14. The calculations yielded $\hat{\sigma}_Z \approx 9m$ and $\hat{\sigma}_Y \approx 13m$.

Assume that random errors in estimates of the coordinates $\{Z_{L1}(t_k), Y_{L1}(t_k)\}$ and $\{Z_{TR}(t_k), Y_{TR}(t_k)\}$ are statistically independent, stationary, and homogeneous. In this case, the variance $\sigma_Z^2 = \langle \hat{\sigma}_Z^2 \rangle$ $(\sigma_Y^2 = \langle \hat{\sigma}_Y^2 \rangle)$ is a sum of the variances $\sigma_Z^2\,|_{L1} = \langle [Z_{L1} - \langle Z_{L1}\rangle]^2 \rangle$ $(\sigma_Y^2\,|_{L1} = \langle [Y_{L1} - \langle Y_{L1}\rangle]^2 \rangle)$ and $\sigma_Z^2\,|_{TR} = \langle [Z_{TR} - \langle Z_{TR}\rangle]^2 \rangle$

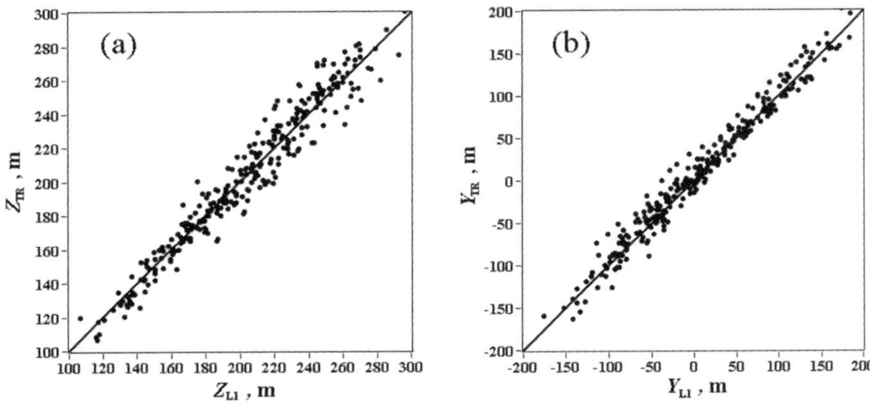

Figure 5.14 Comparison of the (a) vertical and (b) horizontal coordinates of vortex axes obtained through triangulation and from measurements by pulsed lidar. (© 2005 American Institute of Aeronautics and Astronautics. From [36].)

$(\sigma_Y^2 \mid_{TR} = \langle [Y_{TR} - \langle Y_{TR} \rangle]^2 \rangle)$. If we also assume that the errors of estimation of vortex axis coordinates from data of the pulsed lidar (L1) and through triangulation with the use of the L2 and L3 lidar data are identical, then $\sigma_Z \mid_{L1} = \sigma_Z \mid_{TR} \approx \hat{\sigma}_Z / \sqrt{2} \approx 6m$ and $\sigma_Y \mid_{L1} = \sigma_Y \mid_{TR} \approx \hat{\sigma}_Y / \sqrt{2} \approx 9m$. These values of the errors (6m for the vertical wake coordinate and 9m for the horizontal one) are overstated, because the measurements by lidar L1 and lidars L2 and L3 were carried out in different vertical planes (see Figure 5.12), as well as due to the nonideal temporal synchronization between lidars (error ~1s). The analysis carried out in [36] with allowance made for these factors has revealed that the lidar error in estimation of the vortex axis coordinates is approximately $E_Z = 4.5m$ for the vertical coordinate and $E_Y = 6.5m$ for the horizontal one.

During measurements with the ONERA lidar (L3), the focal length F was changed according to the prediction of descent of the wake vortices and their transport by the lateral wind. This allowed the sensing volume to be located near the vortex core, and, as a consequence, more accurate data have been obtained for the velocity envelopes used in (5.12) and (5.13) for the estimation of vortex circulation. At the same time, the results of triangulation were used for R_{Ci}, for which the accuracy of measurement of the angles φ_{Ci} by cw lidars L2 and L3 is quite sufficient [36].

Figure 5.15 exemplifies the velocity envelopes $V_{E+}[\varphi(t)]$ and $V_{E-}[\varphi(t)]$ obtained from the data of simultaneous measurements by the ONERA cw lidar (L3,

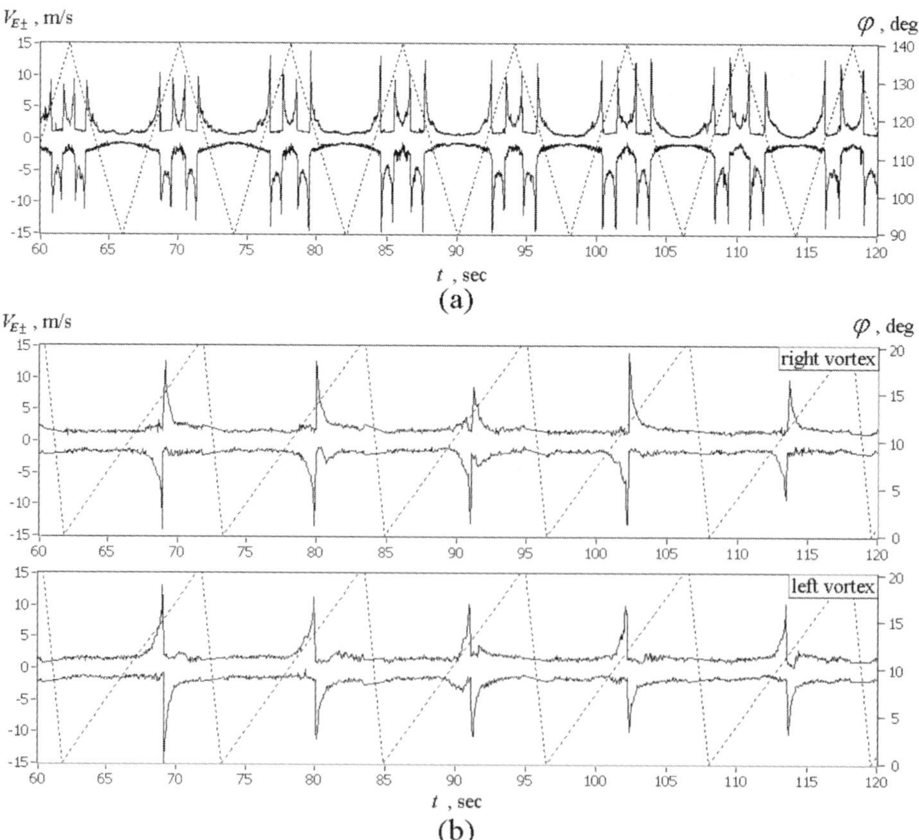

Figure 5.15　Velocity envelopes measured by (a) cw and (b) pulsed CDLs.

heterodyning system) and DLR pulsed lidar (L1). Here t is time after overflight (that is, t is the vortex age). In the case of the pulsed lidar, the velocity envelopes are obtained separately for the right and left wakes with the use of the iterative procedure for calculation of the floating threshold $q_{th}(\varphi)$ (see Section 5.4) and are suitable for estimation of the vortex circulation by (5.12) and (5.13). The depicted envelopes obtained from the cw lidar data are also suitable for the calculation of vortex circulation by these equations. For every scan, the circulation can be estimated for the right and left wakes. The arrays of these estimates represent the wake circulation as a function of time, that is, $\Gamma_i(t_{in})$, where t_{in} is the time of intersection of the ith ($i = 1, 2$) vortex axis by the probing beam, and n is the scan number ($n = 1$ corresponds to the first appearance of the vortex in the scanning sector).

Figure 5.16 depicts three examples of measurement of the normalized circulation of the right vortex Γ/Γ_0 as a function of the normalized time t/t_0 with the DLR pulsed lidar and ONERA cw lidar [hereinafter, the theoretical values of Γ_0 and t_0

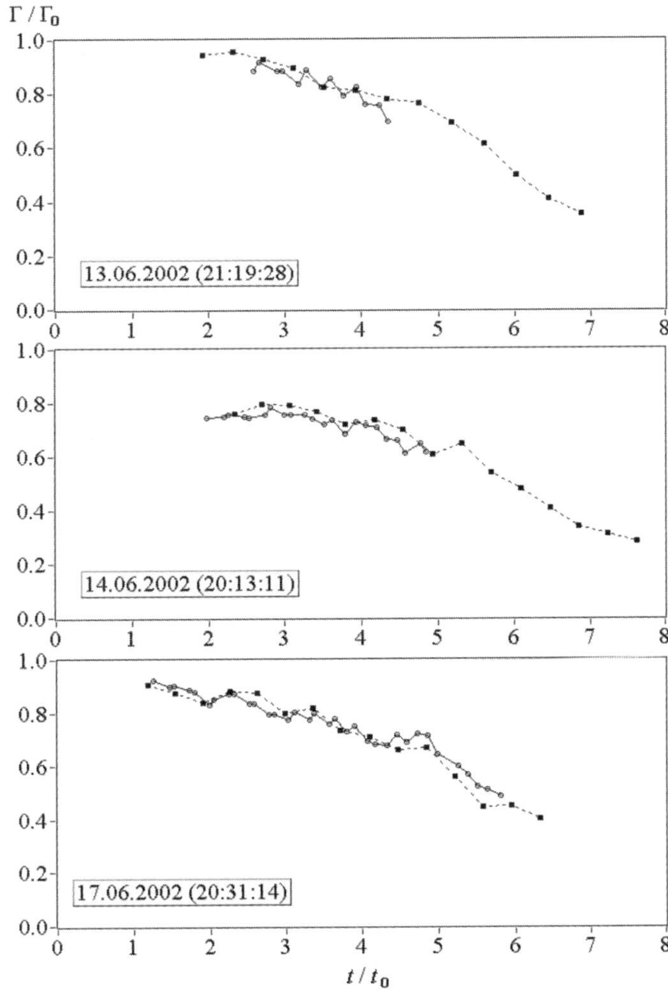

Figure 5.16 Dependences of the normalized vortex circulation on the normalized time measured by a pulsed CDL (squares) and an ONERA cw lidar (circles) in Tarbes in 2002. Local time is shown in parenthesis. (© 2005 American Institute of Aeronautics and Astronautics. From [36].)

are calculated by (5.2) and (5.3)] [36]. One can see good agreement between the data measured by these lidars in the overlapping time intervals.

As already mentioned, in view of the measurement geometry, pulsed lidar L1 allowed for longer observation of wakes than did cw lidars L2 and L3. All estimates of the normalized circulation Γ/Γ_0 obtained from the data of the pulsed lidar and the ONERA cw lidar in the overlapping time intervals are shown by symbols in Figure 5.17 [36]. These data were used to calculate the time dependence of the mean circulation with a moving average. As can be seen from the figure, the calculated results shown by the dot-and-dash (L3) and solid (L1) curves are quite close. The standard deviations of individual estimates of circulation normalized to their average values for lidars L1 and L3 nearly coincide and are equal to about 0.06 (in absolute values, 22 m²/s). This suggests that the errors of estimation of the wake circulation from L1 and L3 lidar data are identical, since the measurements by these lidars were conducted under the same atmospheric conditions.

For comparison, Figure 5.18 depicts individual estimates of the vortex circulation obtained from the data of simultaneous measurements by the DLR pulsed lidar (Γ_{L1}) and ONERA cw lidar (Γ_{L3}) [36]. Using the data of this figure, by analogy with (4.55), we can calculate the dimensionless parameter $\hat{\sigma}_{\Gamma}$ characterizing the level of discrepancy in the circulation estimates Γ_{L1} and Γ_{L3} as

$$d_{\Gamma} = \sqrt{\frac{4}{K} \sum_{k=1}^{K} \left[\frac{\Gamma_{L1}(t_k) - \Gamma_{L3}(t_k)}{\Gamma_{L1}(t_k) + \Gamma_{L3}(t_k)} \right]^2} \qquad (5.18)$$

where $K = 95$. As a result, we have $d_{\Gamma} = 0.0528$. Random errors of the estimates Γ_{L1} and Γ_{L3} are statistically independent. Then, assuming these errors to be stationary, we can present $\langle [\Gamma_{L1}(t_k) - \Gamma_{L3}(t_k)]^2 \rangle$ as a sum $\hat{\sigma}_{\Gamma}^2\big|_{L1} + \sigma_{\Gamma}^2\big|_{L3}$, where the first and second terms are the variances of random errors of vortex circulation estimation from the

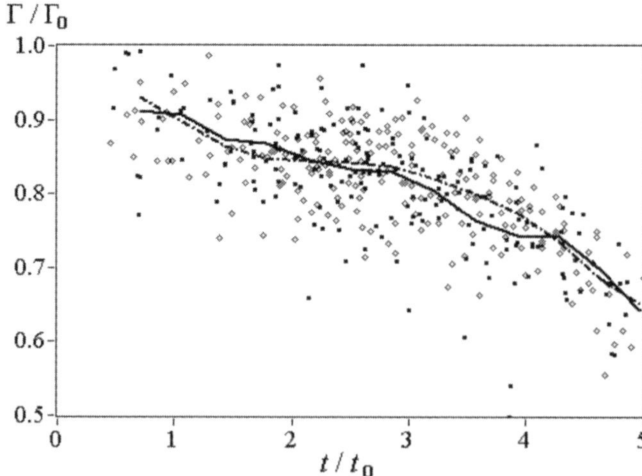

Figure 5.17 Results of measurement of vortex circulation by pulsed (squares are individual estimates; solid curve represents the mean value) and cw (rhombs are individual estimates; dot-and-dash curve represents the mean value) CDLs. (© 2005 American Institute of Aeronautics and Astronautics. From [36].)

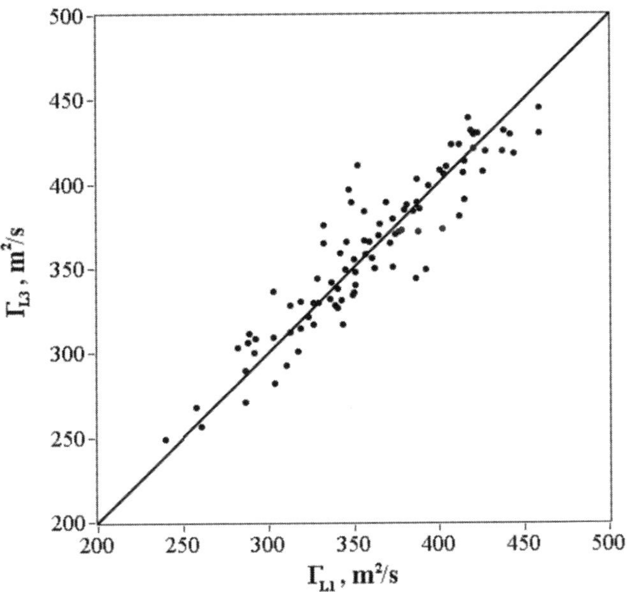

Figure 5.18 Vortex circulation measured simultaneously by pulsed (Γ_{L1}) and (Γ_{L3}) CDLs. (© 2005 American Institute of Aeronautics and Astronautics. From [36].)

data of the L1 and L3 lidars, respectively. Because these variances are equal, the error E_Γ of lidar (L1 or L3) estimation of the vortex circulation can be calculated by the approximate equation

$$E_\Gamma = \Gamma d_\Gamma / \sqrt{2} \qquad (5.19)$$

where Γ is the mean value of the vortex circulation. According to the data of Figure 5.18, the mean circulation equals 350 m²/s. Given $\Gamma = 350$ m²/s and $d_\Gamma = 0.0528$ in (5.19), we obtain the mean estimate of the error: $E_\Gamma = 13$ m²/s, which coincides with the corresponding value reported in [36]. The relative error of circulation estimation $\varepsilon_\Gamma = (E_\Gamma / \Gamma) \times 100\%$ is 3.7%.

We now calculate the error of lidar estimates of the vortex axis coordinates and circulation for the conditions of the field experiment (SNR value, vortex circulation, and distance from the lidar to the axis) using the numerical simulation of the operation of a scanning pulsed lidar (see Section 2.6) for wake vortex measurements. [It is assumed that there are no turbulent variations of wind velocity, and the radial velocity is calculated by (5.6) and (5.1).] We also use the procedures for data processing described in Section 5.4. For this purpose, we use 500 independent realizations of estimates of the vortex axis coordinates $\{\hat{R}_C, \hat{\varphi}_C\}$ and circulation $\hat{\Gamma}$ at SNR = 2, 10, and the given (true) parameters $R_C = 800$m, 400m, $\Gamma = 500$ m²/s, 200 m²/s. The errors of estimation of the radial E_R and angular E_φ coordinates of the vortex axis and circulation E_Γ are calculated as $E_R = \sqrt{(\hat{R}_C - R_C)^2}$, $E_\varphi = \sqrt{(\hat{\varphi}_C - \varphi_C)^2}$, and $E_\Gamma = \sqrt{(\hat{\Gamma}_C - \Gamma_C)^2}$.

Table 5.1 summarizes the results of theoretical calculations of the errors E_R, E_φ, and E_Γ from the simulated data [38]. It follows from Table 5.1 that the error E_Γ is influenced to a greater degree than the other errors by the SNR value. On average, $E_\Gamma = 11.2$ m²/s, which differs only a little from the mean estimate of the error $E_\Gamma = 13$ m^2/s obtained in the field experiment. The calculations by (5.19), where $d_\Gamma = 0.0528$, yield the following results: $E_\Gamma = 18.7$ m^2/s at $\Gamma = 500$ m^2/s and $E_\Gamma = 7.5$ m^2/s at $\Gamma = 200$ m^2/s, which fall within the range of the E_Γ values given in Table 5.1.

To compare the errors of lidar measurements of wake axis coordinates E_Z and E_Y calculated from the data of the numerical and field experiments, we use the equations

$$E_Z = \sqrt{(E_R \sin\varphi_C)^2 + (R_C E_\varphi \cos\varphi_C)^2} \qquad (5.20)$$

$$E_Y = \sqrt{(E_R \cos\varphi_C)^2 + (R_C E_\varphi \sin\varphi_C)^2} \qquad (5.21)$$

where the error E_φ is in radians. If we take $R_C = 600$m, $E_R = 7$m (value averaged over the data of Table 5.1), $\varphi_C = 20$ deg, and $E_\varphi = 0.1$ deg in these equations, then we obtain $E_Z = 2.6$m and $E_Y = 6.6$m. As shown earlier, in the field experiment $E_Z = 4.5$m and $E_Y = 6.5$m.

Note that the results for the vortex circulation given in Table 5.1 were obtained with the use of the integration method (see Section 5.2). For the case where $\Gamma = 500$ m^2/s, the integration limits were specified as $r_{min} = 5$m and $r_{max} = 15$m. To avoid the bias of the circulation estimate at $\Gamma = 200$ m^2/s, the upper boundary of the integration interval r_{max} was decreased to 10m. Another solution of the problem of bias at the small vortex circulation was proposed in [60].

Thus, the comparative analysis of the results of simultaneous measurements by the pulsed and cw CDLs has allowed us to determine the accuracy of estimation of the wake vortex parameters, which is confirmed by the results of numerical experiments and is quite acceptable (since the error of estimation of the wake axis coordinates does not exceed the core diameter of the wake generated by a large aircraft, and the relative error of estimation of the wake circulation is ~4%) for lidar studies of the atmospheric influence on the aircraft wake vortices. It is preferable to carry out these studies with pulsed CDLs, because, in contrast to cw lidars, in this case one lidar is sufficient for the experiment. In addition, the geometry of measurements by a pulsed lidar allows, for a not very strong lateral wind, the evolution of the wake vortex—from its formation to its destruction—to be observed. The numerical

Table 5.1   Results of Theoretical Calculations of E_R, E_φ, and E_Γ from Simulated Data

SNR	Γ (m^2/s)	R_C (m)	E_R (m)	E_φ (deg)	E_Γ (m^2/s)
2	500	800	8.6	0.08	18.6
2	500	400	7.2	0.10	16.0
2	200	800	7.5	0.08	8.5
2	200	400	5.9	0.11	8.1
10	500	800	8.3	0.05	12.5
10	500	400	6.5	0.06	11.7
10	200	800	7.1	0.03	7.3
10	200	400	5.8	0.05	7.0

experiments have shown that with a decrease in the SNR, the error in lidar estimates of the wake vortex parameters increases, and at SNR < 1 the method of estimation of these parameters from measured velocity envelopes $V_{E\pm}(R, \varphi)$ becomes inefficient.

5.6   Measurements of Wake Vortex Parameters in the Atmospheric Surface Layer

In 2004 at the airfield of the Frankfurt am Main Airport, the DLR Lidar Group studied the influence of the Earth's underlying surface on wake vortices with a 2-μm pulsed CDL [39]. The experimental campaign consisted of two stages. First, for 9 measurement days in September, parameters of aircraft wake vortices were measured at low altitudes of the atmospheric boundary layer, where the Earth's surface is expected to affect wake vortices. Then in October through December, for 11 measurement days, measurements were conducted at high altitudes, where the influence of the Earth's surface on wake vortices is negligibly small. The measurements of wake vortices that are influenced by the Earth's surface are referred to as *in-ground effect* (IGE), whereas those beyond the influence of the Earth's surface are referred to as *out-of-ground effect* (OGE). In this section, we present the results of IGE measurements.

For the IGE measurements, the container with the lidar was set on the ground. Due to the conditions for practical realization of the experiment, scanning by the probing beam (the elevation angle φ varied from 0° to 15°, and the scan angular rate was 2 deg/s) was carried out in the vertical plane forming the angle $\theta_g = 28°$ with the vertical plane perpendicular to the two runways separated by 500m. The distance to the nearest runway was about 450m. In processing of raw measurement data (for example, when estimating the circulation or the separation between axes of the vortices) the angle θ_g should be properly taken into account. The lidar was used to measure wake vortices behind landing aircraft. The measurements usually started between 5:00 and 6:00 LT, when the incoming traffic increased quickly, and ended by noontime.

The period of continuous operation of the lidar when taking measurements at the airfield of the Frankfurt am Main airport was 5 to 6 hours. During this time period, a large number of various aircraft intersected the scanning plane. For example, for the 2 hours after 6:00 LT, aircraft landed on two runways with an average interval of 2 min. The information about the actual landing time of aircraft, their wingspan, mass, true speed, and flight height in the scanning plane ZY allowed for calculation of the vortex age t, as well as the parameters Γ_0 and t_0 by (5.2) and (5.3). Next we present the analysis of results of lidar measurements for the case of a LTA.

Figure 5.19 shows four examples of lidar measurements of vortex core trajectories $Z_i(t) = R_{Ci}(t)\sin\varphi_{Ci}(t)$, $Y_i(t) = R_{Ci}(t)\cos\varphi_{Ci}(t)$, and dependences of the normalized vortex circulation Γ_i/Γ_0 on the normalized time t/t_0, where, as before, the subscript $i = 1$ for the right vortex and $i = 2$ for the left one. The curves for the trajectories of the right and left vortices are connected by dashed lines at the points corresponding to one scan by the probing beam. The data plotted in Figures 5.19(a), a′ and 5.19(b), b′ were obtained from measurements taken during calm weather, whereas Figures 5.19(c), c′ and 5.19(d), d′ were obtained at moderate lateral wind. It can be seen

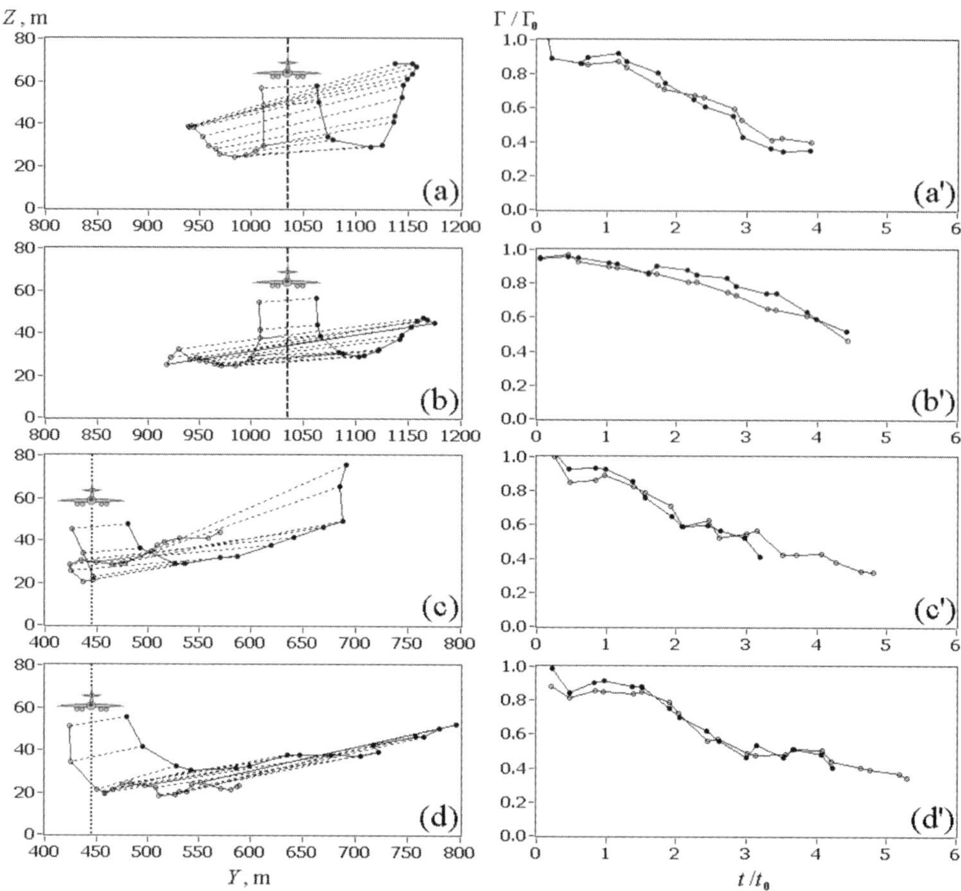

Figure 5.19 (a–d) Vortex core trajectories and dependences of the normalized vortex circulation on (a′–d′) normalized time measured by lidar in Frankfurt am Main on (a, a′, b, b′) 16 September and (c, c′, d, d′) 28 September 2004. Local time of intersection of the scanning plane by aircraft: (a, a′) 06:54:06, (b, b′) 08:21:31, (c, c′) 04:53:41, and (d, d′) 06:44:52. Here, Z is the altitude and Y is the distance from the lidar. Closed circles: data for the right vortex; open circles: data for the left vortex.

that in the first case (Figures 5.19(a) and 5.19(b)) the couple of vortices formed at a height of 60m descended to the height of vortex axes ~25m and then, due to the repelling effect of the Earth's surface, the vortices moved upward and the distance between the vortex axes increased. Later on, the upward motion again alternates with the downward motion. The lateral wind transported the vortices (see Figures 5.19(c) and 5.19(d)). The height of the vortex core at the late stage of the vortex evolution can exceed the flight height in the scanning plane, while its circulation is 2 to 3 times smaller than the initial circulation.

Figure 5.20 shows the results of IGE measurements of the relative distance between the vortex axes b/b_0, tilt angle of the vortex pair θ_p, height of the vortex axis Z, and relative circulation Γ/Γ_0 as functions of the normalized time t/t_0 [39]. The results include 100 measurements of individual realizations of the time dependence of vortex axis coordinates and circulation. Individual estimates of the parameters are shown by dots in the figure. It can be seen that at $t/t_0 < 0.2$, the distance b and circulation Γ measured by lidar agree, on average, with the theoretical values of b_0

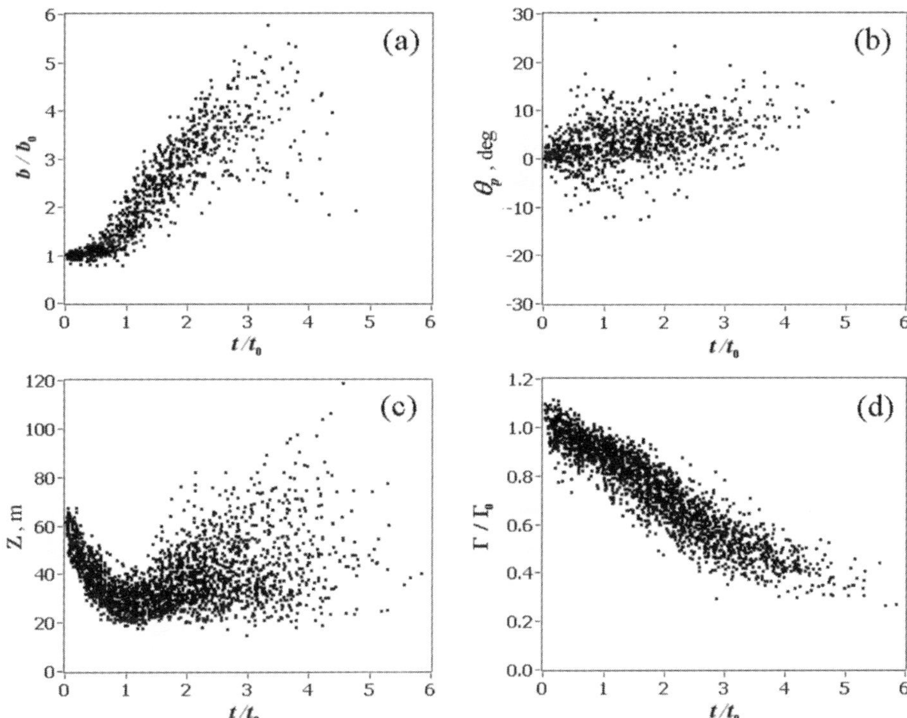

Figure 5.20   Results of lidar IGE measurements of (a) the relative distance between vortex axes b/b_0, (b) tilt angle of the vortex pair θ_p, (c) height of the vortex axis Z, and (d) relative circulation Γ/Γ_0 as functions of the normalized time t/t_0.

and Γ_0. With time, the distance between the vortex axes increases, and, as follows from Figure 5.20(a), the value of b can be almost 6 times as large as the initial distance b_0. The angles θ_p vary within $\pm 15°$. The number of dots in the plots of Figure 5.10(c) and 5.20(d) is more than twice as large as that in Figures 5.20(a) and 5.20(b) because, first, the data for Z and Γ are given for the right and left vortices and, second, the situation was often (at IGE measurements) observed that one vortex was fully destructed (correspondingly, there were no data for b and θ_p), while the other had a close to ideal structure.

According to Figure 5.20(c), in the initial time interval $t/t_0 \leq 1$ the vortices go down. Then, on average, the height of the wake axis increases due to the effect of the Earth's surface. As this takes place, the height of the wake axis can grow to be twice as large as the flight height (~60m). We can see from Figure 5.20(d) that with time the vortex circulation decreases, and the analysis has shown that the circulation, on average, halves at $t/t_0 \approx 3$ as compared to the initial value.

5.7   Lidar Investigations of the Influence of Wind and Atmospheric Turbulence on Wake Vortices in the Atmospheric Boundary Layer

In this section, we analyze the results of measurements of wake vortex parameters taken with a 2-μm pulsed CDL in the atmospheric boundary layer at heights where we can neglect the effect of the Earth's surface (OGE measurements). The experiments were conducted at airfields of the Tarbes airport (June 2002 and August 2003) [36, 43] and Frankfurt am Main airport (October–December 2004) [39, 40, 50]. The geometry of measurements at Tarbes is shown in Figure 5.12 (lidar L1). For OGE measurements in Frankfurt am Main, the container with the lidar was set on the roof of a 20m-high building. The measurements involved scanning by the probing beam in the vertical plane perpendicular to two runways, and the height of landing aircraft in the scanning plane was about 280m.

Four examples of lidar measurements of vortex axis trajectories are shown in Figure 5.21 [40, 50]. We can see that, in general, the vortex pair descended, since the vortices were in the field of each other (that is, the wakes "pushed" each other down). The errors of estimation of the vertical and horizontal coordinates of the vortex core are 4.5m and 6.5m, respectively (see Section 5.5). Therefore, the effects observed in these examples, such as the horizontal motion and the upward vertical motion of the vortex cores, as well as the increase (decrease) of the distance between the vortex axis b and change of the tilt angle θ_p, can be generally explained by the effect of wind on the wake vortices, rather than by measurement errors.

From the data measured during the experimental campaign in Frankfurt am Main in October-December 2004, the results for LTA were selected in [40]. The category of LTA incorporates various types of large passenger aircrafts. In addition to the wake vortex parameters such as the vertical $Z_{Ci}(t_n) = R_{Ci}(t_n)\sin[\phi_{Ci}(t_n)] + h_L$ and horizontal $Y_{Ci}(t_n) = R_{Ci}(t_n)\cos[\varphi_{Ci}(t_n)]$ coordinates of the axis of the ith vortex ($i =1, 2$; t_n is the measurement time, $n = 1, 2, 3, \ldots$ is the scan number, and h_L is the height of lidar location), the separation between the vortex axes $b = |r|$, and the tilt angle $\theta_p = \arg(r)$ ($r = (Y_{C1} - Y_{C2}) + j(Z_{C1} - Z_{C2})$, $j = \sqrt{-1}$), the following parameters were calculated: velocity of the vertical motion of the vortex core (axis)

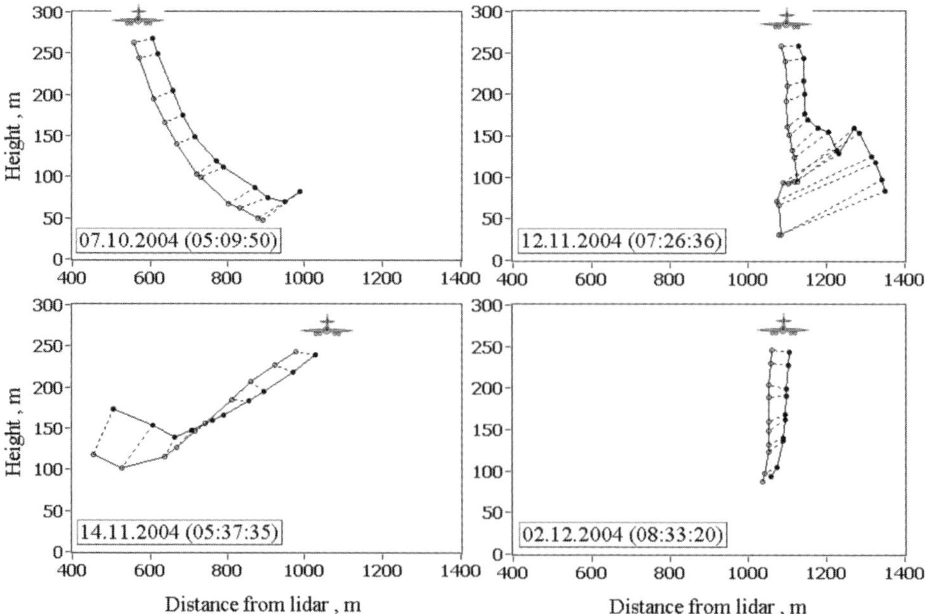

Figure 5.21   Vortex core trajectories measured by lidar on the airfield of the Frankfurt am Main airport. Local time is given in parentheses.

$$w_i(t_{n+1/2}) = [Z_{Ci}(t_{n+1}) - Z_{Ci}(t_n)]/(t_{n+1} - t_n) \qquad (5.22)$$

and velocity of the horizontal motion

$$u_i(t_{n+1/2}) = [Y_{Ci}(t_{n+1}) - Y_{Ci}(t_n)]/(t_{n+1} - t_n) \qquad (5.23)$$

where $t_{n+1/2} = (t_{n+1} + t_n)/2$.

Figure 5.22 depicts the results of lidar measurements of the normalized distance between the vortex axes b/b_0, tilt angle θ_p, height of the vortex axis Z_C, and normalized vertical component of the vortex core velocity w/w_0 as functions of the normalized time t/t_0 in Frankfurt am Main [40]. Dots show individual estimates of these parameters. To obtain the time dependences of the measured parameters, the average of individual estimates inside the temporal window $\Delta t/t_0 = 1$ was used. These dependences are shown by solid curves in the figure. It can be seen from Figure 5.22(a) that at $t < t_0$ the measured values of b are close to the theoretical values of the initial separation between the vortex axes b_0. The separation b can both decrease and increase with time. The decrease of b can be caused by the Crow instability, when variations of b along the wake have an oscillating character [9]. As follows from our measurements, in many cases as the vortices approach each other too closely $(b/b_0 < 0.5)$, they destruct rapidly. An increase in the separation between vortex axes is mainly caused by wind shears [17]. One can see that b can increase almost threefold compared to its initial value. On average, the increase of b starts from $t/t_0 \approx 2$ (curve in Figure 5.22(a)). The inhomogeneity in the vertical distribution of wind velocity causes a tilt of the vortex pair with respect to the horizontal plane, and the angle θ_p can vary within a wide range (see Figure 5.22(b)). As a rule, $\theta_p > 0$, when the lateral wind is directed from the lidar (projection of the wind velocity vector onto the scanning plane $V_T > 0$), and $\theta_p < 0$, when it has the opposite direction. The curve showing the variation of the mean angle θ_p with time in Figure 5.22(b) deviates a little from the zero level, which is connected with the larger number of cases with $V_T > 0$ as compared to $V_T < 0$ in the experiment [40].

As follows from Figure 5.22(c), the vortex descent is described, on average, by the linear time dependence $(Z_C - Z_a)/b_0 = -t/t_0$, where Z_a is the flight height in the measurement plane at $t/t_0 \le 2.5$. The slowing of vortex descent is connected with, first of all, a decrease in the vortex circulation with time after overflight. The scatter of points in this figure far exceeds the error of measurement of the vertical coordinate of the vortex axis ($E_Z = 4.5$m), which can be explained by the effect of large-scale turbulent inhomogeneities of the vertical component of wind and wind shears on the wake vortices. We can see from Figure 5.22(d) that at small values of t/t_0 the ratio w/w_0 is close, on average, to unity, and with time it decreases. The standard deviation of single velocity estimates w from the average velocity value exceeds by approximately twofold the rough estimate of the measurement error $E_w = E_Z/T = 0.3$ m/s, where $T = 15$s is the period of up-down scanning by the probing beam [40].

The wake vortices are transported by the lateral wind. To know how the horizontal velocity of the vortex core u corresponds to the lateral wind velocity V_T, the vertical profiles of $V_T(h)$ were retrieved along with the velocities u for every scan from the same lidar data with the use of (4.43). Figure 5.23 compares the results of

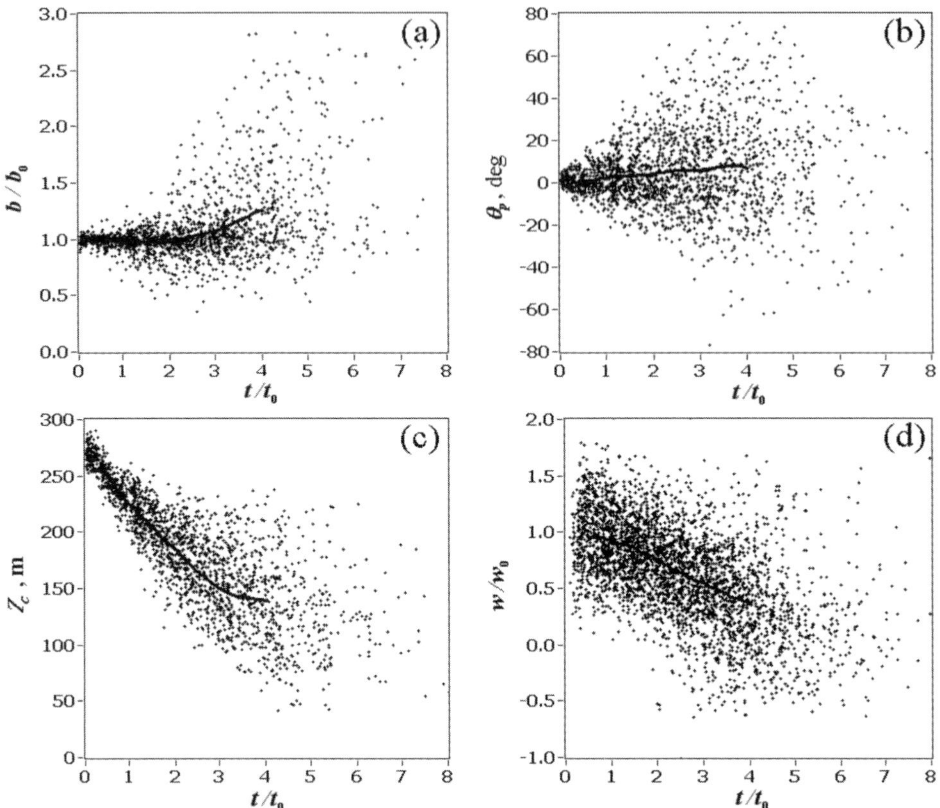

Figure 5.22 Results of lidar measurements of the normalized separation between the (a) vortex axes, (b) tilt angle, (c) height of the vortex axis, and (d) normalized vertical component of the vortex core velocity as functions of the normalized time after overflight. Dots: individual estimates; solid curves: averaged values.

simultaneous (within two scans) measurements of the velocities $V_T(h_C)$ and u, where the height $h_C = [Z_C(t_{n+1}) + Z_C(t_n)]/2$ [40]. We can see that the horizontal component of the vortex core velocity generally corresponds to the transverse wind velocity. The histogram of velocity difference $V_T - u$ calculated from the data shown in this figure has the Gaussian form with zero mean and standard deviation equal to 0.84 m/s. The observed regular excess of $|u|$ over $|V_t|$ can be explained by the acceleration of the horizontal motion of vortices due to an increase of the angle $|\theta_p|$ (see Figure 5.22(b)).

Examples of lidar measurements of the wake vortex circulation are shown in Figure 5.24 [40]. One can easily see two phases in the behavior of this parameter: initially slow and then fast wake vortex decay, which is in agreement with the theory developed in [18]. Through the least-squares fitting of the circulation estimates in each of these phases to the linear functions, we can find the time t_s of transition from one phase to another as a point of intersection of the obtained lines (see lines in Figure 5.24). Arrows in the figure show the normalized time t_s/t_0 for every measured realization $\Gamma(t)$. At $t < t_s$ the vortex has an almost ideal shape, and at longer times the vortex destructs fast under the effect of small-scale atmospheric turbulence. That is why the time t_s is referred to in the following as the vortex (with ideal shape) lifetime. According to the theory of [18], the time t_s is determined by the dissipation rate of the kinetic energy of atmospheric turbulence ε. Sections 4.6 and 5.5 show

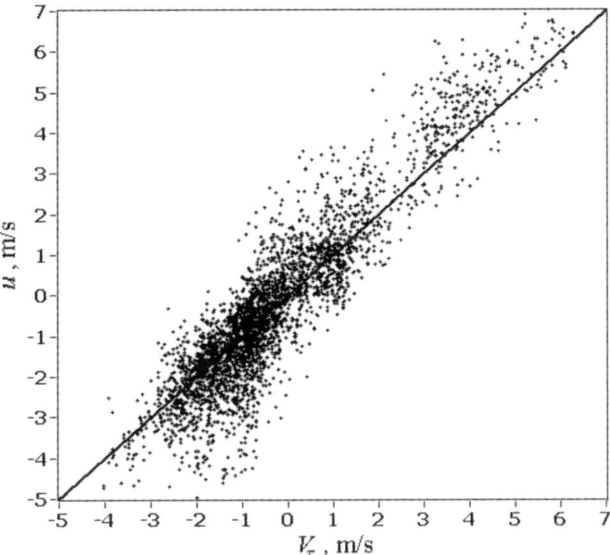

Figure 5.23 Comparison of results of simultaneous lidar measurements of the transverse component of wind velocity (abscissa) and the horizontal component of vortex core velocity (ordinate).

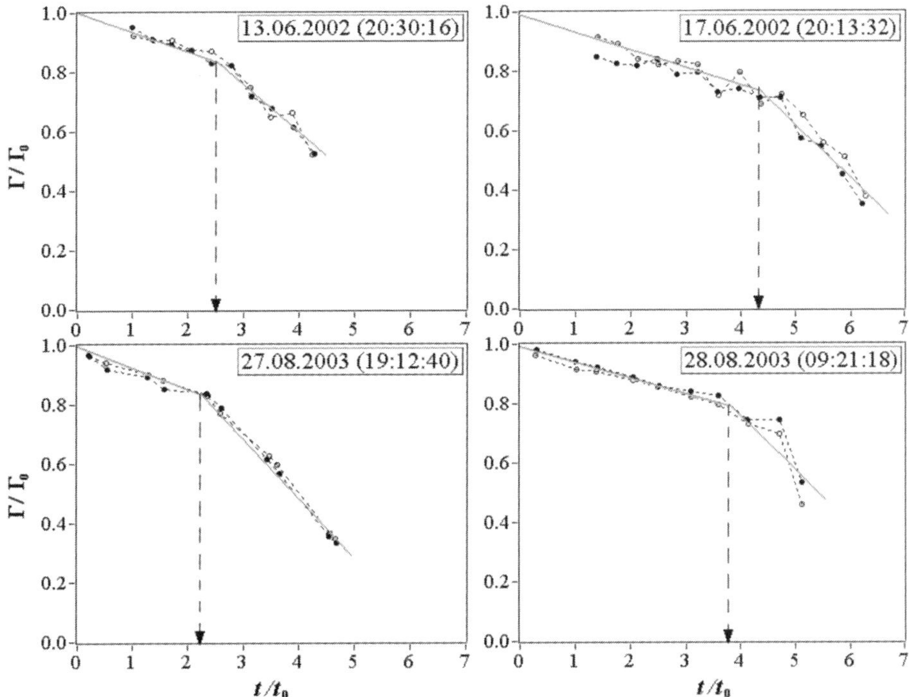

Figure 5.24 Normalized circulation of the right (closed circles) and left (open circles) vortices as functions of the normalized time obtained from lidar measurements in Tarbes. Arrows show the time of transition from the slow phase to the fast phase of circulation decay. Local time of aircraft passage through the scanning plane is given in parentheses.

that the relative error of lidar measurements of the turbulent energy dissipation rate does not exceed 25%, and the accuracy of lidar estimates of the vortex circulation is, on average, ±13 m^2/s, which is quite acceptable for our analysis of the influence of atmospheric turbulence on the wake vortex decay process.

The data from lidar measurements in Tarbes in 2002 and 2003 were used to retrieve the vertical profiles $\varepsilon(h)$ for every obtained time dependence of the circulation Γ. Then the values of $\varepsilon(h)$ were averaged over the heights of wake vortex observation Z_C. The range of the averaged values was $5 \cdot 10^{-5} - 2 \cdot 10^{-3}$ m^2/s^3. It was divided into three parts: (1) $5 \cdot 10^{-5} - 2 \cdot 10^{-4}$ m^2/s^3, (2) $2 \cdot 10^{-4} - 5 \cdot 10^{-4}$ m^2/s^3, and (3) $5 \cdot 10^{-4} - 2 \cdot 10^{-3}$ m^2/s^3. The estimates of vortex circulation were sorted into the corresponding three groups [36, 40].

Figure 5.25 shows the normalized vortex circulation Γ/Γ_0 as a function of the normalized time t/t_0 [40]. Symbols are for individual estimates of the circulation, whereas curves demonstrate the variation of the average circulation (averaging of individual estimates) with time at different levels of turbulence. One can see that the stronger the turbulence, the faster the vortex decay. Arrows show the average vortex lifetime t_s for each of the considered ranges of ε: (1) $t_s/t_0 \approx 3.8$, (2) $t_s/t_0 \approx 2.9$, and (3) $t_s/t_0 \approx 2$.

To obtain the empirical wake vortex lifetime dependence on the turbulent energy dissipation rate $t_s(\varepsilon)$, the data of lidar measurements in Frankfurt am Main in 2004 were used [40]. For every individual measured dependence of the vortex circulation Γ on time t, the wake vortex lifetime t_s was determined. The vertical profile of the turbulent energy dissipation rate was determined from the same raw lidar data, and the obtained values of ε were averaged over heights of wake location during the measurement of $\Gamma(t)$.

Figure 5.26 depicts individual estimates (dots) of the normalized time t_s/t_0 (ordinate, linear scale) obtained at different values of the dissipation rate ε (abscissa, log

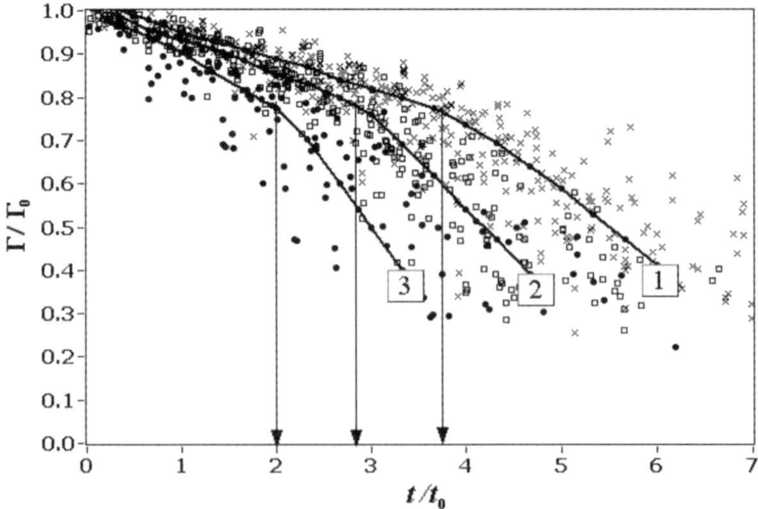

Figure 5.25 Normalized wake vortex circulation measured by lidar at (curve 1) $\varepsilon \in [5 \cdot 10^{-5}, 2 \cdot 10^{-4}]$ m^2/s^3 (crosses), (curve 2) $\varepsilon \in [2 \cdot 10^{-4}, 5 \cdot 10^{-4}]$ m^2/s^3 (squares), and (curve 3) $\varepsilon \in [5 \cdot 10^{-4}, 5 \cdot 10^{-3}]$ m^2/s^3 (circles). Curves 1, 2, and 3 show the time dependence of the average circulation for these ranges of ε.

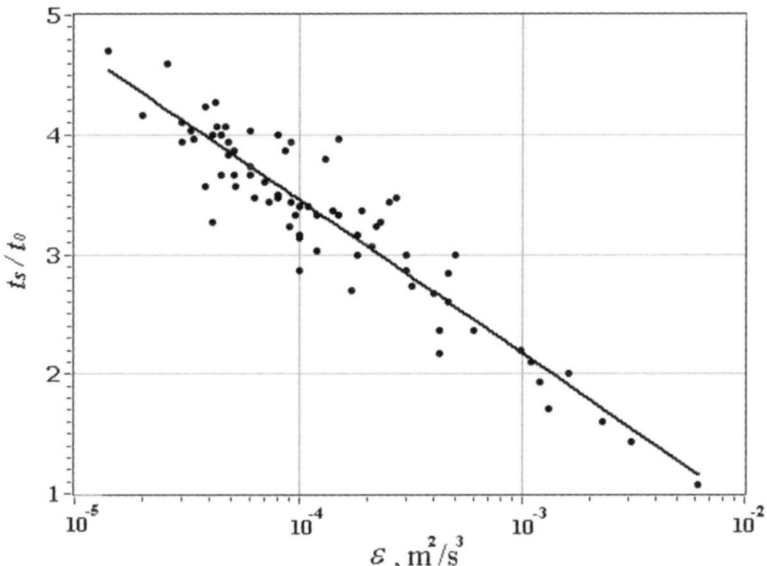

Figure 5.26 Normalized wake vortex lifetime t_s/t_0 as a function of the dissipation rate of kinetic energy of atmospheric turbulence. Dots: individual estimates; line: calculation by empirical formula (5.24).

scale). It can be seen that in the range of 10^{-5} m^2/s$^3 < \varepsilon < 10^{-2}$ m^2/s^3 the wake vortex lifetime t_s varies roughly from t_0 to $5t_0$. In general, the ratio t_s/t_0 decreases with an increase in ε. The depicted individual estimates of t_s/t_0 as a function of $\log_{10} \varepsilon$ were used in [40] to obtain, by the least-squares method, the linear dependence

$$t_s/t_0 = c_1 \log_{10} \varepsilon + c_2 \tag{5.24}$$

where ε is measured in (m^2/s^3), $c_1 = -1.282$, and $c_2 = -1.676$.

The result of calculation by (5.24) is shown as a solid line in Figure 5.26. The standard (rms) deviation of individual estimates of t_s/t_0 (dots) from this line is 0.27. Having the information about parameters of aircraft (t_0) and atmospheric turbulence (ε), one can predict the wake vortex lifetime with an accuracy of ~0.27t_0 using (5.24). The relative error in this case does not exceed 10%.

5.8   Measurement of Wake Vortex Parameters Using an Airborne Lidar in the Free Atmosphere

Ground-based measurements of aircraft wake vortices by a Doppler lidar are possible only for heights within the atmospheric boundary layer. Wake vortices formed at heights of the free atmosphere can be measured by a CDL installed aboard another aircraft. Studies of wake vortices in the free atmosphere were carried out by the DLR Lidar Group in 2005 in Bavaria and in 2006 in southern France. A 2-μm pulsed CDL was installed aboard the Falcon-20 aircraft. Another aircraft involved in the experiment (whose wake vortices were measured) was the ATTAS in 2005 [37] and a LTA in 2006 [38].

Figure 5.27 depicts the idealized geometry of measurements by an airborne lidar used in these experiments. The Falcon (A1) aircraft with CDL flew along a closed trajectory at a fixed altitude. The flight altitude in different experiments was from 3 to 12 km. Another aircraft (ATTAS, LTA), denoted as A2 in Figure 5.27, flew in the opposite direction at a lower altitude. The difference in the flight altitudes of A1 and A2 was usually 900m. The measurements involved scanning by the probing beam (see Figure 5.27) with the angular rate of $\omega_s = 10$ deg/s in the angular range of $\varphi \in [-15°, +15°]$ about the vertical axis. The duration of one scan was 3s. Smoke generators were installed on the wings of the A2 aircraft. Due to this, the power of the lidar echo signal was sufficient to measure the wake vortices at different ages. In addition, the Falcon pilot could see the smoke plume and correct the flight trajectory so that the plume was always within the scanning sector. During the measurements in March and June 2006 (A1 flight altitude of ~11 to 12 km), smoke generators were not used, because the wake was visible due to condensation of moisture.

Let the trajectory of the A1 aircraft coincide with the X axis of the Cartesian coordinate system $\{Z, X, Y\}$, whose origin lies at a fixed point at the A1 flight altitude, where Z is the vertical coordinate, and Y is the axis in the horizontal plane transverse to the flight trajectory. When scanning by the probing beam was perpendicular to the X axis, the coordinates of the position of the sensing volume at the time t (in the horizontal plane, in which the A2 aircraft wake is located and whose altitude is Δh lower than the A1 flight altitude) are described by the equations [37]

$$X(t) = V_a \cdot (t - T/2)$$
$$Y(t) = \Delta h \cdot \tan[\omega_s \cdot (t - T/2)]$$

(5.25)

where V_a is the aircraft true speed, $T = 3$s is the time of one scan, and $t \in [0, T]$. For the time Δt between $t = T/2$ and $t = T/2 + \Delta t$, the probing beam intersecting the wake moves by the distances $\Delta Y = \Delta h \tan(\omega_s \Delta t)$ (along the Y axis) and $\Delta X =$

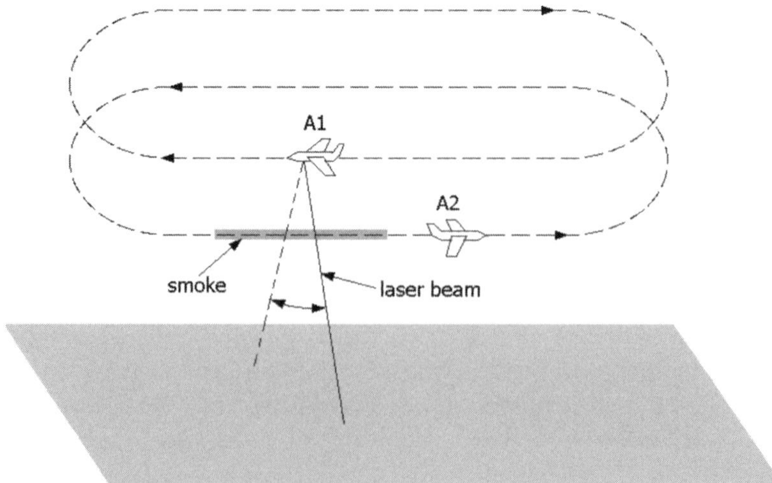

Figure 5.27 Geometry of measurement of wake vortices generated by aircraft A2 (ATTAS or LTA) with coherent Doppler lidar installed aboard aircraft A1 (Falcon). (© 2007 American Institute of Aeronautics and Astronautics. From [37].)

$V_a \Delta t$ (along the X axis). From measured velocity envelopes the circulation of the vortex is estimated around its axis within ± 15 m. Therefore, we take $\Delta Y = 30$m. Then, the shift ΔX equals 19m at $\Delta h = 1$ km, $\omega_s = 10$ deg/s, and $V_a = 111$ m/s. This value of the shift along the X axis can appear to be too large for the case in which an "old" wake vortex is measured, when the coordinates $\{Z_C, Y_C\}$ of the vortex axis have a strongly peaked profile (along the X axis) due to wind flow turbulence, which causes significant distortions of the measured velocity envelope $V_E(\varphi)$.

To decrease ΔX, the proposal was made [37] to have the probing beam scan at some angle $\theta_L \neq 0$ to the Y axis rather than perpendicularly to the X axis. In this case,

$$X(t) = V_a \cdot (t - T/2) - \Delta h \cdot \sin\theta_L \cdot \tan[\omega_s \cdot (t - T/2)]$$
$$Y(t) = \Delta h \cdot \cos\theta_L \cdot \tan[\omega_s \cdot (t - T/2)]$$

$$(5.26)$$

It is necessary for the value of $X(t)$ to be close to zero during scanning by the probing beam in the direction opposite to the A1 aircraft flight direction. Because the scanning limits are $\pm 15°$, in (5.26) we can take $\tan[\omega_s \cdot (t - T/2)] \approx \omega_s \cdot (t - T/2)$. ω_s Then, from the solution of the equation for $X(t) = 0$, we obtain the equation for the angle θ_L in the form [37]

$$\theta_L = \arcsin\left(\frac{V_a}{\omega_s \Delta h}\right)$$

$$(5.27)$$

where ω_s is measured in radians per second. In the experiment, the initial difference between the altitudes of the A1 aircraft and the wake of the A2 aircraft is 900m. With time, the wake can go down by 200m; that is, $\Delta h = 1,100$m. Upon the substitution of $V_a = 111$ m/s, $\omega_s = 0.1745$ rad/s, and $\Delta h = 1,100$m into (5.27), we obtain $\theta_L \approx 35°$.

Figure 5.28 depicts the trajectories of the center of the sensing volume calculated by (5.26) at $\Delta h = 900$m (solid curve 1), 1,000m (solid curve 2), and 1,100m (solid

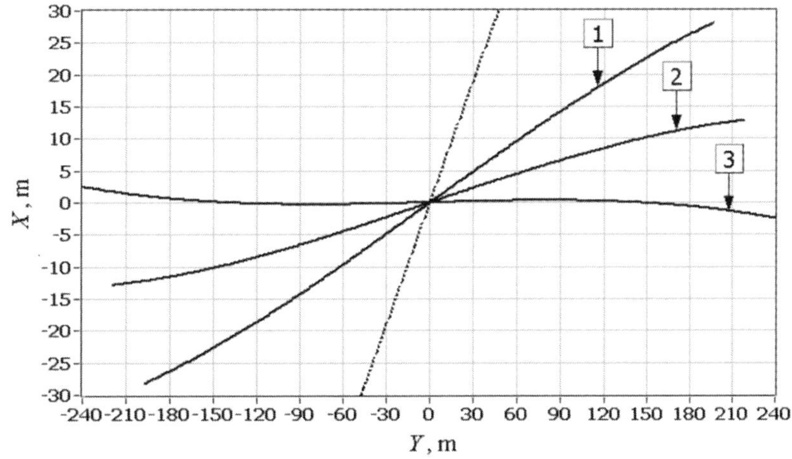

Figure 5.28 Trajectories of the center of the sensing volume in the horizontal plane below Falcon by 900m (curve 1), 1,000m (curve 2), and 1,100m (curve 3) at $\theta_L = 35°$ (solid curves); trajectory at $\theta_L = 0°$ and $\Delta h = 1000$m (dashed curve). (© 2007 American Institute of Aeronautics and Astronautics. From [37].)

curve 3) and $\theta_L = 35°$ and at $\theta_L = 0°$ and $\Delta h = 1,000m$ (dashed curve). In contrast to the case $\theta_L = 0°$ ($\Delta X = 19m$), at $\Delta Y = 30m$ the shift ΔX is $5m$ ($\Delta h = 900m$), $2m$ ($\Delta h = 1,000m$), and 0 ($\Delta h = 1,100m$). Thus, the use of the angle $\theta_L = 35°$ allows the shift ΔX to be decreased significantly.

In the case of ground-based measurements, the angular scanning rate by the probing beam was 5 times lower than in the case of airborne measurements. Since the pulse repetition frequency is the same in the both cases (500 Hz), to obtain data with the scanning angle resolution $\Delta\varphi = 0.1°$ by an airborne lidar, it is necessary to accumulate 5 Doppler spectra, rather than 25 as in the case of ground-based measurements at $\omega_s = 2$ deg/s. To obtain velocity envelopes from Doppler spectra at $L = 5$, the optimal threshold is $q_{th} = 3.5$ (for ground-based measurements, $q_{th} = 2.5$).

The threshold $q_{th} = 3.5$ allows the velocity envelopes $V_{E\pm}(R, \varphi)$ and then coordinates of wake axes $\{R_{Ci}, \varphi_{Ci}\}$ to be obtained from measured Doppler spectra. The floating threshold $q_{th}^{(i)}(\varphi)$ depends significantly on the SNR. Due to the operation of smoke generators, a strong local inhomogeneity of scattering properties of the medium takes place in the wake behind the A2 aircraft. Therefore, through generalization of (5.5), we can write the theoretical model of the Doppler spectrum in the form [38]

$$S_D(V) = \frac{B_V}{\sqrt{2\pi}\sigma_{VI}} \int_{-\infty}^{+\infty} dz' Q_s(z')[S_a(R + z', \varphi) + S_b]\exp\left\{-\frac{[V - V_r(R + z')]^2}{2\sigma_{VI}^2}\right\} + 1 \quad (5.28)$$

where the sum $S_a(R + z', \varphi) + S_b$ is proportional to the backscattering coefficient with contributions from the generated smoke $[S_a(R + z', \varphi)]$ and atmospheric aerosol (S_b). In the absence of smoke, SNR $= S_b$. Using the equation

$$\text{SNR}(R, \varphi) = B_V^{-1} \int_{-\infty}^{+\infty} dV[S_D(V) - 1] \quad (5.29)$$

we have the following equation for the signal-to-noise ratio from (5.28):

$$\text{SNR}(R, \varphi) = S_b + \int_{-\infty}^{+\infty} dz' Q_s(z') S_a(R + z', \varphi) \quad (5.30)$$

In [38], the following model of the smoke plume was proposed:

$$S_a(R, \varphi) = \sum_{i=1}^{2} S_{mi} \exp\left[-\frac{(R\cos\varphi - R_{Ci}\cos\varphi_{Ci})^2 - (R\sin\varphi - R_{Ci}\sin\varphi_{Ci})^2}{L_{si}^2}\right] \quad (5.31)$$

where S_{mi} and L_{si} are model parameters. Substituting (2.25) and (5.31) into (5.30) and integrating with respect to z', we derive the equations

$$\text{SNR}(R,\varphi) = S_b + \sum_{i=1}^{2} \frac{S_{mi} L_{si}}{\sqrt{L_{si}^2 + (\Delta z)^2/\pi}} \exp\left\{ -\frac{R^2 - 2RR_{Ci}\cos(\varphi - \varphi_{Ci}) + R_{Ci}^2}{L_{si}^2 + (\Delta z)^2/\pi} - \right.$$

$$\left. -\frac{(\Delta z)^2/\pi}{L_{si}^2 + (\Delta z)^2/\pi} \frac{R_{Ci}^2 \sin^2(\varphi - \varphi_{Ci})}{L_{si}^2} \right\} \tag{5.32}$$

After estimation of $\hat{\text{SNR}}(R,\varphi)$ and the wake axis coordinates $\{R_{Ci}, \varphi_{Ci}\}$, the parameters S_{mi} and L_{si} are determined through the fitting of the results calculated by (5.32) to $\hat{\text{SNR}}(R,\varphi)$ values by the least-squares method. Then, the threshold $q_{th}^{(i)}(\varphi)$ is calculated with the use of (5.31) and (5.28), and the vortex circulation Γ_i is ultimately determined by the integration method (see Section 5.2).

The first measurements of wake parameters by an airborne lidar were conducted in south Bavaria on 21 and 22 April 2005. ATTAS (weight of 20t and wing spread of 21.1m) with a smoke generator on the left wing was used as the A2 aircraft. The results of these measurements published in [37] surpassed all expectations of the authors in the practical implementation of this complex and unique experiment.

The following experiment was conducted in southern France on 4 March and 23 and 25 June 2006. In this experiment, the wake vortices of different types of LTAs were studied. Smoke generators were not used because, according to the weather forecast, the weather conditions were expected to favor the condensation of moisture at the exhaust emission from the aircraft engine nozzle. In measurements on 25 June 2006, the wake behind the A2 was invisible most of the time (no condensation) and only about 20% of the raw lidar data for this day was suitable for processing.

In an experiment [38] conducted on 9 and 15 November 2006 in the region between Toulouse and the Pyrenees, smoke generators were set on the wings of the LTA used. Measurements using the 2-μm CDL from aboard the Falcon aircraft were carried out in clear weather, with north wind (first measurement day) and south wind (second measurement day). This experiment included one flight on the first day (flight N1) and three flights (flights N2, N3, and N4) on the second day. The flight altitude of the LTA was about 2,700m (flights 1 and 2) and 3,600m (flights 3 and 4). During flights N1 and N2, the true speed of the LTA was 90 m/s, while the Falcon speed was 115 m/s. For flights N3 and N4, $V_a = 95$ m/s (LTA) and $V_a = 125$ m/s (Falcon). As in other experiments, the Falcon flight altitude was about 900m higher than that of the LTA. The Falcon and LTA flight times (UTC) are tabulated in Table 5.2. Each flight included eight legs, in which the Falcon flew over the LTA and the lidar could measure the smoke-seeded wake. The smoke generators installed on the wings of the LTA were switched on after every turn approximately 140s before the head-on intersection of the vertical plane by the A1 and A2 aircraft and then switched off after the intersection. The off-cycle periods of the smoke generators were from 7 to 20 min (the average for all flights was 9 min).

The trajectories of all flights of the Falcon and LTA (GPS data) are shown in Figure 5.29. In flights N1, N3, and N4, the lidar measurements of the wake vortices were mostly conducted when the Falcon flew predominantly west to east and in the opposite direction (east to west), whereas in flight N2 it flew almost along the north-to-south (south-to-north) line.

Table 5.2 Flight Times for November 2006 Experiment

Flight N	Date	Time of Flight
1	09 November 2006	13:54–15:24
2	15 November 2006	08:54–10:00
3	15 November 2006	12:20–13:20
4	15 November 2006	15:36–16:30

Figure 5.30 shows the trajectories of the Falcon and LTA flights in the horizontal plane in the $\{X, Y\}$ coordinate system with its origin at 0.834086° longitude and 43.075592° latitude or at the point where the Falcon was located at time UTC = 12:20:00 (see Table 5.2, flight N3) [38]. The corresponding legs of the LTA trajectory (with the smoke generators switched on) and the Falcon trajectory (when the lidar measured the smoke-seeded wake) do not coincide except for a small zone around the point of location of the LTA at the time of the smoke generator was switched off. This is connected with the fact that the south wind transported the smoke-seeded wake, and the Falcon pilot had to correct the flight trajectory for the wake vortices generated by the LTA to remain within the scanning angle (±15°).

The measured raw lidar data (UTC, scanning angle, echo signal samples for every shot) and the Falcon and LTA flight data (latitude, longitude, altitude, roll

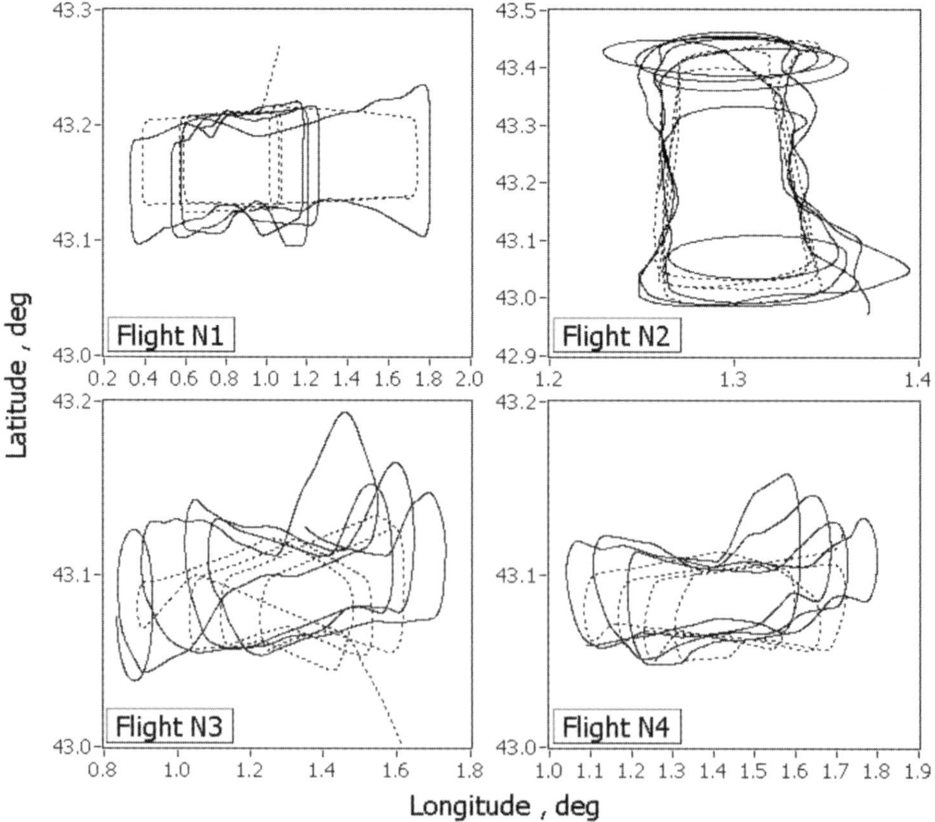

Figure 5.29 Trajectories of Falcon (solid curves) and LTA (dashed curves) flights. (© 2007 American Institute of Aeronautics and Astronautics. From [38].)

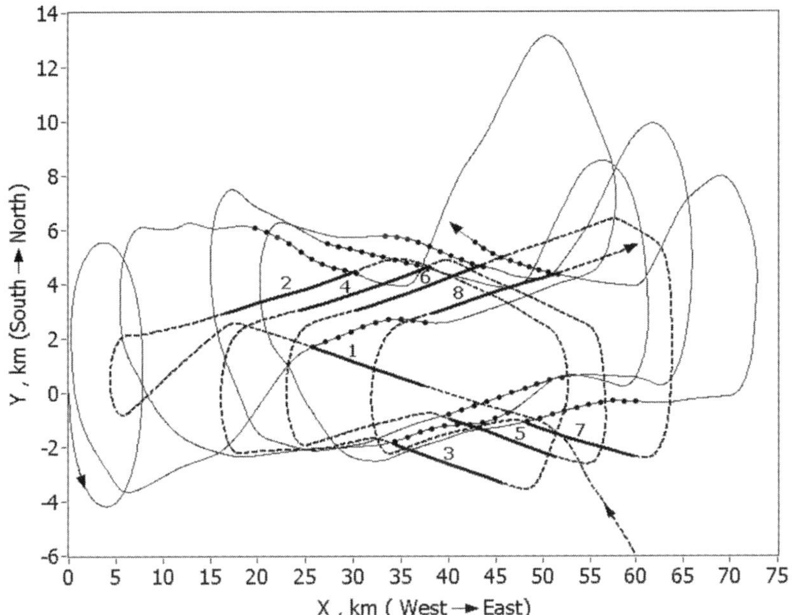

Figure 5.30   Trajectories of Falcon (solid curve) and LTA (dashed curve) (flight N3). Heavy curves are areas with smoke seeding (legs 1–8). Solid curves with dots show areas where wake vortices were measured. (© 2007 American Institute of Aeronautics and Astronautics. From [38].)

angle, pitch angle, mass, and true speed) were used for analysis of the wake vortices generated by the LTA.

The processing of the raw lidar data yielded estimates of the Doppler spectra $\hat{S}_D(V, R_l, \varphi_{i'}; n)$, where $R_l = R_0 + l\Delta R$ and where $l = 0, 1, 2, ..., L_R$ and $L_R = 60$, $\Delta R = 12$m, $R_0 = 740$m, $\varphi_{i'} = \varphi_0 + i'\Delta\varphi$, $\varphi_0 = -15°$, $i' = 0, 1, 2, ..., I, I = 300$, $n = 0, 1, 2, 3, ...$ is the scan number, and $V \in [-25$ m/s, $+25$ m/s]. With the threshold $q_{th} = 3.5$, the velocity envelopes were found from these spectra, and then the vortex axis coordinates $\{R_{Ci}(t_n), \varphi_{Ci}(t_n)\}$ were determined ($i = 1$ for the right vortex and $i = 2$ for the left vortex) in the Falcon-fixed coordinate system with the origin at the point of lidar location.

Figure 5.31 illustrates the evolution of the Doppler spectra $\hat{S}_D(V, R_{C1}(t_n), \varphi_{i'}; n)$ with the scan number [38]. The values of the spectra are given in the gray scale, where the white and black colors correspond, respectively, to the minimum and to the maximum. Due to the smoke seeding, the maximum values of the spectrum are concentrated near the wake vortex cores.

The coordinates of the axis of the ith vortex $\{Z_{Ci}, Y_{Ci}, X_{Ci}\}$ in the $\{Z, Y, X\}$ coordinate system (Z is the vertical axis in the upward direction, Y is the horizontal south–north axis, and X is the horizontal west–east axis), whose origin coincides with the spatial position of the A2 aircraft (LTA), were calculated as follows:

$$Z_{ci} = (h_1 - h_2) - Z' \tag{5.33}$$

$$Y_{ci} = Y'\cos(\alpha_1 - \alpha_2) + X'\sin(\alpha_1 - \alpha_2) + (Y_1 - Y_2)\cos\alpha_2 + (X_1 - X_2)\sin\alpha_2 \tag{5.34}$$

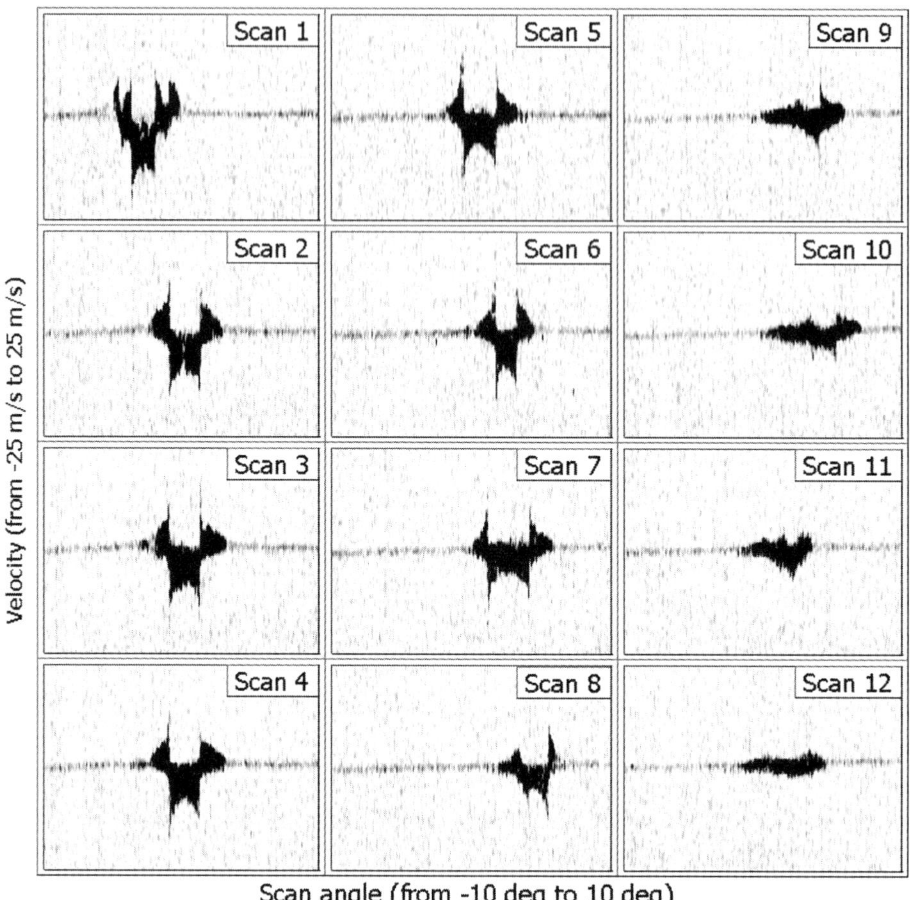

Figure 5.31   Measured Doppler spectra (flight N1, leg 1). (© 2008 American Institute of Aeronautics and Astronautics. From [38].)

$$X_{ci} = X' \cos(\alpha_1 - \alpha_2) - Y' \sin(\alpha_1 - \alpha_2) + (X_1 - X_2)\cos\alpha_2 + (Y_1 - Y_2)\sin\alpha_2 \quad (5.35)$$

where h_1 is the A1 (Falcon) flight altitude; h_2 is the A2 flight altitude;

$$Z' = R_{Ci}\cos\varphi', \quad Y' = R_{Ci}\sin\varphi'\cos\theta_L, \quad X' = R_{Ci}\sin\varphi'\sin\theta_L \quad (5.36)$$

$$\varphi' = \arctan\left(\sqrt{\tan^2[\varphi_{\mathrm{roll}} + \arctan(\cos\theta_L\,\tan\varphi_{Ci})] + \tan^2[\varphi_{\mathrm{pitch}} + \arctan(\sin\theta_L\,\tan\varphi_{Ci})]}\right)$$

$$(5.37)$$

φ_{roll} is the A1 roll angle; φ_{pitch} is the A1 pitch angle; $\{Y_1, X_1\}$ and $\{Y_2, X_2\}$ are the horizontal coordinates of, respectively, the A1 and A2 aircraft at the time t' of wake intersection by the probing beam; $\alpha_1 = \arg\{\mathbf{V}_1\}$ and $\alpha_2 = \arg\{\mathbf{V}_2\}$ are the A1 and A2 aircraft flight direction angles (with respect to the west-east axis); and $\mathbf{V} = V_x + jV_y$ is the aircraft velocity vector in the complex form. These vectors at the time $t'_m = t'_0 + m\Delta t'$, where $m = 1, 2, 3, \ldots$ and $\Delta t' = 0.5$s, can be calculated as

$$V_k = [X_k(t'_{m+1}) - X_k(t'_{m-1})] + j[Y_k(t'_{m+1}) - Y_k(t'_{m-1})] \qquad (5.38)$$

where $k = 1$ for A1 flight trajectories and $k = 2$ for A2 flight trajectories.

When the coordinates X_{Ci} and Y_{Ci} are known, the age t of the ith vortex is calculated by the simple equation

$$t = \frac{\sqrt{Y_{Ci}^2 + X_{Ci}^2}}{V_a} \qquad (5.39)$$

where V_a is the true speed of A2, which differs from the aircraft speed with respect to the Earth's surface due to the wind.

Figure 5.32 exemplifies the lidar measurements of vortex core trajectories (Figure 5.32(a)), as well as the normalized vertical coordinate of axes of the vortex pair (Figure 5.32(b)) and the normalized vortex circulation (Figure 5.32(c)) as functions of the normalized vortex time (age), where the parameters Γ_0 and t_0 are calculated by (5.2) and (5.3) [38]. [The theoretical initial distance between the vortex axes $b_0 = (\pi/4)B_a$, where B_a is the wingspan of the A2 aircraft.] These results were obtained from the raw data measured during leg 1 of flight N1. Due to the lateral wind, the wake shifted along the Y axis with the distance ($L_{Ci} = \sqrt{Y_{Ci}^2 + X_{Ci}^2}$) between the axis of the ith vortex and the LTA (see Figure 5.32(a)). In this example, the maximum shift was equal to about $10b_0$, and the maximal distance was $L_{Ci} \approx 400b_0$. As this took place, the wake vortex cores descended by about $6b_0$ (see Figure 5.32(b)). We can see from Figure 5.32(c) that with an increase of the vortex age, after $t/t_0 \approx 2$, the wake vortex decayed rapidly, and already at $t/t_0 \approx 4.5$ the circulation was halved compared to the initial value.

Using the data of this experiment (four flights or 32 legs), we obtained 768 estimates for every wake vortex parameter. From estimates of the vortex axis

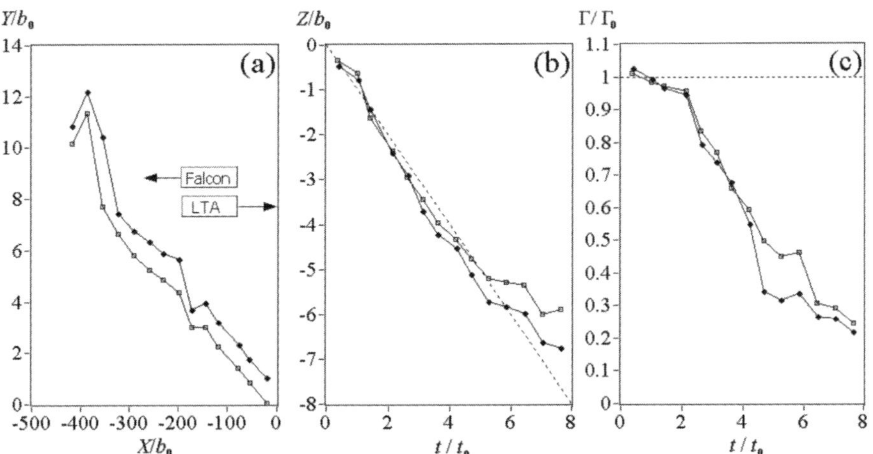

Figure 5.32 Normalized (a) horizontal and (b) vertical coordinates of the vortex axis, and (c) vortex circulation as functions of the normalized vortex age (flight N1, leg 1). Squares: right vortex; closed circles: left vortex. (© 2008 American Institute of Aeronautics and Astronautics. From [38].)

coordinates, we can calculate the separation of the vortex axes as $b = |\{Z_{C1}, Y_{C1}, X_{C1}\} - \{Z_{C2}, Y_{C2}, X_{C2}\}|$. By averaging the b values calculated for $t/t_0 < 1$ (73 estimates), we found that $\langle b \rangle \approx b_0$; that is, the distance b measured by lidar is in agreement, on average, with the theoretical value [7] for the initial part of the wake vortex evolution. In this case, the standard deviation was $[\langle b^2 \rangle - \langle b \rangle^2]^{1/2} \approx 0.14 b_0$.

Figures 5.33 and 5.34 show the results of lidar measurements of the normalized vertical coordinate of the vortex axis and the vortex circulation on 9 and 15 November 2006 [38]. The curves in these figures describe the dependence Z_{Ci}/b_0 (Γ_i/Γ_0) on t/t_0 for the right ($i = 1$) and left ($i = 2$) vortices obtained at one leg. For every flight, we have 16 curves for Z_{Ci}/b_0 and for Γ_i/Γ_0. It can be seen that in flight N1 the Z_{Ci}/b_0 curves are smoother compared to the corresponding data of other flights. Flight N1 took place on the first measurement day, when a north wind was observed, whereas flights N2 through N4 occurred on the second day during a south wind. Reference [38] proposed that during the experiment on 15 November 2006, the mesoscale structure of the wind field with large absolute values of the vertical component of the wind velocity vector was formed in the free atmosphere due to the air mass transport by the south wind over the Pyrenees. As a result, the aircraft wake was transported upward in some legs and downward in other legs. That is why the vertical coordinate of the wake vortex axis measured on this day has such

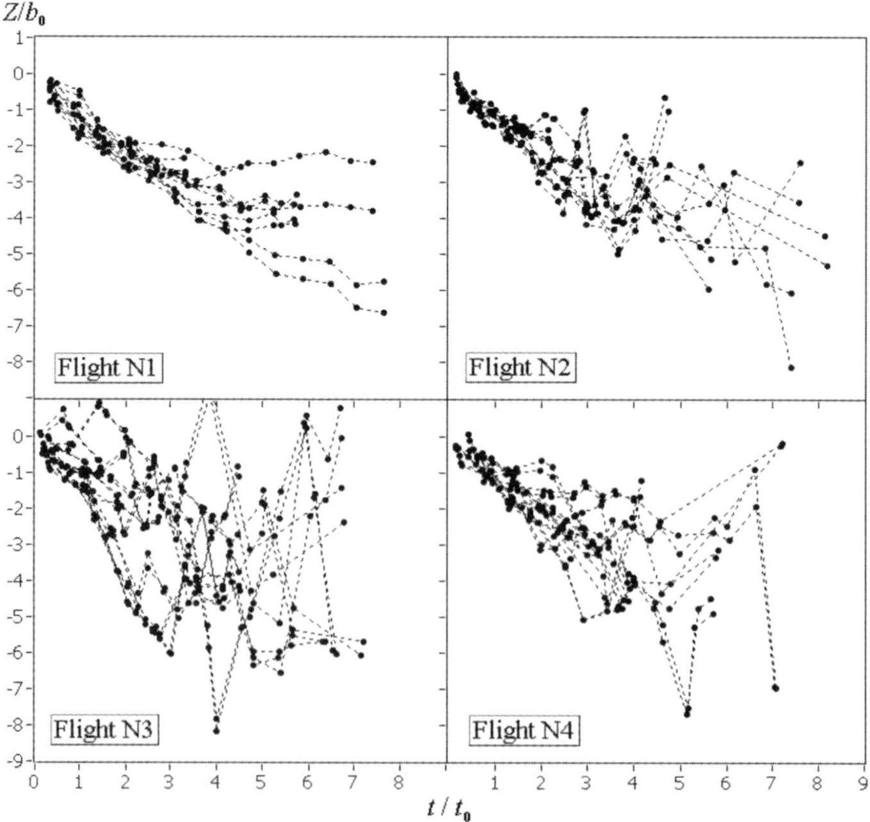

Figure 5.33 Normalized vertical coordinate of the vortex axis vs normalized vortex age, measured at different flights. (© 2008 American Institute of Aeronautics and Astronautics. From [38].)

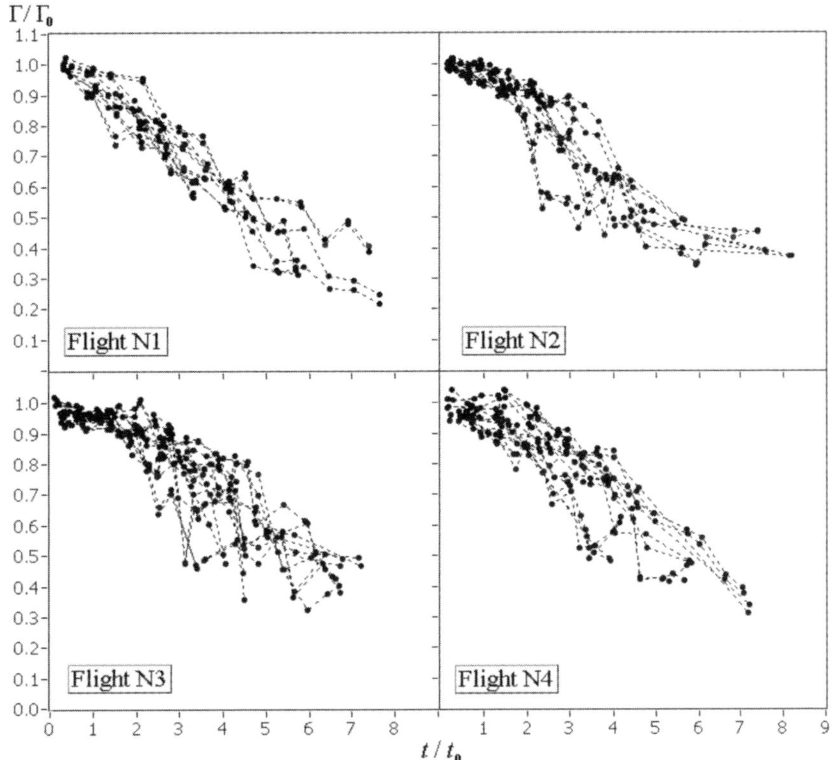

Figure 5.34   Normalized vortex circulation versus normalized vortex age, measured at different flights. (© 2008 American Institute of Aeronautics and Astronautics. From [38].)

a complex dependence on the wake vortex age t (or on the distance L_{Ci} to the aircraft generating the wake).

All estimates of the vertical coordinate of the vortex axis and the vortex circulation obtained in the experiment on 9 and 15 November 2006 are shown as dots in Figure 5.35. In this experiment, the wake of the LTA was observable up to $t/t_0 \approx 8$. The averaging of individual estimates of the considered wake vortex parameters in the range $0 < t/t_0 < 5$ was used in [38] to calculate the average values (curves in Figure 5.35). One can see from Figure 5.35(a) that, on average, the vortex core descent ($\langle Z_{Ci}\rangle/b_0$, curve in the figure) is described by the linear time dependence; that is, $\langle Z_{Ci}\rangle/b_0 = -t/t_0$, up to $t/t_0 \approx 3$. According to the estimates (with the use of numerical simulation), the accuracy of determination of the coordinates Z_{Ci} is only a little worse than the accuracy of measurement by a ground-based lidar [38]. Therefore, the scatter of points about the curve in Figure 5.35(a) should mostly be connected with the influence of the vertical component of the wind velocity on the vortex axis height. For example, during flight N3 at the eighth leg when $t/t_0 < 2$, the vertical component of the wind $\tilde{V}_z$ was positive (the estimate $\tilde{V}_z$ is obtained from the same experimental data; that is, from the radial wind velocity measured by lidar [35] with regard for the data on the Falcon flight velocity and trajectory) and higher than the theoretical value of the vortex descent velocity b_0/t_0. Consequently, in Figure 5.35(a) for this case we see the excess over the zero level. In the same flight N3, but at the third and seventh legs, the value of $\tilde{V}_z$ was negative (vertical component of the wind velocity is directed downward) and $\left|\tilde{V}_z\right| > b_0/t_0$. Here, in the range $2 \le t/t_0 \le 3$, we can see the far larger descent of the wake vortex cores than the average values (curve in Figure 5.35(a)).

In contrast to Figure 5.35(a), the scatter of points about the curve in Figure 5.35(b) has a far smaller amplitude. The point is that large-scale inhomogeneities of wind can determine, to a significant extent, the spatial position of a wake, practically not affecting the wake structure and, consequently, wake vortex circulation. As in the case of lidar measurements of wake vortex circulation in the atmospheric boundary layer (see Section 5.7), the curve in Figure 5.35(b) can be divided into two parts in terms of the slow decay and fast decay of the vortices. This result, as noted earlier, is in agreement with the theoretical model of two phases of wake vortex decay [7, 18, 20]. Since the SNR for lidar measurements from the Falcon outside the smoke plume was very low (SNR < −10 dB), we failed to estimate the turbulent energy dissipation rate from lidar data. It follows from Figure 5.35(b) that, on average, the time of transition from the slow phase to the fast phase of vortex decay was $t_s \approx 2.7t_0$. According to the empirical model (5.24), this lifetime of the wake vortex (of ideal shape) t_s corresponds to the dissipation rate $\varepsilon = 3.8 \cdot 10^{-4}$ m²/s³.

5.9 Lidar Investigations of a Wind Turbine Wake

As a wind turbine operates, a fraction of the wind flow energy is transferred to rotate the turbine blades; therefore, a wind-velocity deficit is generated downwind of the turbine. Studies on the influence of atmospheric conditions (in particular, wind velocity and wind turbulence intensity) on the length of a wind wake and the velocity deficit inside the wake are needed because of the increasing number of wind farms and the need to optimize the turbine arrangement in a wind farm.

Experimental investigations of the wind field in the vicinity of a wind turbine with the use of a sodar have been carried out in [61]. Results of a study of the wake generated by a wind turbine with the aid of a continuous-wave CDL are presented in [62, 63]. However, pulsed CDL opens up a wide range of possibilities to investigate

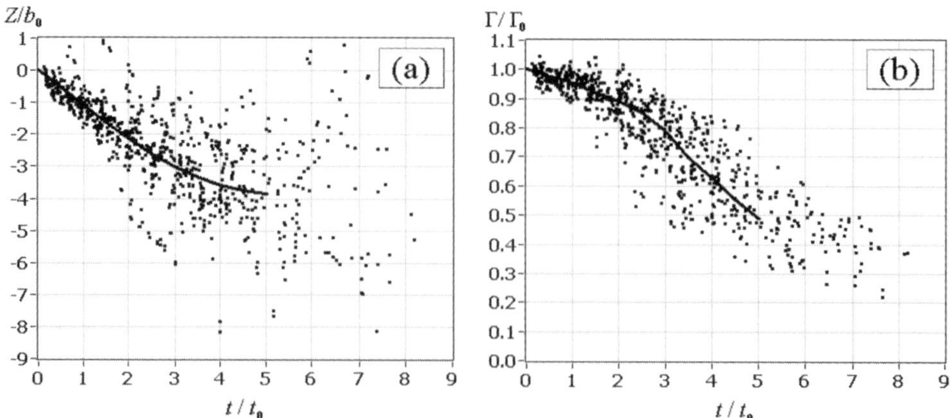

Figure 5.35 Normalized vertical coordinate of the (a) vortex axis and (b) vortex circulation versus normalized vortex age. Dots: individual estimates; curves: averaged values. (© 2008 American Institute of Aeronautics and Astronautics. From [38].)

wind turbine wakes by using the geometry of scanning by the sensing beam during the measurement time, as demonstrated in [64]. This section presents the results of experimental investigations into the spatial structure of a turbulent wind field in the vicinity of a wind turbine with the use of data measured by a 2-μm pulsed CDL under various atmospheric conditions [51–54].

A deficit of wind velocity takes place inside the wake generated behind a wind turbine on its leeward side. At some distance from the turbine, this deficit fully disappears. The most characteristic wake parameters are the maximum value of the wind velocity deficit and the effective transverse and longitudinal dimensions of the wake. The transverse wake dimension is initially determined by the diameter D of the circle described by the outer end of the turbine blade. The maximum velocity deficit and the longitudinal dimension of the wake depend on the turbine type and atmospheric conditions. To investigate these parameters with the aid of a pulsed CDL, different geometries of scanning can be used. In this case, the lidar should be at a sufficient distance from the turbine, and the wind direction should be nearly aligned with the lidar–turbine line. As seen in Figure 5.36, the angle of wind direction θ_V should, if possible, be close to the azimuth angle θ_T between the direction to the north and the line running from the lidar position to the turbine position (angle between the axis OY and the line OT in Figure 5.36).

We studied the wind field in the vicinity of the wind turbine using conical-sector scanning. The elevation angle φ should be set so that the sensing beam intersects with the largest area of the turbine wake, including the point of the maximum velocity deficit, if possible. Figure 5.36 shows the geometry of the lidar measurement for conical-sector scanning.

One of the aims of this study was to obtain information about the velocity deficit at a distance R from a turbine along the wind direction from data measured by a scanning lidar. We defined the velocity deficit VD as

$$VD = (1 - U/U_A) \times 100\% \qquad (5.40)$$

where U_A is the mean ambient wind velocity outside of the wake and U is the mean wind velocity within the wake downstream of the turbine.

We assume that the velocity U_A is horizontally homogeneous. Then, we can choose the point A (see Figure 5.36) on the scanning plane with the coordinates $\{X_A, Y_A\}$ that lies at the same radial distance from the lidar as the point with the turbine coordinates $\{X_T, Y_T\}$, where $X_A = b_T \sin\theta_A$, $Y_A = b_T \cos\theta_A$, $X_T = b_T \sin\theta_T$, $Y_T = b_T \cos\theta_T$, and b_T is the distance between the lidar and the turbine. The angle θ_A between the OY axis and the line OA can be set as $\theta_A = \theta_T \pm d/b_T$, where the arc length $d \ll b_T$. In this case, point A can be located either to the right (+) or to the left (−) of the turbine.

As a result of the multiple repetitions of sector scans during the lidar measurement and data processing, we obtained an array of estimated radial velocities $\hat{V}_r(z_i, \theta_m; n)$, where $n = 1, 2, \ldots, N$ is the scan number, for the shaded area in Figure 5.36. Then we averaged these estimates as

$$\langle \hat{V}_r(z_i, \theta_m) \rangle = N^{-1} \sum_{n=1}^{N} \hat{V}_r(z_i, \theta_m; n) \qquad (5.41)$$

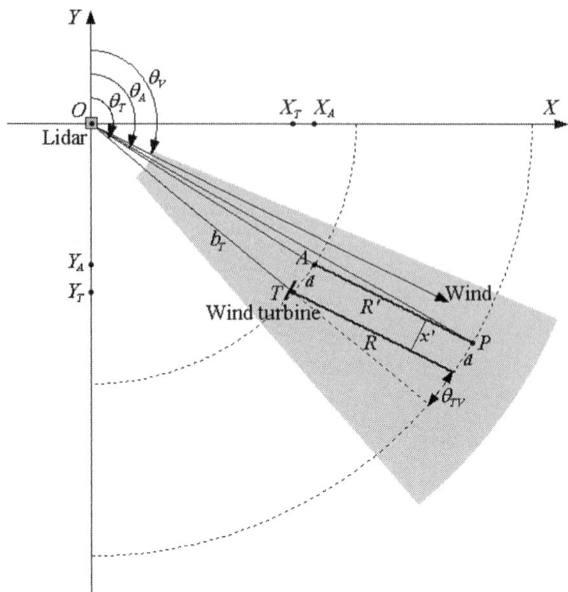

Figure 5.36 Geometry of lidar measurements for conical-sector scanning by the sensing beam in the vicinity of the wind turbine (top view). (© 2013 American Meteorological Society. Used with permission. From [54].)

and transformed from polar $\{z', \theta\}$ to the rectangular coordinates $\{X, Y\}$ by interpolating the data to a computational grid with a fine mesh $(\langle \hat{V}_r(z',\theta)\rangle \rightarrow \langle \hat{V}_r(X,Y)\rangle)$.

According to the measurement geometry (see Figure 5.36), the dependence of the mean wind velocity U on the distance $R \in [0, R_{\max}]$ along the wind direction and on the arc length d can be calculated as:

$$U(R,d) = \langle \hat{V}_r(X_A + R'\sin\theta_V, Y_A + R'\cos\theta_V)\rangle / \cos(\theta_{AV} - \theta_R) \qquad (5.42)$$

where $\theta_{AV} = \theta_A - \theta_V$, $\theta_R = \arctan[R'\sin\theta_{AV}/(b + R'\cos\theta_{AV})]$ is the angle between the lines OP and OA, $R' = (R^2 + 2Rb_T\cos\theta_{TV} + b_T^2\cos^2\theta_{AV})^{1/2} - b_T\cos\theta_{AV}$, and $\theta_{TV} = \theta_T - \theta_V$. This derivation assumes that the vertical variation of the wind direction angle θ_V can be neglected over the interval $[0, R_{\max}]$, and that the mean radial velocity, measured at the azimuth angle θ and at very small elevation angle φ ($\cos\varphi \approx 1$), can be described by the equation $\langle \hat{V}_r \rangle = U\cos(\theta - \theta_V)$. The measurement height $h(R) = h_L + (R^2 + 2Rb\cos\theta_{TV} + b^2)^{1/2}\sin\varphi$ does not depend on the arc length d. The wind direction angle θ_V can be found through minimization of the measured mean radial velocity as a function of the azimuth angle at the fixed measurement range [54].

Based on (5.40) through (5.42) and in accordance with Figure 5.36, we estimated the velocity deficit $VD(R, x')$ as

$$VD(R,x') = [1 - U(R - x'\tan\theta_{TV}, x'/\cos\theta_{TV})/U_A(R - x'\tan\theta_{TV})] \times 100\% \qquad (5.43)$$

where $U_A(R) = U(R, 1.2D)$ [53, 54] and the axis x' is perpendicular to the wake axis. Then the obtained dependences $VD(R, 0)$ on the distance R are used to estimate the maximum velocity deficit $VD_{\max}$ and the wake length L_W determined from the decrease of $VD(R, 0)$ down to 10% with an increase of R.

The investigation of the wind field in the vicinity of a 2.3-MW wind turbine located at the National Wind Technology Center (NWTC) 10 km south of Boulder, Colorado, USA, was conducted in April 2011. The measurements were carried out with a NOAA 2-μm pulsed CDL, whose main parameters can be found in [65]. The wind turbine had a rotor diameter of $D = 100$m and hub height of 85m above ground level. The angle θ_T was 130.55° and the distance between the lidar and the wind turbine was $b = 891$m.

During the lidar measurements, a sequence of different scanning geometries was employed. The geometries included both conical-sector scanning in azimuth at different elevation angles and scanning in elevation in the vertical plane at fixed azimuth angles close to θ_T. Full conical scanning was used roughly every half-hour. This sequence of scanning allowed us to estimate the wind direction angle, which we used to set minimum and maximum azimuth scanning angles for the sector scans.

In this case, sector scanning at elevation angles of 3° to 3.5° was optimal for obtaining information about the wake structure. At a range of 890m (the location of the turbine), heights of the laser beam relative to the wind turbine base equaled 60m (25m below the turbine hub) at $\varphi = 3°$, and 67m (18m below the turbine hub) at $\varphi = 3.5°$. At a range of 1,890m, the heights of the laser beam were 110m ($\varphi = 3°$) and 128m ($\varphi = 3.5°$) above the turbine base elevation. Only these elevation angles (3° and 3.5° alternately after each scan) were used for lidar measurements from 19:20 LT on 14 April to 17:30 LT on 15 April 2011. The duration of individual scan sequences was, as a rule, 7 min (57% of all cases) and 12 min (41%). For each case, all data measured alternately at elevation angles of 3° and 3.5° were used for averaging. We present the data processing results of these measurements next.

Figure 5.37(a) shows an example of the 2D distribution of the radial velocity $\langle \hat{V}_r(X, Y) \rangle$ (Figure 5.37(a)). Figure 5.37(b) shows wind velocity $U_T(R) = U(R, 0)$ (curve 1) and $U_A(R)$ (curve 2); Figure 5.37(c) shows distribution of the wind velocity deficit along the wake axis $VD(R, 0)$, and Figure 5.37(d) shows distributions of the wind-velocity deficit perpendicular to the wake $VD(R, x')$ at different distances from the wind turbine. We estimated the transverse size of the wake σ_T by fitting a Gaussian model $\exp[-(x'/\sigma_T)^2]$ to the measured value of $VD(R, x')/VD(R, 0)$ by the least-squares method. The estimates of σ_{TW} from the data of Figure 5.37(d) vary from 60m to 76m.

In Figure 5.38, curves 1 through 4 show the dependency of velocity deficit $VD(R, 0)$ on the normalized distance on the leeward side of the turbine, as obtained from lidar measurements. For comparison, curve 5 shows the velocity deficit estimated from the data measured by a CDL near Bremerhaven (Germany) during nighttime at stable thermal stratification and very weak turbulence [64]. It can be seen that curves 1 and 2, obtained from nighttime lidar data, are closest to curve 5. The values of the normalized turbine wake length L_W/D are shown in the R/D axis as closed circles. It can be seen that L_W/D can change nearly 10-fold for different realizations. For the data shown in Figure 5.38, the maximum velocity deficit $VD_{\max}$ varied from 32% to 74%.

We used the data measured by full conical scanning and an elevation angle of 10° every half-hour for 24 hours, starting from 18:00 LT on 14 April, to retrieve

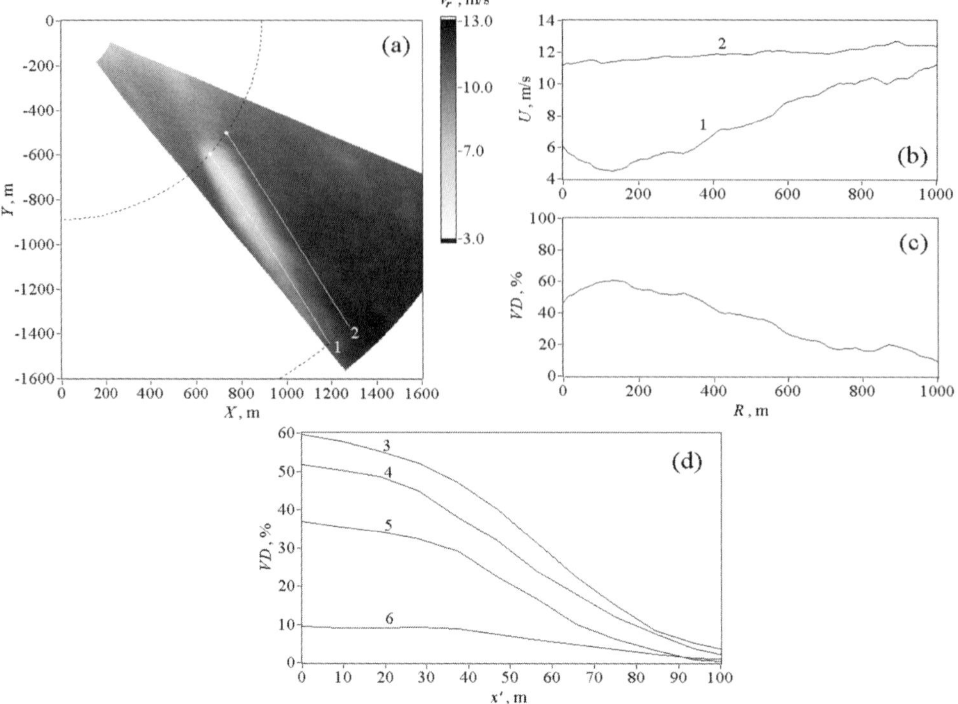

Figure 5.37 (a) Distribution of the radial wind velocity in the scanning plane {X, Y} as obtained from HRDL measurements from 00:18 to 00:23 LT on 15 April 2011; (b) profiles of the wind velocity along dashed lines 1 and 2 starting from points marked by open circles; (c) profiles of wind velocity deficit $VD(R, 0)$ along the dashed line 1 and (d) $VD(R, x')$ along the line perpendicular to the dashed lines 1 and 2 at $R = 100$m (curve 3, $\sigma_T = 60$m), 300m (curve 4, $\sigma_T = 63$m), 500m (curve 5, $\sigma_T = 61$m), and 1,000m (curve 6, $\sigma_T = 76$m). (© 2013 American Meteorological Society. Used with permission. From [54].)

the vertical profiles of the ambient wind velocity U_A and direction θ_V. The same lidar data were also used to retrieve the vertical profiles of the turbulent energy dissipation rate ε and the wind velocity variance σ_V^2 obtained using the transverse structure function method described in Section 4.7. The resulting temporal profiles of U, θ_V, ε (indicated as TEDR), and σ_V (indicated as SDWV) at a height $h = 85$m (height of the wind turbine hub) are shown in Figures 5.39(a) through 5.39(c) as squares connected by lines. It can be seen that during the period considered (from 18:00 LT on 14 April to 18:00 LT on 15 April), the wind velocity varied from 2 to 18 m/s. Most of the time, the wind velocity exceeded 10 m/s. Typically, the deviations of the angle θ_V from θ_T did not exceed 30° in absolute value, which allowed the turbine wake information to be obtained from the lidar data and main wake parameters to be monitored for almost the entire period when the turbine operated. (From 03:10 to 09:30 on 15 April, the research 2.3-MW wind turbine was shut down, the turbine blades did not rotate, and no wake with velocity deficit was generated behind the turbine.)

The processing procedures described in this section were used to determine the ambient wind velocity U_A ($h = 85$m) and direction θ_V, the maximum velocity deficit $VD_{\max}$, and the wake length L_W, which are shown as circles in Figures 5.39(a), 5.39(b),

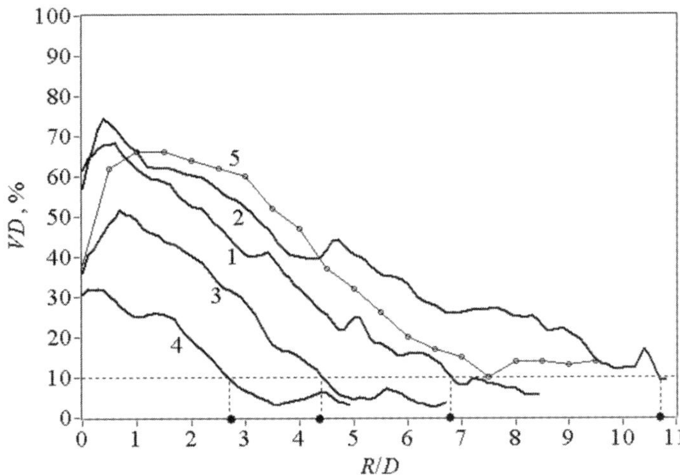

Figure 5.38 Dependences of the wind velocity deficit $VD(R, 0)$ on distance R, normalized to the rotor diameter D, downwind from the wind turbine along the wind flow, as obtained from lidar measurements from 2339 to 2346 LT on 14 April 2011 (curve 1), from 0023 to 0030 LT (curve 2), from 1339 to 1346 LT (curve 3), and from 1548 to 1600 (curve 4) on 15 April 2011. Curve 5 represents the result from [64]. Closed circles show normalized distances (lengths of the turbine wake) determined from the velocity deficit drop down to 10% of ambient. (© 2013 American Meteorological Society. Used with permission. From [54].)

5.39(d), and 5.39(e), from the lidar data measured by conical-sector scanning across the wind turbine location. The same data were also used to obtain the temporal profile of ε at a height of 85m by using the longitudinal structure function method (see Section 4.5), which is shown in Figure 5.39(c). The results that are shown as open squares and circles in Figures 5.39(a) through 5.39(c) are in good agreement for most of the measurement time.

According to the data of Figure 5.39, at wind velocities higher than 5 m/s, the 2.3-MW wind turbine generates a wake with a maximal velocity deficit VD_{max} from 27% to 74%, and the wake length L_W varies from 120m to 1,180m depending on the wind velocity and turbulence.

Analysis of the results depicted in Figure 5.39 has shown that as the wind velocity increases, the wake length L_W first increases, achieves its maximum at $U_A \approx 12$ m/s, and then decreases rapidly. This behavior is characteristic both during the daytime and at nighttime at, respectively, the stable and unstable temperature stratifications of the atmosphere.

For the analysis of the influence of wind turbulence on the wake length L_W, the estimates of ε and L_W obtained from nighttime (from 23:50 on 14 April to 03:00 on 15 April) and day (from 12:20 to 15:25 on 15 April) measurements were selected, and the estimates for each of these time periods were averaged [53, 54]. It turned out that the average wake length was 680m for the night and 340m for the day. These estimates correspond to averaged values of the dissipation rate of 6.6×10^{-3} m²/s³ and $1.3 \cdot 10^{-2}$ m²/s³. Thus, taking into account that the values of the wind velocity averaged for these periods are close and exceed 10 m/s, we can conclude that at the strong wind the twofold increase of the turbulent energy dissipation rate leads to a twofold decrease in the length of the wake generated by the wind turbine.

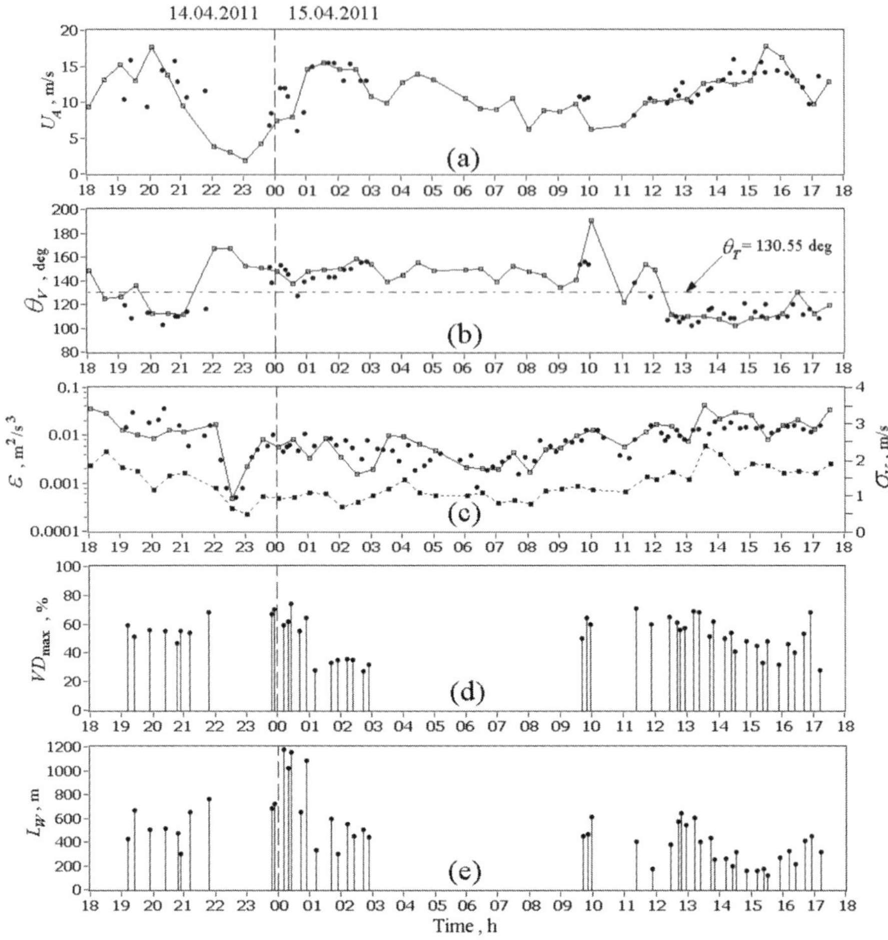

Figure 5.39   Diurnal profiles of the (a) ambient wind velocity, (b) wind direction, (c) turbulent energy dissipation rate (TEDR), and standard deviation of wind velocity (SDWV) at a height of 85m as obtained from the data measured by a 2-μm pulsed CDL using full conical scanning (open squares connected by solid lines are U_A, θ_V, and ε, and closed squares connected by dashed lines are σ_V). ε is estimated from the transverse structure function of the radial velocity (open squares) and from the longitudinal structure function of the radial velocity calculated from lidar measurements using sector scanning (circles). Estimates of the (a) wind velocity and (b) direction, (d) maximum wind velocity deficit, and (e) turbine wake length, determined as the velocity deficit drop down to 10%, were obtained from lidar sector-scan measurements (shown as circles). (© 2013 American Meteorological Society. Used with permission. From [54].)

5.10   Conclusions

1. The method of using velocity envelopes for estimation of aircraft wake vortex parameters from pulsed CDL data has an accuracy of ±4.5m for the vertical coordinate of the vortex axis, ±6.5m for the horizontal coordinate (which is comparable with the core diameter of a wake vortex generated by LTA), and ±13 m²/s (relative error of about 4%) for the wake vortex circulation. The method is applicable at SNR ≥ 1.

2. Due to the effect of the Earth's surface on the wake under IGE conditions, the distance between the vortex axes increases with time and can exceed by

6 times the initial one. Initially, aircraft wake vortices descend and then, once the height above the Earth's surface equal to about a half distance between the vortex axes is achieved, the wake usually begins to ascend. The height of the core of a vortex reflected from the Earth's surface with time can exceed the flight height of the aircraft generating this wake.

3. For aircraft wake vortices in the atmospheric boundary layer under OGE conditions, the following is valid:
   - The separation between the vortex axes b can both increase and decrease with time. On average, at $t/t_0 > 2$ the value of b increases with time.
   - Due to the effect of the atmosphere (regular vertical wind inhomogeneity, large-scale turbulence) on the wake, the tilt angle of the vortex pair with respect to the horizontal plane can vary within $\pm 80°$.
   - The difference of the horizontal velocity of the aircraft wake vortices from the lateral wind velocity is mostly connected with the tilt of the vortex pair.
   - Due to the effect on each other, the aircraft wake vortices move down. Until $t \approx 2.7t_0$ the height of the vortex axis decreases, on average, by the linear law. Then descent of the vortex pair slows down due to a decrease of the vortex circulation.
   - The process of the wake vortex evolution consists of two phases. In the first phase, the slow decay of the vortex circulation occurs due to the diffusion of the wake energy into the environment. In the second phase, under the effect of small-scale atmospheric turbulence, the vortex undergoes deformation and its circulation decreases quickly. Ultimately, the vortex fully collapses, transforming into a local turbulent formation.
   - Without the influence of the Earth's surface and wind shears, the time of transition from one phase of wake decay into another is determined by the turbulent energy dissipation rate ε. The higher the value of ε, the faster the wake decay. The empirical dependence of the time of transition from one phase of vortex decay to another on the turbulent energy dissipation rate is described by (5.24).

4. The wake measurement strategy and the procedure for processing raw data presented in the chapter provide accurate lidar measurements in the free atmosphere that are comparable to the accuracy of ground-based measurements in the atmospheric boundary layer (up to heights of ~400m) if smoke generators are used for the increase in the SNR.

5. The mesoscale structure of the wind field with strong variations of the vertical component formed near mountains can be a reason for sharp variations in the vortex axis height with a very large amplitude. However, the decay of vortex circulation in the free atmosphere, as in the boundary layer, has two phases: the slow and fast ones.

6. The results presented in Section 5.9 show that the use of pulsed coherent Doppler lidars allows not only estimation of the velocity deficit in a wake behind a wind turbine, but also obtaining of information about wind turbulence parameters. This allows us to study the dependence of wake vortex parameters on the turbulent state of the atmosphere. In particular, it follows from [53, 54] that the increase of the turbulent energy dissipation rate (turbulent intensity) can lead to a decrease in the length of a wake behind a wind turbine.

References

[1]   Shur, G.N., "Spectrum of continuous turbulence and coherent structures," *TsAGI Science Journal*, Vol. XXIV, Issue 3, 1993.

[2]   Shur, G.N., and Volkov, V.V., "Coherent clusters in zones of intense atmospheric turbulence," *Russian Meteorology and Hydrology*, No. 1, 1999.

[3]   Shur, G.N., et al., "Scales and energy of wind mesostructures in the low-latitude atmosphere," *Russian Meteorology and Hydrology*, No. 6, 2001, pp. 22–30.

[4]   Shur, G.N., Lepukhov, B.N., and Sokolov, L.A., "Mesoscale structure of wind and temperature fields in the stratosphere of the Southern Hemisphere high latitudes," *Russian Meteorology and Hydrology*, No. 5, 2003, pp. 40–46.

[5]   Shur, G.N., "Chaotic and ordered structures of atmospheric turbulence (analysis of aircraft data)," *Russian Meteorology and Hydrology*, No. 1, 1997.

[6]   Babkin, V.I., et al., *Aircraft Flight Safety Systems*, Nauka, Moscow, 2008, p. 373.

[7]   Gerz, T., Holzäpfel, F., and Darracq, D., "Commercial aircraft wake vortices," *Progress in Aerospace Sciences*, Vol. 38, 2002, pp. 181–208.

[8]   Ginevskii, A.S., and Zhelannikov, A.I., *Aircraft Wakes*, Fizmatlit, Moscow, 2008, p. 170.

[9]   Crow, S.C., "Stability theory for a pair of trailing vortices," *AIAA Journal*, Vol. 8, No. 12, 1970, pp. 2172–2179.

[10]   Brashears, M.R., and Hallock, J.N., "Aircraft wake vortex transport model," *Journal of Aircraft*, Vol. 11, No. 5, 1974, pp. 256–272.

[11]   Crow, S.C., and Bate, Jr., E.R., "Lifespan of trailing vortices in a turbulent atmosphere," *Journal of Aircraft*, Vol. 13, No. 7, 1976, pp. 476–482.

[12]   Hecht, A.M., et al., "Turbulent vortices in stratified fluids," *AIAA Journal*, Vol. 18, No. 7, 1980, pp. 738–746.

[13]   Greene, G.C., "An approximate model of vortex decay in the atmosphere," *Journal of Aircraft*, Vol. 23, No. 7, 1986, pp. 566–573.

[14]   Sarpkaya, T., and Daly, J.J., "Effect of ambient turbulence on trailing vortices," *Journal of Aircraft*, Vol. 24, No. 6, 1987, pp. 399–403.

[15]   Robins, R. E., and Delisi, D.P., "Numerical study of vertical shear and stratification effect on the evolution of a vortex pair," *AIAA Journal*, Vol. 28, No. 4, 1990, pp. 661–669.

[16]   Schilling, V., Siano, S., and Elting, D., "Dispersion of aircraft emissions due to wake vortices in stratified shear flows: A two-dimensional numerical study," *Journal of Geophysical Research*, Vol. 101, No. D15, 1996, pp. 20,965–20,974.

[17]   Hofbauer, T., and Gerz, T., "Effect of nonlinear shear on the dynamics of a counter-rotating vortex pair," *Proc. First International Symposium for Turbulence and Shear Flow Phenomena*, Santa Barbara, CA, USA, 12–15 September 1999, p. 6.

[18]   Holzäpfel, F., "Probabilistic two-phase wake vortex decay and transport model," *Journal of Aircraft*, Vol. 40. No. 2, 2003, pp. 323–331.

[19]   Holzäpfel, F., et al., "Analysis of wake vortex decay mechanisms in the atmosphere," *Aerospace Science and technology*, Vol. 7, No. 4, 2003, pp. 263–275.

[20]   Holzäpfel, F., and Robins, R.E., "Probabilistic two-phase aircraft wake-vortex model: Application and assessment," *Journal of Aircraft*, Vol. 41, No. 1, 2004, pp. 1–10.

[21]   Voevodin, A.V., et al., "Evolution of jet-vortical wake of passenger aircraft," *Aeromekhanika i Gazovaya Dinamika*, No. 4, 2004, pp. 23–31.

[22]   Vyshinskii, V.V., and Sudakov, G.G., *Aircraft Wake in the Turbulent Atmosphere*, TsAGI Proceedings, Moscow, Issue 2667, 2005, p. 155.

[23]   Henderson, S.W., et al., "Eye-safe coherent laser radar system at 2 μm using Tm. Ho: YAG lasers," *Optics Letters*, Vol. 16, 1991, pp. 773–775.

[24] Constant, G., et al., "Coherent laser radar and the problem of aircraft wakes," *Journal of Modern Optics*, Vol. 41, No. 11, 1994, pp. 2153–2173.

[25] Werner, Ch., "Fast sector scan and pattern recognition for a cw laser Doppler anemometer," *Applied Optics*, Vol. 24, No. 21, 1985, pp. 3557–3564.

[26] Huffaker, R.M., et al., Development of a Laser Doppler System for Detection, Tracking, and Measurement of Aircraft Wake Vortices, Report FAA-RD-74-231, 1974.

[27] Harris, M., et al., "Aircraft wake vortices: a comparison of wind-tunnel data with field-trial measurements by laser radar," *Aerospace Science and Technology*, Vol. 4, No. 5, 2000, pp. 363–370.

[28] Vaughan, J.M., and Harris, M., "Lidar measurement of B747 wakes: Observation of a vortex within a vortex," *Aerospace Science and Technology*, Vol. 5, No. 6, 2001, pp. 409–411.

[29] Köpp, F., "Doppler lidar investigation of wake vortex transport between closely spaced runways," *AIAA Journal*, Vol. 32, No. 4, 1994, pp. 805–810.

[30] Köpp, F., "Wake-vortex characteristics of military-type aircraft measured at airport Oberpfaffenhofen using the DLR laser Doppler anemometer," *Aerospace Science and Technology*, Vol. 3, 1999, pp. 191–199.

[31] Harris, M., et al., "Wake vortex detection and monitoring," *Aerospace Science and Technology*, Vol. 6, 2002, pp. 325–331.

[32] Hannon, S.M., and Thomson, J.A., "Aircraft wake vortex detection and measurement with pulsed solid-state coherent laser radar," *Journal of Modern Optics*, Vol. 41, 1994, pp. 2175–2196.

[33] Brockman, P. B., et al., "Coherent pulsed lidar sensing of wake vortex position and strength, winds and turbulence in the terminal area," *Proc. 10th Coherent Laser Radar Conference*, Mount Hood, OR, USA, 28 June–2 July 1999, pp. 12–15.

[34] Köpp, F., et al., "Characterization of aircraft wake vortices by multiple-lidar triangulation," *AIAA Journal*, Vol. 41, No. 6, 2003, pp. 1081–1088.

[35] Köpp, F., Rahm, S., and Smalikho, I.N., "Characterization of aircraft wake vortices by 2-μm pulsed Doppler lidar," *Journal of Atmospheric and Oceanic Technology*, Vol. 21, No. 2, 2004, pp. 194–206.

[36] Köpp, F., et al., "Comparison of wake-vortex parameters measured by pulsed and continuous-wave lidars," *Journal of Aircraft*, Vol. 42, No. 4, 2005, pp. 916–923.

[37] Rahm S., Smalikho, I. N., and Köpp, F. "Characterization of aircraft wake vortices by airborne coherent Doppler lidar," *Journal of Aircraft*, Vol. 44, No. 3, 2007, pp. 799–805.

[38] Rahm, S., and Smalikho, I.N., "Aircraft wake vortex measurement with airborne coherent Doppler lidar," *Journal of Aircraft*, Vol. 45, No. 4, 2008, pp. 1148–1155.

[39] Smalikho, I.N., and Rahm, S., "Measurements of aircraft wake vortex parameters with a coherent Doppler lidar," *Atmos. Oceanic Opt.*, Vol. 21, No. 11, 2008, pp. 854–868.

[40] Smalikho, I.N., and Rahm, S., "Lidar investigation of the effect of wind and atmospheric turbulence on aircraft wake vortices," *Atmos. Oceanic Opt.*, Vol. 22, No. 12, 2009, pp. 1160–1169.

[41] Holzäpfel, F., et al., "The wake vortex prediction and monitoring system WSVBS—Part I: Design," *Air Traffic Control Quarterly*, Vol. 17, No. 4, 2009, pp. 301–322.

[42] Köpp, F., et al., "Wake-vortex characterization by pulsed Doppler lidar," *Proc. 12th Coherent Laser Radar Conference*, Bar Harbor, ME, USA, 15–20 June 2003, pp. 152–158.

[43] Köpp, F., et al., "Comparison of wake-vortex parameters measured by 2-μm pulsed and 10-μm cw lidars," *Proc. Workshop Aircraft Vortices and Atmospheric Turbulence*, Institute of Atmospheric Physics, DLR, Oberpfaffenhofen, Germany, September 2004, pp. 3–19.

[44] Smalikho, I.N., Köpp, F., and Rahm, S., "Measurement of atmospheric turbulence by 2-μ Doppler lidar," *Proc. Workshop Aircraft Vortices and Atmospheric Turbulence*, Institute of Atmospheric Physics, DLR, Oberpfaffenhofen, Germany, September 2004, pp. 20–31.

[45] Rahm, S., Smalikho, I.N., and Simmet, R., "Recent lidar-based wake vortex measurements at DLR," *Proc. 14th Coherent Laser Radar Conference*, Snowmass, CO, USA, 8–13 July 2007, pp. 62–65.

[46] Holzäpfel, F., et al., "The wake vortex prediction and monitoring system WSVBS—Part I: Design," *Proc. 1st European Air and Space Conference (CEAS 2007)*, Deutscher Luft- und Raumfahrtkongress 2007, CEAS-2007-177, Berlin, Germany, 10–13 September 2007, pp. 3383–3390.

[47] Rahm, S., and Smalikho, I.N., *Aircraft Wake Vortex Measurement with Airborne Coherent Doppler Lidar*, Report No. 223, DLR, Oberpfaffenhofen, July 2007, p. 15.

[48] Baumann, R., et al., *Assessment of Weather Effect on the Wake-Vortex Behavior*, AWIATOR Technical Report No. D 1.14-10, DLR, Oberpfaffenhofen, June 2007, p. 64.

[49] Smalikho, I.N., and Rahm, S., "Doppler lidar measurements of parameters of aircraft wake vortices and atmospheric turbulence," *Proc. XV International Symposium Atmospheric and Ocean Optics and Atmospheric Physics*, Krasnoyarsk, Russia, 22–28 June 2008, CI-07, p. 97.

[50] Smalikho, I.N., and Rahm, S., "Lidar investigations of the influence of wind and atmospheric turbulence on aircraft wake," *Proc. XVI International Symposium on Atmospheric and Ocean Optics and Atmospheric Physics*, Tomsk, Russia, 12–15 October 2009, IAO SB RASP, pp. 533–536.

[51] Smalikho, I.N., et al., "Pulsed coherent lidar measurements of parameters of the wake generated by a wind turbine under different atmospheric conditions," *Izvestiya vuz. Fizika*, Vol. 55, No. 8, 2012, pp. 91–95 (in Russian).

[52] Smalikho I.N., et al., "Lidar investigation of the wake generated by a wind turbine under different atmospheric conditions," *Proc. XVIII International Symposium Atmospheric and Oceanic Optics and Atmospheric Physics*, Irkutsk, 2–6 July 2012, pp. C-230–C-233 (in Russian).

[53] Smalikho, I.N., et al., "Lidar investigation of atmosphere effect on a wind turbine wake," *Proc. 26th International Laser Radar Conference*, 25–29 June 2012, Porto Heli, Greece, S5p-02, p. 4.

[54] Smalikho, I.N., et al., "Lidar investigation of atmosphere effect on a wind turbine wake," *Journal of Atmospheric and Oceanic Technology* (accepted on 12 February 2013). http://dx.doi.org/10.1175/JTECH-D-12-00108.1.

[55] Köpp, F., et al., *Comparison of Wake-Vortex Parameters Measured by 2 mm Pulsed and 10 mm CW Lidars*, Report No. 194, DLR, Oberpfaffenhofen, March 2004, p. 18.

[56] Lamb, H., *Hydrodynamics*, 6th ed., Dover, New York, 1932, p. 592.

[57] Holzäpfel, F., et al., "Strategies for circulation evaluation of aircraft wake vortices measured by lidar," *Journal of Atmospheric and Oceanic Technology*, Vol. 20, No. 8, 2003, pp. 1183–1195.

[58] Jazwinski, A.Y., *Stochastic Processes and Filtering Theory*, Academic Press, New York, 1970, p. 376.

[59] Smalikho, I.N., "Techniques of wind vector estimation from data measured with a scanning coherent Doppler lidar," *Journal of Atmospheric and Oceanic Technology*, Vol. 20, No. 2, 2003, pp. 276–291.

[60] Wassaf, H.S., Burnham, D.C., and Wang, F.Y., "Wake vortex tangential velocity adaptive spectral (TVAS) algorithm for pulsed lidar systems," *Proc. 16th Bi-annual Coherent Laser Radar Conference*, Long Beach, CA, USA, 20–24 June 2011, Session 9: Wind Measurement Systems II, p. 4.

[61] Barthelme, R.J., et al., "Offshore wind turbine wakes measured by sodar," *Journal of Atmospheric and Oceanic Technology*, Vol. 20, 2003, pp. 466–477.

[62] Bingöl, F., Mann, J., and Larsen, G.C., "Light detection and ranging measurements of wake dynamics. Part I: One-dimensional scanning," *Wind Energy*, Vol. 13, 2010, pp. 51–61.

[63] Trujillo, J.-J., et al., "Light detection and ranging measurements of wake dynamics. Part II: two-dimensional scanning," *Wind Energy*, Vol. 14, 2011, pp. 61–75.

[64] Käsler, Y., Rahm, S., and Simmet, R., "Wake measurements of a multi-MW wind turbine with coherent long-range pulsed Doppler wind lidar," *Journal of Atmospheric and Oceanic Technology*, 2010, Vol. 27, pp. 1529–1532.

[65] Grund, C.J., et al., "High-resolution Doppler lidar for boundary layer and cloud research," *Journal of Atmospheric and Oceanic Technology*, Vol. 18, No. 3, 2001, pp. 376–393.

List of Acronyms

ALIENS	Atmospheric Lidar End-to-End Simulator
ATTAS	Advanced Technologies Testing Aircraft System
CDL	Coherent Doppler lidar
CRLB	Cramer-Rao lower bound
cw	Continuous wave
DLR	German Aerospace Center (Deutsches Zentrum für Luft- und Raumfahrt)
DSW	Doppler spectrum width
DSWF	Direct sine wave fitting (wind vector estimation by the least square technique)
DWD	German Weather Service (Deutsche Wetter Dienst)
FFT	Fast Fourier transform
FSWF	Filtered sine wave fitting
IAO	Institute of Atmospheric Optics
IGE	In-ground effect
LSF	Longitudinal structure function of wind velocity
LTA	Large transport aircraft
MFAS	Maximum of the function of accumulated spectra
ML	Levin's maximum likelihood estimator
ML DSWF	Direct sine wave fitting, where radial velocities are estimated using ML
ML FSWF	Filtered sine wave fitting, where radial velocities are estimated using ML
OGE	Out-of-ground effect
ONERA	French Aerospace Lab (Office National d'Etudes et de Recherches Aerospatiales)
PM	Periodogram maximum estimator
PM DSWF	Direct sine wave fitting, where radial velocities are estimated using PM
PM FSWF	Filtered sine wave fitting, where radial velocities are estimated using PM
RAS	Russian Academy of Science
SB	Siberian Branch
TEDR	Turbulent energy dissipation rate
TSF	Transverse structure function of wind velocity
WV ML	Maximum likelihood for wind vector estimation

Nomenclature

a	Radius of aerosol particle
a_0	Initial radius of the probing beam
a_R	Smallest radius of the probing beam
b	Distance between aircraft wake vortex axes
B	Bias of lidar estimate of the wind velocity
b_0	Initial distance between aircraft wake vortex axes (calculated theoretically)
B_a	Aircraft wingspan
B_D	Bandwidth determined by the Nyquist frequency
b_e	Fraction of bad estimates of the radial velocity
B_F	Frequency bandwidth
b_r	Bias of lidar estimate of the radial velocity
b_T	Distance between the lidar and the wind turbine
$B_T(\tau)$	Temporal correlation function of the normalized echo signal power $\bar{P}_S / <P_S>$
B_U	Bias of lidar estimate of the wind velocity
B_V	Bandwidth in velocity units
$B_{\bar{V}}(\tau)$	Temporal correlation function of the radial velocity averaged over the sensing volume
b_ε	Relative bias of lidar estimate of the turbulent energy dissipation rate
c	Light speed
$C_K \approx 2$	Kolmogorov constant
$C_S(\tau)$	Covariance function of the normalized complex echo signal
C_n^2	Structure characteristic of the refractive index
D	Wind turbine diameter
$D_e(r)$	Structure function of random error of lidar estimate of the radial velocity in the case of a pulsed CDL
$D_L(\theta, R)$	Azimuth (transverse) structure function of the radial velocity measured by conically scanning pulsed CDL at range R
$\bar{D}_L(\theta; R)$	Azimuth (transverse) structure function of the radial velocity averaged over the sensing volume
$D_V(r)$	Longitudinal structure function of the radial velocity
$D_V(y)$	Transverse structure function of the radial velocity
$D_{\bar{V}}(x)$	Structure function of the radial velocity averaged over the sensing volume
$D_{\hat{V}}(x)$	Structure function of the radial velocity measured by a CDL
d_ε	Discrepancy parameter of turbulent energy dissipation rate estimates by LSF and DSW methods

$D_\Psi(\boldsymbol{\rho})$	Structure function of phase fluctuations of the plane wave propagating in the turbulent atmosphere
e	Electron charge
E	Error of lidar estimate of the wind velocity
E_P	Probing pulse energy
E_U	Error of lidar estimate of the wind velocity U
E_ε	Relative error of lidar estimate of the turbulent energy dissipation rate
E_σ	Random error of estimate of the squared Doppler spectrum width
F	Focal length of the probing beam
$F_a(V)$	Function of accumulated Doppler spectra
f_I	Intermediate frequency (difference of frequencies of probing and reference beams)
f_L	Focal length of the received telescope
$\hat{f}_r$	Doppler frequency estimate
$f_s(a)$	Aerosol particle size distribution function
g	Gravitation acceleration
$G(0,\boldsymbol{\rho}_a; z,\boldsymbol{\rho}_b; t)$	Green's function describing the propagation of a spherical wave from the point $\{0, \boldsymbol{\rho}_a\}$ to the point $\{z, \boldsymbol{\rho}_b\}$
h	Plank's constant
h_B	Height of the atmospheric boundary layer
$H_s(\kappa_z)$	Transfer function of the low-frequency spatial filter in the case of cw CDL
$H_p(\kappa_z)$	Transfer function of the low-frequency spatial filter in the case of pulsed CDL
$H_t(f)$	Transfer function of the temporal low-frequency filter in the case of cw CDL
$H_\perp(\kappa_y)$	Transfer function of the low-frequency filter along the axis y' perpendicular to the axis of probing beam
$I_L(z, \boldsymbol{\rho})$	Intensity of the equivalent reference beam propagating from the lidar to the atmosphere
$I_P(z, \boldsymbol{\rho}, t)$	Intensity of the probing beam at point $\{z, \boldsymbol{\rho}\}$ and time t
$I_{PN}(z, \boldsymbol{\rho}, t)$	Normalized intensity of the probing beam
$j = \sqrt{-1}$	Imaginary unit
J	Photocurrent
$J_C(t)$	Coherent component of the photocurrent
J_N	Noise component of the photocurrent (shot noise)
$J_{NF}(t)$	Noise component of the photocurrent in the frequency band B_F
J_S	Signal component of the photocurrent
$K_e(r)$	Correlation coefficient of random error of lidar estimate of the radial velocity in the case of pulsed CDL

$K_n(\tau)$	Correlation function of fluctuations of the lidar estimate of radial velocity caused by refractive turbulence
$K_{PS}(\tau)$	Correlation coefficient of echo signal power fluctuations
$K_T(\tau)$	Temporal correlation coefficient of fluctuations of the echo signal power averaged over microphysical parameters of the scattering medium
$K_V(r)$	Correlation coefficient of wind velocity fluctuations
L	Number of pulses used for spectral accumulation
$L_d = 2\pi a_0^2/\lambda$	Diffraction length
L_{MO}	Monin-Obukhov scale
L_n	Integral (outer) scale of refractive turbulence
l_n	Inner scale of refractive turbulence
L_V	Integral correlation scale of wind velocity fluctuations (outer scale of turbulence)
l_V	Inner scale of wind turbulence
L_W	Length of the wind turbine wake
M_a	Aircraft mass
N_{eff}	Mean number of efficiently scattering particles (mean number of particles, 90% of which determine the mean echo signal power)
N_P	Mean number of aerosol particles in the sensing volume
$p(x)$	Probability density function
$P(t)$	Normalized signal power
$P_B(t)$	Incoherent component of the power of detected backscatter radiation
$P_C(t)$	Coherent component of the power, which is connected with the interference of the reference and backscattered waves
$P_D(t)$	Power of radiation within the sensitive plate of the detector
P_{JN}	Noise power
P_L	Reference beam power
P_P	Power of the probing beam
P_n	Probability of the number of photoelectrons
$P_N(t)$	Normalized noise power
$P_S(t)$	Normalized power of the echo signal
$\bar{P}_S(t)$	Normalized power of the echo signal averaged over microphysical parameters of the scattering medium
$Q(V)$	Function for wind vector estimation by FSWF technique
$Q_s(z)$	Function characterizing the spatial resolution along the axis of the probing beam (weighting function of averaging over the sensing volume) for a cw CDL
$Q_s(z')$	Function characterizing the spatial resolution along the axis of the probing beam (weighting function of averaging over the sensing volume) for a pulsed CDL

q_{th}	Spectral threshold
r_C	Radius of the aircraft wake vortex core
R_{Ci}	Distance between lidar and axis of ith (left or right) wake vortex signal-to-noise ratio (defined as the ratio of the echo signal power to the noise power in the receiver passband)
$\hat{S}(f_k)$	Estimate of the normalized Doppler spectrum
$S_D(V)$	Doppler spectrum (normalized spectrum of lidar signal power)
S_e	Noise component of the Doppler spectrum measured by CDL
$S_S(f)$	Spectral power density of the lidar echo signal
$\overline{S}_S(f)$	Spectral density of the echo signal power obtained using the conditional averaging over an ensemble of realizations of microstructure parameters of the scattering medium
$S_V(f)$	Temporal spectrum of the radial wind velocity at fixed point ($z = R$)
$S_{\overline{V}}(f)$	Temporal spectrum of the radial wind velocity averaged over the sensing volume
$S_{\hat{V}}(f)$	Temporal spectrum of the radial wind velocity measured by CDL
$S_V(\kappa_z)$	One-dimensional spatial spectrum of turbulent fluctuations of the radial wind velocity
$S_V^{(2)}(\kappa_z, \kappa_y)$	Two-dimensional spatial spectrum of turbulent fluctuations of the radial wind velocity
$S_W(f)$	Spectrum of the time window function
$S_z(f)$	Spectrum of random displacements of the sensing volume (formed in the case of a cw CDL) along the optical axis
$S_\Psi(\kappa)$	Spectrum of phase fluctuations of the plane wave propagating in the turbulent atmosphere
t_0	Theoretically calculated time during which the aircraft wake vortices descend at a distance equal to the initial distance between the vortex axes
T_A	Atmospheric transmission
T_P	Time interval between lidar shots
t_s	Wake vortex lifetime
$T_s = 1/B_F$	Sampling interval for complex lidar signal
T_W	Effective width of the time window
U	Mean wind velocity
U_A	Ambient mean wind velocity
$U_{0L}(\boldsymbol{\rho}'')$	Complex amplitude of the field of the reference beam
$U_{0P}(\boldsymbol{\rho}, t)$	Complex amplitude of the probing beam at the telescope aperture
$U_B(\boldsymbol{\rho}'', t)$	Complex amplitude of the backscattered wave field at the plane $\boldsymbol{\rho}''$ of the telescope aperture
$\tilde{U}_L(z_i, \boldsymbol{\rho}_i, t)$	Complex amplitude of the equivalent reference beam propagating from the lidar to the atmosphere (or the backpropagated local oscillator field)

$\tilde{U}_P(z_i, \boldsymbol{\rho}_i, t)$	Complex amplitude of the probing beam propagating in the atmosphere
u_*	Friction velocity
V_a	Aircraft speed
VD	Velocity deficit
V_e	Random error of lidar estimate of the radial velocity
V_{E-}	Negative velocity envelope
V_{E+}	Positive velocity envelope
V_{eff}	Effective lidar sensing volume
V_r	Radial velocity at point $\{z, \boldsymbol{\rho}\}$ and time t
$\hat{V}_r$	Lidar estimate of the radial velocity
$\overline{V}_r$	Radial velocity averaged over the sensing volume
V_T	Projection of the wind vector to the vertical plane of scanning by the probing beam
$V_\perp$	Transverse velocity vector
$W(t)$	Function of time window
w_0	Theoretically calculated initial speed of descent of the wake vortex core
$Z(t)$	Normalized complex lidar signal
Z_C	Height of wake vortex axis
$Z_N(t)$	Normalized noise
$Z_S(t)$	Normalized echo signal
α_i	Backscatter amplitude of the ith particle
β_t	Radiation extinction coefficient
β_π	Backscatter coefficient
Γ	Wake vortex circulation
Γ_0	Theoretically calculated initial wake vortex circulation
γ'	Angle between axis of the probing beam and wind direction
δ_i	Kronecker delta
δf	Width of a single frequency bin
Δf	Frequency resolution
$\Delta p = \sigma_P c/2$	Longitudinal "instantaneous" dimension of the sensing volume in the case of a pulsed CDL
Δt	Measurement duration of one Doppler spectrum
ΔV	Velocity resolution
Δz	Effective longitudinal dimension of the sensing volume
ε	Dissipation rate of the kinetic energy of turbulence
ε_{xy}	Relative error of mean wind velocity measurement by a cw CDL
η	Quantum efficiency of the detector
θ	Azimuth angle

θ_p	Tilt angle
θ_V	Wind direction angle
λ	Optical wavelength
$\nu = c/\lambda$	Optical wave frequency
$\Pi(\boldsymbol{\rho})$	Pupil function of the transmit/receive aperture of the telescope
ρ_0	Concentration of aerosol particles
ρ_a	Density of air
$\rho_s(a)$	Concentration of particles with radii larger than a
$\sigma_{CR,WV}^2$	Cramer-Rao lower bound for lidar estimate of the wind velocity
σ_e^2	Variance of the random error of estimation of the radial velocity
σ_{en}^2	Variance of random error of lidar estimate of the radial velocity, calculated by assuming Gaussian statistics for the echo signal
$\hat{\sigma}_f^2$	Estimate of the squared Doppler spectrum width (second spectral moment)
σ_{fl}^2	Instrumental broadening of the echo signal power spectrum
σ_g^2	Variance of good estimate of the radial velocity
σ_P	Probing pulse duration determined from the power drop to the e^{-1} level from the point of maximum
σ_{PS}^2	Relative variance of echo signal power fluctuations
σ_S	Doppler spectrum width in velocity units
σ_T^2	Variance of the echo signal power $\bar{P}_S$ averaged over microphysical parameters of the scattering medium
σ_t^2	Turbulent broadening of the Doppler spectrum
σ_V^2	Wind velocity variance
$\sigma_{\bar{V}}^2$	Variance of the radial velocity averaged over the sensing volume
$\sigma_{\hat{V}}^2$	Variance of lidar estimate of the radial velocity
$\sigma_{<V>}^2$	Doppler spectrum broadening due to inhomogeneity of the mean wind
σ_{VI}^2	Instrumental broadening of the echo signal power spectrum in velocity units
$\bar{\sigma}_V^2$	Broadening of the Doppler spectrum due to inhomogeneity of the radial velocity inside the sensing volume (in velocity units)
σ_W	Half-width of the Gaussian time window determined by the $e^{-1/2}$ level
σ_z^2	Variance of random displacements of the sensing volume (formed in the case of cw CDL) along the optical axis
σ_π	Differential backscatter cross section of an aerosol particle
τ_C	Correlation time of fluctuations of the echo signal power
τ_P	Probing pulse duration defined by the power drop down to the half-maximum level to the right and to the left from the point of maximum
τ_V	Correlation time of wind velocity fluctuations

τ_z	Integral timescale of random displacements of the sensing volume (formed in the case of cw CDL) along the optical axis
φ	Elevation angle
φ_{Ci}	Angle of intersection of the ith (left or right) vortex axis by the scanning probing beam
$\Phi(V)$	Log-likelihood function
ψ_i	Wave phase with the uniform distribution of the probability density function over the interval $[0, 2\pi]$.
ω_0	Angular speed of scanning by the probing beam
$<...>$	Operator for averaging over ensemble of all random variables
$<...>_m$	Operator for averaging over an ensemble of realizations of microstructure parameters of the scattering medium
$\hat{X}$	Estimate of X
$<...>_E$	Experimental estimate of mean value

About the Authors

Viktor A. Banakh, Doctor of Science, is a specialist in wave propagation in random media, atmospheric turbulence, and laser probing the atmosphere. He is a head of the Wave Propagation Laboratory of the Institute of Atmospheric Optics of the Siberian Branch of the Russian Academy of Sciences, the author and coauthor of more than 150 papers in the peer reviewed journals and few monographs on laser propagation in a turbulent atmosphere, techniques of wind and turbulence parameters estimation from the coherent Doppler lidar data, and relative problems.

Igor Smalikho graduated from the Tomsk State University (TSU) in 1983 with a diploma in optics and radiophysics. From 1983 to 2001 he worked as a research scientist in the Institute of Atmospheric Optics (IAO) of Siberian Branch of the Russian Academy of Sciences, where he concentrated on research of optical wave propagation in the atmosphere, including effects of refractive turbulence, scattering, and thermal blooming on laser radiation. In 1989 he received his Ph.D. from TSU. From 2001 to 2008 he worked as a research scientist at the Institute of Atmospheric Physics of DLR in Oberpfaffenhofen, where he developed methods of coherent Doppler lidar measurements of wind, atmospheric turbulence, and aircraft wake vortices and processed experimental data. He has worked at the IAO since 2008.

Index

Recent Titles in the Artech House
Applied Photonics Series

Brian Culshaw and Alan Rogers, Series Editors

Advanced Optical Communication Systems and Networks, Milorad Cvijetic and Ivan B. Djordjevic

Chemical and Biochemical Sensing with Optical Fibers and Waveguides, Gilbert Boisdé and Alan Harmer

Coherent and Nonlinear Lightwave Communications, Milorad Cvijetic

Coherent Doppler Wind Lidars in a Turbulent Atmosphere, Viktor Banakh and Igor Smalikho

Coherent Lightwave Communication Systems, Shiro Ryu

DWDM Fundamentals, Components, and Applications, Jean-Pierre Laude

Fiber Bragg Gratings: Fundamentals and Applications in Telecommunications and Sensing, Andrea Othonos and Kyriacos Kalli

Finite Element Modeling Methods for Photonics, B. M. Azizur Rahman and Arti Agrawal

Frequency Stabilization of Semiconductor Laser Diodes, Tetsuhiko Ikegami, Shoichi Sudo, and Yoshihisa Sakai

Handbook of Distributed Feedback Laser Diodes, Second Edition, Geert Morthier and Patrick Vankwikelberge

Helmet-Mounted Displays and Sights, Mordekhai Velger

Introduction to Infrared and Electro-Optical Systems, Second Edition, Ronald G. Driggers, Melvin H. Friedman, and Jonathan Nichols

Introduction to Lightwave Communication Systems, Rajappa Papannareddy

Introduction to Semiconductor Integrated Optics, Hans P. Zappe

Liquid Crystal Devices: Physics and Applications, Vladimir G. Chigrinov

New Photonics Technologies for the Information Age: The Dream of Ubiquitous Services, Shoichi Sudo and Katsunari Okamoto, editors

Optical Document Security, Third Edition, Rudolf L. van Renesse

Optical FDM Network Technologies, Kiyoshi Nosu

Optical Fiber Amplifiers: Materials, Devices, and Applications, Shoichi Sudo, editor

Optical Fiber Communication Systems, Leonid Kazovsky, Sergio Benedetto and Alan Willner

Optical Fiber Sensors, Volume Three: Components and Subsystems, John Dakin and Brian Culshaw, editors

Optical Fiber Sensors, Volume Four: Applications, Analysis, and Future Trends, John Dakin and Brian Culshaw, editors

Optical Measurement Techniques and Applications, Pramod Rastogi

Optical Transmission Systems Engineering, Milorad Cvijetic

Optoelectronic Techniques for Microwave and Millimeter-Wave Engineering, William M. Robertson

Reliability and Degradation of III-V Optical Devices, Osamu Ueda

Signal Processing and Performance Analysis for Imaging Systems, S. Susan Young, Ronald G. Driggers, and Eddie L. Jacobs

Smart Structures and Materials, Brian Culshaw

Substrate Surface Preparation Handbook, Max Robertson

Surveillance and Reconnaissance Imaging Systems: Modeling and Performance Prediction, Jon C. Leachtenauer and Ronald G. Driggers

Wavelength Division Multiple Access Optical Networks, Andrea Borella, Giovanni Cancellieri, and Franco Chiaraluce

For further information on these and other Artech House titles, including previously considered out-of-print books now available through our In-Print-Forever® (IPF®) program, contact:

<table>
<tr><td>Artech House</td><td>Artech House</td></tr>
<tr><td>685 Canton Street</td><td>16 Sussex Street</td></tr>
<tr><td>Norwood, MA 02062</td><td>London SW1V 4RW UK</td></tr>
<tr><td>Phone: 781-769-9750</td><td>Phone: +44 (0)20 7596-8750</td></tr>
<tr><td>Fax: 781-769-6334</td><td>Fax: +44 (0)20 7630-0166</td></tr>
<tr><td>e-mail: artech@artechhouse.com</td><td>e-mail: artech-uk@artechhouse.com</td></tr>
</table>

Find us on the World Wide Web at: www.artechhouse.com